Coupled Processes in Subsurface Deformation, Flow, and Transport

Mao Bai, Ph.D.

Derek Elsworth, Ph.D.

ASCE PRESS

American Society of Civil Engineers
1801 Alexander Bell Drive
Reston, VA 20191-4400

Abstract: This book presents the fundamental concepts and analytical and numerical approaches available in representing deformation, flow, and transport behavior in geologic media as relevant to many engineering disciplines – civil, mining, petroleum, environmental, chemical, process – and the geological sciences. The individual processes governing deformation, flow, and transport are presented, with emphasis on the coupling and feedbacks present where solid deformation, fluid flow, and solute transport combine, and in the representation of heterogeneous media through multi-porosity approaches. Analytical and numerical solutions for subsurface systems subjected to varying mechanical, thermal, and chemical disturbances are presented. The implications of the theory and solutions presented are reflected in the example applications included throughout the text and in the final chapter.

Library of Congress Cataloging-in-Publication Data

Pai, Miao, 1952-

Coupled processes in subsurface deformation, flow and transport / Mao Bai and Derek Elsworth.

p. cm.

Includes bibliographical references and index.

ISBN 0-7844-0460-7

1. Engineering geology. 2. Engineering geology–Mathematical models. 3. Hydrogeology. 4. Geochemistry. 5. Soils–Solute movement. I. Elsworth, Derek. II. Title.

TA705 .P34 2000

624.1′51–dc21

00-024569

Library of Congress Catalog Card No: 00-024569
ISBN 0-7844-0460-7
Manufactured in the United States of America.

To our parents: Demao and Yongzhi, and Jack and Rosalind.

Contents

ACKNOWLEDGEMENTS

The authors thank the following individuals who have contributed to this book: Younane Abousleiman, who both offered encouragement to write this book and assisted in understanding the bases and pertinent applications associated with the theory of poroelasticity, particularly in the validation of the numerical codes in the book; Fanhong Meng, who undertook many useful tasks embedded in the book, especially in many numerical investigations; Sheik Ahseek, who helped in the theoretical development of anisotropic dual-porosity poroelasticity; Ashene Bouhroum, who provided experimental data; and Faruk Civan, who added to the comprehension of transport phenomena in heterogeneous porous media. Other individuals who have contributed to the book include: Mian Chen, Huaxing Zhang, Qinggang Ma, Musharraf Zaman, Zhengying Shu, and Jinggang Cao. The authors are especially grateful to Jean-Claude Roegiers for his unfailing technical and administrative support, and to Hilary Inyang for his enduring assistance throughout the study. Mao Bai is indebted to Tianquan Liu for the initial encouragement that resulted in the publication of the book. The sacrifice and understanding of the authors' families during the writing of this book, are greatly appreciated. The forbearance of Nai and Susan, and of Andi, Geneviève, and Cooper are warmly appreciated. Finally, this book is dedicated to our parents: Demao and Yongzhi, and Jack and Rosalind, whose unfailing encouragement and understanding have made this possible.

PREFACE

This book presents the mathematical underpinnings that represent the deformation, flow, and transport behavior of porous and porous-fractured media, as an analog for geologic media. This grouping of behaviors addresses the pressing issues of resource recovery of oil, gas, and water; of environmental protection; and in describing the progress of natural geologic processes. Of particular interest are the couplings and feedbacks present in the attendant processes of deformation, flow, and transport. The book also provides fundamental concepts and analytical and numerical approaches in representing subsurface flows relevant to many disciplines in engineering and science, notably civil, mining, petroleum, environmental, chemical, process and geological engineering, and the geological sciences. The component topics of deformation, flow, and transport are included in separate chapters, with emphasis given to the coupled processes that result when solid deformation, fluid flow, and solute transport combine.

The predominant theme that permeates the text is the utility of describing behavior at different length and time scales through the concept of a multi-porous medium. This representation follows directly from the continuum theory of mixtures, and enables the observed diffusive and dispersive behavior of fractured rocks to be straightforwardly described. Analytical and numerical solutions are presented for subsurface systems subjected to varying mechanical, thermal, and chemical disturbances. The implications of the presented theory and solutions are reflected in the example applications included throughout the book and especially in the final chapter.

The book addresses deformation, flow, and transport from a common standpoint. The conservation and constitutive laws are sequentially defined and combined, with other constraints levied on continuity of the dependent variables. This arrangement is natural and illustrates the commonality in the mathematical development of the governing equations. The approach is also convenient where the coupling between processes is also included, and the individual component behaviors can be represented, as presented, with the coupling terms becoming self-evident.

The book is based largely on the authors' past and current research interests. It can be used as a reference text for those who are interested in the study of coupled subsurface processes, or as a teaching text for an advanced graduate course in the mechanics of coupled processes. Instructors may wish to develop their own set of "homework" problems in conjunction with appropriate readings from the book to help students digest its contents. The solutions presented in the text are specific to the chosen physical conceptualizations and specific assumptions; they must be modified if the conceptualizations or assumptions are different.

NOMENCLATURE

Symbol	Definition
A	Skempton coefficient A
a	acceleration
$a^{\#}$	cross-sectional area of transport
A^*	effective flow cross-sectional area
a_{ij}^*	solute exchange intensity factor
A_a	analytically calculated cross-sectional area
A_c	flow cross-sectional area
A_e	area of integration
A_e^*	uniform cross-sectional area of bar element
A_f	flow area
A_t	numerically calculated cross-sectional area
b	fracture aperture
B	Skempton coefficient B
$\mathbf{B}$	strain-displacement matrix
b_i^*	flow velocity ratio
b_0	original hydraulic aperture
b_h	fracture hydraulic aperture
B_i	formation volume factor
b_m	fracture mechanical aperture
b_m^*	dimensionless fracture mechanical aperture
Δb_s	perturbed aperture due to the stress change
c	solute concentration
c_0	constant concentration at the inlet
c^0	initial concentration
c^*	lumped compressibility
$\mathbf{c}^*$	lumped compressibility matrix
$\tilde{C} = C_{ijkl}$	compliance tensor
c_d	diffusivity coefficient
c_f	fluid compressibility

c_g	constant related to grain packing and shape
$c_g^{\#}$	constant related to mean grain size and shape
c_h^*	specific heat
c_h^f	fluid heat capacity
$\tilde{C}^s$	compliance of solid constituent
c_v	coefficient of consolidation
$\mathbf{D}$	elastic modulus matrix
$\tilde{D}^h = D_{ij}$	hydrodynamic dispersion tensor (2nd order)
$\tilde{D} = D_{ijkl}$	elastic modulus tensor (4th order)
$D^{\#}$	equivalent dispersion coefficient
D^m	mechanical dispersion coefficient
$\tilde{D}_{12}$	elastic stiffness tensor of dual-porosity medium
D^d	effective diffusion coefficient
D_h	fracture hydraulic diameter
d_m	hydraulic radius or size factor
D_r	radial dispersion coefficient
E	elastic modulus
E_0	effective grain modulus
$E^{\#}$	elastic modulus of fractured porous medium
erf	error function
erfc	complementary error function
$\mathbf{F}$	boundary tractions
$F = F_0$	force, e.g., external load
$\mathbf{f}$	vector of applied nodal boundary tractions
f^*	pore fraction contained within the macropores
$f^{\#}$	corrective factor
F_l	constant related to rock mechanical properties
$F(n)$	porosity factor
f_s	fracture length per unit flow area
$f(s)$	system modifier
g	gravitational acceleration
G	shear modulus
g_c	effective grain contact area
G_e	equivalent geometric factor
G_f	geometric factor
h	height of consolidating column
H	Biot constant
h_a	domain thickness
h_j^*	elevation head

h_r	reservoir thickness
I_0	modified Bessel function of zero order
l_a	flow geometry
L	linear length of 1-D domain
L^*	length of a laboratory sample
$L^\#$	characteristic length of micropores
L_e	domain length in numerical analysis
L_f	linear flow length
L_h	compatibility factor
$\vec{J}$	hydraulic gradient
$\mathbf{J}_t$	total mass flux
$\mid \mathbf{J} \mid$	Jacobian determinant
JRC	joint roughness coefficient
k	permeability
$\mathbf{k}$	permeability tensor
K	bulk modulus of porous medium
k_0	initial or original permeability
K_0	modified Bessel function of zero order
K_1	modified Bessel function of first order
$K^\#$	bulk modulus of fractured porous medium
K^*	bulk modulus of fractured medium
$\mathbf{K}_c$	conductance matrix
$\mathbf{K}_e$	stiffness matrix
K_f	fluid bulk modulus
K_{fr}	bulk grain modulus of fractured medium
K_h	hydraulic conductivity
K_h^*	thermal conductivity
K_h^{*f}	thermal conductivity for fluid
k_n	absolute permeability of nonwetting fluid
K_n	fracture normal stiffness
K_{nw}	bulk modulus of nonwetting fluid
$K_p^\#$	sorption intensity factor
k_{rw}	relative permeability of wetting fluid
k_{rn}	relative permeability of nonwetting fluid
K_s	bulk modulus of solid grains
K_h^*	thermal conductivity
K_h^{*s}	thermal conductivity of solid
K_{sh}	fracture shear stiffness
k_w	absolute permeability of wetting fluid

K_w	bulk modulus of wetting fluid
m	mass
$\mathbf{m}$	one-dimensional vector
m^*	Fourier parameter
m_d	average 'walking' distance
m_ω	mass of species, ω
M	Biot modulus
M_i	shape function (fluid)
M_s	total mass for solid
M_f	total mass for fluid
n	porosity
$N^\#$	total number of steps
N_g	dimensionless number
N_i	shape function (solid)
p	fluid pressure
p_0	initial fluid pressure
p_c	capillary pressure or interfacial tension
Pe	Peclet number
q	flow rate
$\mathbf{q}$	vector of prescribed nodal discharge
Q	volumetric flow rate
$\mathbf{Q}$	boundary discharge
q_s	specific discharge
R	Biot constant
$R^\#$	radius of spherical matrix block
$r_0 = k_1/k_2$	permeability ratio
$r_1 = K_f/K_{fr}$	fracture compressibility ratio
$r_2 = K_f/K_s$	grain compressibility ratio
$r_3 = K_f/E$	matrix compressibility ratio
r_b	radial distance from center of micropore block
R_b	radius of spherical micropore block
r_c	arbitrary distance from matrix center
r_{De}	dimensionless reservoir radius
r_e	reservoir outer boundary radius
Re	Reynolds number
r^c	local coordinate
$\mathbf{R}_d$	coupling matrix
R_k	ratio between matrix and fracture permeabilities
R_l	absolute fracture roughness
$R_m = E_{mass}/E_{intact}$	modulus reduction ratio

R_r	retardation factor
r_w	wellbore radius
s	Laplace parameter
s^*	fracture spacing
s^c	local coordinate
S	saturation
$\mathbf{S}$	vector of nodal saturations
S_e	surface of the calculated domain
s_h^*	lumped intrinsic heat capacity
t	time
T	temperature
T_0	initial temperature
t^c	local coordinate
t_d	dimensionless time or pore volume injected
u	displacement
$\mathbf{u}$	vector of nodal displacements
u_0	initial displacement
u_i^s	vector of solid displacements
u_i^f	vector of fluid displacements
u_n	normal displacement
v	average flow velocity
$v^* = nv$	intrinsic flow velocity
$v^\#$	velocity of the plume front
V	total volume
v_a	average flow velocity
$\mathbf{v}_d$	Darcy's velocity
v_{xi}	fluid velocity in x_i direction
V_p	pore volume
V_s	solid volume
W	strain energy
w_i^c	weight in numerical integration
z	elevation of the control volume

<u>Greek</u>	<u>Definition</u>
α	Biot coefficient
α_2	pressure ratio factor of fracture
$\alpha^\#$	heterogeneous dispersivity

α_f	fluid thermal expansion coefficient
α_g	geometric constant related to matrix-fracture pattern
α_h^*	thermal expansion coefficient
α_l	longitudinal dispersivity
α_r	radial dispersivity
α_s	solid thermal expansion coefficient
α_t	transverse dispersivity
β_f	volume fraction of fractures
β_h	thermal expansion factor
β_s^*	site parameter
γ	shear strain
Γ	geometric leakage (interporosity flow) factor
γ^*	equivalent Peclet number
$\Gamma^{\#}$	mass exchange rate between macropores and micropores
Γ_c	concentration exchange coefficient
γ_f	unit weight of the fluid
$\delta(t)$	*delta* function
δ_{ij}	Kronecker delta
ε_{ij}	strain tensor
$\tilde{\varepsilon}^p$	strain in fluid constituent
$\tilde{\varepsilon}^s$	strain in solid constituent
ε_{kk}	volumetric strain
$\tilde{\varepsilon}_v$	volumetric strain tensor
ζ	fluid content
η_i^*	equivalent sorption intensity factor
$\theta^{\#}$	relative saturation of macropores
ι^*	sorption process factor
$\kappa = \dfrac{k}{\mu}$	intrinsic permeability (mobility)
λ	*Lamé* constant
λ^*	rate constant or interporosity coefficient
λ_d	first-order decay coefficient
μ	dynamic viscosity
μ_{nw}	dynamic viscosity of nonwetting fluid
μ_w	dynamic viscosity of wetting fluid
ν	Poisson ratio
$\nu^{\#}$	Poisson ratio of fractured porous medium
ν_u	undrained Poisson ratio
ξ	concentration exchange coefficient
ξ_l	parameter determined from experimental data

ξ_w	characteristic dimension of the matrix blocks
ρ	density
ρ_s	solid density
$\rho_f = \rho_h^f$	fluid density
$(\rho_i)_{std}$	density at standard condition
ρ_ω	density of species ω
$\bar{\sigma}$	mean stress
σ_c	confining or mean stress
$\tilde{\sigma}^e$	effective stress tensor
$\sigma_e = \sigma^e$	effective stress
σ_e^0	initial effective stress
σ_i	equivalent cementing pressure
σ_n	normal stress
σ_{kk}	total stress (summation of stresses)
$\tilde{\sigma}^s$	stress in solid constituents
$\tilde{\sigma}^t$	total stress tensor
ς	standard deviation
ς^*	sorption isotherm
$\tilde{\tau}$	shear stress tensor
$\tau^* = t + \Delta t$	desired time level
$\tau_{ij}^{\#}$	tortuosity tensor
ϕ_d	fracture dilatation angle
φ	storage coefficient
$\tilde{\Psi}$	elastic modulus tensor
ψ_{fd}	time discretization schematic constant
ω	ratio of fracture storativity
Ω	mass transfer rate between stagnant and flowing fluids
∇	divergence
$\vec{\nabla}$	gradient

Subscript	Definition
0	original or absolute or initial
1	matrix or macropore or component x
2	fracture or mesopore or component y
3	micro-fracture or micropore or component z
a	analytical quantity
c	confining
D	dimensionless quantity
e	effective

t	total or numerical quantity
fr	fracture
ma	matrix or macropores
mi	micropores
$nw = n$	non-wetting fluid phase
w	wetting fluid phase

Superscript	Definition
0	initial
e	effective
T	matrix or vectorial transposition

Coordinate	Definition
x, y, z	Cartesian (global)
r, θ, z	cylindrical (global)
r, θ	axisymmetric (global)
$\bar{r}^c, \bar{s}^c, \bar{t}^c$	local (3D)

Conversion (English)	Conversion (SI)
1 in	2.54 cm
1 md	9.87×10^{-6} cm^2
1 cp	1×10^{-3} Pa $\cdot$ s

Chapter 1

INTRODUCTION

This book emphasizes the couplings present in describing the motion of fluids in the subsurface. This view is different from the many texts that emphasize flow behavior in isolation from significant thermal, chemical and mechanical effects. The emphasis is on flow within fractured geologic media with the behavior of uncoupled processes viewed as a special case of the coupled focus; hence, no generality is lost from the chosen approach. Traditional methods dealing with uncoupled behavior are used to introduce the more complex issues.

1.1 STATE OF THE ART

The processes of deformation, flow, and transport are typically viewed in isolation, although a coherent thread links the three processes. In common with all three processes, the governing equations are defined by enforcing conservation, ensuring continuity of the dependent variable, applying a constitutive relationship, and satisfying boundary and initial conditions. Momentum must be conserved for deformation, and mass or energy must be conserved for flow or transport. Where these systems are coupled, the individual conservation laws must be jointly and simultaneously satisfied. Isolated description of individual events (e.g., deformation) can be justified only when the single event becomes relatively dominant, and when the fundamental basis of this single process is probed, in isolation.

1.1.1 Individual Process

The state-of-the-art with respect to deformation, flow, and transport is discussed first, with the aim of reviewing the fundamentals that apply to each process. Within each process, discussion is divided between homogeneous and heterogeneous media, with emphasis on the methods that may be applied for this description.

1.1.1.1 Deformation

Mechanical influence is an important issue if the impact of the external or internal loading is the primary concern, since the consequence of the loading may result in structural instability or system failure.

Homogeneous Media

Classical analytical tools applied in the theory of elasticity are documented by Love (1927), Timoshenko (1934), and Timoshenko and Goodier (1970). These principles may be applied to geologic materials, accounting for the presence of defects and heterogeneities that may influence structural stability. The discipline of rock mechanics bridges the gap between continuum mechanics and the applied endeavors of mining engineering and petroleum engineering, among others (Jaeger and Cook 1979; Hoek and Brown 1980). Extension of the precepts of linear elasticity has been made in various areas beyond the linear limitation, such as in large deformation (Atkin and Fox 1980), elastoplasticity (Coussy 1995), and viscoelasticity (Lai et al. 1993). This trend is compatible with the increasing use of multi-disciplinary approaches to tackle problems involving coupled processes, heralded, for example, by the application of rock mechanics to solve important problems in the petroleum industry.

Heterogeneous Media

The recognition that natural fractures are pervasive in geologic media places geological engineering apart from other branches of continuum engineering. As a result, the study of geological media has been complicated by the presence of these fractures, which show complex nonlinear and hysteretic response to deformation and closure. The application of linear elastic fracture mechanics and other continuum approaches to rock mechanics (Rudnicki 1980; Atkinson 1987; Sih et al. 1975) has expanded to include the interpretation of subsurface processes, including reservoir and aquifer stimulation, or waste disposal using hydraulic fracturing (Economides and Nolte 1987). Besides fracture mechanics, the determination of fracture stiffnesses (i.e., joint stiffnesses) has been a key activity due to its practical implications in geotechnical engineering construction, such as tunneling, including rock mass classification, and in underground storage and slope stability (Hoek and Bray 1981; Nash 1987; Hencher 1987; Amadei and Pan 1995). Widely referenced papers include those by Goodman (1976), Barton (1976), Bandis and Barton (1983), and Barton et al. (1985).

1.1.1.2 Flow

Fluid flow in the subsurface is an important subject that links together deformation and transport. Fluid flow may be considered as an uncoupled process, where the release of fluid from storage is approximated by critical assumptions regarding the assumed deformation profile. Specifically, changes in fluid storage are assumed to progress without any change in total stress. This enables flow problems to be solved without recourse to the full coupling anticipated with the mechanical behavior.

Homogeneous Media

Several classic texts describe the fundamental principles of fluid mechanics, including those by Batchelor (1967), Turner (1973), Tritton (1988) and Acheson (1990). Similar to the theory of elasticity, complications associated with classical approaches involve extension to new constitutive laws, including nonlinearities and turbulence. In subsurface flows, the greatest difficulties are involved in the characterization of the porous or fractured medium, defining distributions of important parameters, such as permeability, and determining appropriate scales required to enable behavior to be represented as a continuum. Many excellent texts describe the important characteristics of subsurface flows, varying from the fundamental (Bear 1972) to the practical (Freeze and Cherry 1979). Early works include the classic texts by Lamb (1932) and Muskat (1937) in which attempts were made to apply Navier-Stokes formulations to characterize flow in porous media. Other texts of note include those by Bear and Verruijt (1987), Cushman (1990), and Bear and Bachmat (1990), each addressing a particular niche. Important current developments in subsurface flow relate to multi-phase flows, with references to this topic included in Wallis (1969), Peaceman (1977), Butterworth and Hewitt (1977), Hetsroni (1982) and Greenkorn (1983). Multi-phase flow itself represents a coupled process. As a result, a multi-disciplinary approach to describe this behavior is necessary.

Heterogeneous Media

Two primary approaches have been used in characterizing the behavior of fractured porous media: stochastic and deterministic.

The stochastic approach envisions the flow process as being random, unpredictable unless sufficient data are accumulated and analyzed. Several methods are popular in defining this behavior, including classic Markov processes, uncertainty analyses, reliability analyses, geostatistical analysis, and more recently, fractal analyses (Ghanem and Dham 1998; Bai et al. 1998a).

The deterministic approach considers the flow process to be predictable via theoretical rules, in which single-phase or multi-phase flow in fractured porous media has been modeled based on one of three possible conceptualizations:

- *Discrete fracture networks*, in which the geometric characteristics of each fracture are fully defined. Early discrete models were developed by

Castillo et al. (1972). The percolating fractures usually act as primary flow pathways; hence, matrix replenishment is less important (Long et al. 1982; Smith and Schwartz 1984). A more general model, presented by Shimo and Long (1987), was limited to a single configuration of the fracture network. Andersson and Dverstorp (1987), and Elsworth (1986) investigated flow and transport in three-dimensional systems of disk-like fractures using boundary element methods. Lin and Fairhurst (1991) applied algebraic topological theory to describe flow in a network. Using percolation theory, Mo et al. (1998) investigated fluid flow and solute transport in fracture network characterized by randomly distributed fractures.

- *Dual-porosity media*, in which the fluid in the fractures and in the matrix blocks are considered as separate continua, related interactively through a transfer function (Barenblatt et al. 1960; Warren and Root 1963; Shapiro 1987). The importance of the dual-porosity models rests not only on the significant differences in terms of time and length scales between conductive fractures and storage-rich matrix blocks, but also on the transient interactive mass exchange between the two media. The latter property frequently results in nonlinear flow characteristics. In continuum modeling of flow and transport in fractured porous media, relevant coupled processes have been studied extensively, including partial and comprehensive coupling of fluid flow, solid deformation, heat transfer, and solute transport (Elsworth and Bai 1992; Bai et al. 1993; Bai and Roegiers 1995).

- *Equivalent porous media*, in which the medium is fractured to the extent that it behaves hydraulically as a homogeneous porous medium. In this case, the existence of fractures is reflected in the material coefficients (hydraulic conductivity, storativity, etc.) which may be orders of magnitude different from a homogeneous medium (Grisak and Cherry 1975; Shapiro 1987).

At the microscopic scale, characterizing hydraulic properties for single fractures and fracture networks (i.e., fractured media) has been an important research focus. This has been spurred by our inability to adequately describe the role of fractures on flow and transport. The simplest representation of flow within fractures is based on the parallel plate concept (Snow 1968, 1969), including the interconnectivity of fractures comprising a ubiquitous network. The arrangement of fractures may be shown to yield an equivalently anisotropic permeability. Under the simplest assumption of a smooth-faced fracture, the concept of flow in fractures is analogous to the flow between parallel plates (Bear 1972), and flow rate may be shown proportional to the cube of the fracture aperture (e.g., Witherspoon et al. 1980). This cubic dependency suggests that even small fractures may dominate the hydraulic response, and that small changes in aperture may result in large changes in flow rates. The importance of this observation is reflected in the understanding of fluid flow channeled by the major fractures (Hsieh and Neuman 1985; Hsieh et al. 1985), or the determination of the direction in which a well should be drilled to intercept the most permeable flow features (Vaziri and Byrne 1990). Permeability anisotropies of fractured media and their relation to the statistical configurations of the fractures may also be defined (Oda 1985; Sagar and Runchal 1982).

At the macroscopic scale, fluid flow in fractured porous media has been a subject of intensive study for almost four decades. The seminal discourse by Barenblatt et al. (1960) introduced the phenomenology of dual-porosity behavior. Independent exposition by Warren and Root (1963) suggested a similar reservoir model based on a simplification of the Barenblatt et al. (1960) model. These works stimulated reservoir engineers to simulate naturally fractured reservoirs. These works include Kazemi (1969) and deSwaan's (1976) adoption of transient interporosity flow between fractures and matrix blocks, adoption by Crawford et al. (1976) where the temporal pressure slope changes were identified as the dual-porosity behavior in actual well test results, and adoption by Bourdet et al. (1984) where the dual-porosity response in well tests was further probed using the pressure derivative method.

Those who want to study the subject systematically should consider books by: Auguilera (1980), Reiss (1980), Streltsova (1988), Prat (1990) and Chilingarian et al. (1992). Notable review papers are available by Kamal (1983), Gringarten (1984), Chen (1989), Firoozabadi (1990), and Breitenbach (1991). For a broad understanding of the subject, readers should refer to: Odeh (1965), Duguid and Lee (1977), Mavor and Cinco-Ley (1979), Kucuk and Sawyer (1980), Najurieta (1980), Gilman and Kazemi (1983), Serra et al. (1983), Streltsova (1983), Bourdet et al. (1984), Moench (1984), Braester (1984), Chen et al. (1985), Ershaghi and Aflaki (1985), Reynolds et al. (1985), Abdassah and Ershaghi (1986), Liu and Chen (1987), Dykhuizen (1990), Al-Bemani and I. Ershaghi (1991), Nanba (1991), Kazemi et al. (1992), Jelmert (1993), Zimmerman et al. (1993), Lim and Aziz (1995), and Aly et al. (1996).

1.1.1.3 Transport

Transport processes are becoming increasingly more important in defining the modes and rates that contaminated fluids may migrate and potentially degrade groundwater resources.

Homogeneous Media

Transport phenomena usually are considered to be linked to the flow process. For this reason, flow and transport are generally compiled as a unified subject. Texts in this area include books by Bear (1972), Freeze and Cherry (1979), Bear and Verruijt (1987), Bear and Corapcioglu (1987), and Bear et al. (1993). Transport has become a dominant theme within the water resources literature since the protection of groundwater resources has become an increasing focus. Ogata and Banks (1961) provided insight into the degree of difficulty in deriving analytical solutions to the advection-dispersion transport equation. Ogata (1964, 1970) provided a certain framework for the further development of transport models, although restricted to one-dimensional solutions. The subsequent development of numerical solutions to transport problems has been hindered primarily due to the asymmetric nature of the transport equations, especially when convection processes dominate, since the equation degenerates to a first-order hyperbolic equation which is inherently instable (Huyakorn and Pinder 1983; Bai et al. 1993). A review of transport studies by Gee et al. (1991) documented various technical papers in general areas of common interest.

Heterogeneous Media

Modeling solute transport through heterogeneous porous media has attracted increased interest due to the recognition that the form and rate of plume migration is difficult to predict using the conventional theory of flow and transport through a homogeneous medium. Common abnormalities observed in the flow pattern include premature breakthrough and extensive tailing. These phenomena are traditionally interpreted as due to either one or a combination of the following mechanisms: (a) local flow rectification due to fluid transport between mobile (flowing) and immobile (dead-end pore) regions (Coats and Smith 1964); (b) tortuous flow pathways as a result of heterogeneous grain and pore distributions (Bear 1972); (c) variable flow channels due to particle-pore clogging, size exclusion, and deposition (Joy and Kouwen 1991; Imdakm and Sahimi 1991); (d) directional pollutant migration and spatial storage as a consequence of dominant anisotropy in permeability and variation in porosity distributions (Noltimier 1971; Sardin et al. 1991); (e) regional perturbation of solute concentration due to velocity contrast between layered and fractured media (McKibbin 1985; Houseworth 1988), and (f) fluids and/or formation nonlinear characteristics (Bai and Roegiers 1994a).

Due to the striking similarity in the concentration profiles resulting from these varied solute transport modes, research interest has focused on just one of these influential factors: the impact of two-region flow (factor a). To address this particular mechanism, Coats and Smith (1964) provided an adequate phenomenological model in which solute migration in the mobile region was modified by a "quasi-steady" flow between macropores and dead-end pores, useful for characterizing the heterogeneities of the medium. Using the average-volume theory, Piquemal (1992) derived a slightly different formulation, envisioning a similar scale for the selection of parameters. Coats and Smith's (1964) model has been further improved by Bai and Elsworth (1995) through coupling the complete transport processes (dispersion and convection) within so-called "dead-end" pores.

Fluid flow and solute transport are different in that the transported medium is a type of fluid for the former case, but a component of the fluid for the latter scenario (Bear 1993). Aside from this difference, Coats and Smith's (1964) model is similar, if not identical, to Warren and Root's (1963) model which was primarily used to interpret fluid flow through fractured porous media. In many cases, Coats and Smith's (1964) model has received wide applications in the modeling of fluid flow and contaminant and energy transport through fractured porous media (Tang et al. 1981; Bibby 1981; Huyakorn et al. 1983; Elsworth 1989; Elsworth and Xiang 1989; Nilson and Lie 1990; Rowe and Booker 1990; Sudicky and McLaren 1992; Harrison et al. 1992; Leo and Booker 1993). In contrast, however, Warren and Root's model is rarely cited in the literature of solute transport through micropore-macropore regions (Passioura 1971; Passioura and Rose 1971; Joy and Kouwen 1991; Koenders and Williams 1992; Joy et al. 1993; Piquemal 1992, 1993) with an exception that Sahimi (1993) provided an implicit link between the two models. As a result of this ineffective communication between between disciplines, models based on the same conceptualization have been developed independently (e.g., spherical block model for transport through fractured porous media by Huyakorn et al. 1983; and for transport through micropore-macropore region by Correa et al. 1987). Communication needs to be improved to accelerate information dissemination between these two seemingly different fields.

Due to the existence of physical and mathematical analogies in modeling flow and transport through fractured porous media and through micropore-macropore media, the multi-porosity/mul-tipermeability model proposed by Bai et al. (1993) can be made applicable to solute transport through multi-pore regions. Gwo et al. (1995) made a parallel effort to discretize partially saturated heterogeneous media into three component regions representing micropore, mesopore, and macropore reservoirs, where inter-region flow is activated in accordance with the state of fluid saturation.

Some related reference books provide useful information to obtain an initial understanding and general knowledge on the state of the art, which include those by Evans and Nicholson (1987), Bear and Corapcioglu (1987), Bear et al. (1993), and National Research Council (1996). Relevant review papers are by Sudicky and Huyakorn (1991), Wang (1991), and Wheatcraft and Cushman (1991).

1.1.2 Multiple Processes

The present review focuses on coupled deformation and flow processes in homogeneous and heterogeneous porous media, respectively.

1.1.2.1 Homogeneous Media

The transient flow and deformation behavior in a porous medium may result from changes in either the fluid pressure or total stress boundary conditions applied to the system. It is the admissibility of changes in total stress within the system that describes the essence of coupled deformation-dependent flow behavior within porous media and sets it apart from decoupled diffusive flow systems. Comprehensive coupling between stresses and pore pressures was first rationalized by Biot (1941) and later adopted in many applications to specific deformation flow systems (Ghaboussi and Wilson 1973; Zienkiewicz et al. 1977; Simon et al. 1984; Lewis and Schrefler 1987; Detournay and Cheng 1988).

The most influential papers on this subject were related to the concept of effective stress by Terzaghi (1923), and to a generalized three-dimensional theory of consolidation by Biot (1941). These contributions remain of undiminished importance, presently, and have been augmented by subsequent work. Skempton (1954) quantified the relationship between total stress and fluid pressure under undrained initial loading through the well-known parameters, A and B, credited as *Skempton pore pressure parameters*. The significant implication of the A and B parameters is in their ability to separate the effects of pore pressure from the ambient stress. Mandel (1953) identified a non-diffusive form of temporal pressure evolution in a two-dimensional configuration, providing a strong justification for the importance of poroelastic behavior in the deformation-flow process. Nur and Byerlee (1971) offered theoretical evidence that the effective stress law presented by Biot (1941) was more general and physically sensible than that initially proposed by Terzaghi (1923). Under specific conditions, Cryer (1963) presented analytical solutions for three-dimensional consolidation formulated by Biot (1941). Rice and Cleary (1976) proposed poroelastic solutions using stresses and pressure as primary unknowns, which are different from Biot's displacement based solutions. Based on micromechanical concepts, Carroll (1979), and later Thompson and Willis (1991), extended Biot's poroelasticity to include more general anisotropic media. Other work of note includes applications to consolidation (Schiffman et al. 1969; Ghaboussi and Wilson 1973), mixture theory (Crochet and Naghdi 1966), poroelastic parameters (Green and Wang 1986; Zimmerman et al. 1986; Kumpel 1991), poroelastic applications (Detournay and Cheng 1988) and poroelastic theory (McNamee and Gibson 1960). In addition, readers may be interested in pertinent expositions by Charlez (1991) and Coussy (1995) on theoretical aspects, by Lewis and Schrefler (1987) on numerical developments, and by Detournay and Cheng (1993) on general aspects.

1.1.2.2 Heterogeneous Media

In comparison, fluid flow in fractured poroelastic media has received less attention due to the complications in the mathematical formulation and in parametric determination. Elsworth (1993) extended the dual-porosity poroelastic conceptualization, initially proposed by Aifantis (1977, 1980), to coupled processes of energy and mass transport. Aifantis' proposal may be viewed as a natural extension of Biot's single-porosity poroelasticity (Biot 1941) with the combination of the dual-porosity fluid flow model proposed by Barenblatt et al. (1960). The mathematical basis for this extension can be traced to mixture theory (Crochet and Naghdi 1966; Atkin and Craine 1976) in which any material in a composite medium that shows its physical, thermal, hydraulic, and mechanical characteristics distinctly different from those of other intervening materials deserves separate description; this leads to individual governing equations in the total system. Practically, the necessity of this extension may be attributed to the inclusion of significant mechanical impacts on soil consolidation in fractured or heterogeneous media (Wilson and Aifantis 1982), on groundwater flow in fractured aquifers (Huyakorn et al. 1983), and, in particular, on petroleum production from naturally fractured reservoirs (Bai et al. 1993). Although analytical solutions are occasionally available for dual-porosity poroelastic systems, subject to simplified boundary and initial conditions (Bai et al. 1995a), numerical methods, preferably the finite element methods, appear to be the dominant tools adapted for any sensible utilizations (Khaled et al. 1984; Elsworth and Bai 1992; Bai and Elsworth 1994; Bai et al. 1995b). Compared with single-porosity poroelasticity (Biot 1941) and flow in dual-porosity media (Barenblatt et al. 1960), the development of dual-porosity poroelastic models has been restricted primarily due to the complications in the determination of parameters. Physical conceptualization and laboratory determination of these parameters is difficult, but may be determined from the methods of Wilson and Aifantis (1982), Bai et al. (1993), and Berryman and Wang (1995).

Analytical solutions for flow in deformation-coupled dual-porosity systems provide an important means of distilling the essential behavioral components in the response, albeit for simplified geometric representations of reality. The governing equations representing flow through fractured-porous media were initially proposed by Barenblatt et al. (1960). The mathematical model was further developed as a potential reservoir simulator by Warren and Root (1963). Analytical solutions for flow towards a single well are available for a variety of reservoir conditions. Streltsova-Adams (1978) evaluated many of the possible solutions based on the dual-porosity conceptualization. Raghavan et al. (1985) proposed several approaches for wells intercepting single discrete fractures. Moench (1984) assumed an interactive dual-porosity behavior as that involving fluid flows across a fracture skin. In terms of coupling fluid flow with solid deformation, and using Melan's (1940) solution for a central point dilatation in a semi-elastic space, Segall (1985) solved the coupled flow-deformation problem for a reservoir subjected to fluid extraction. Other advances that extend the traditional dual-porosity approach to encompass coupled processes include: partial or comprehensive coupling of fluid flow, solid deformation, and heat transfer (Bai et al. 1993; Bai and Roegiers 1994a), identifying local influences such as convective flow (Bai and Roegiers 1994b), and nonlinear flow near a well (Bai et al. 1994a) in a dual-porosity medium. Accompanying numerical advances include the development of a three-dimensional finite element model capable of evaluating coupled flow-deformation behavior in poroelastic dual-porosity media for a single-phase fluid (Bai et al. 1995b), triple-porosity media (Bai et al. 1997a), and for two-phase fluid flow (Bai et al. 1998b).

Further information on this subject can be obtained from Bai et al. (1995a, 1995b, 1996), Bai and Elsworth, (1993, 1994), Beskos and Aifantis (1986), Ghafouri and Lewis (1996), Hill and Aifantis (1980), Huyakorn et al. (1983), Valliappan and Khalili-Naghadeh (1990), Swenson and Beikmann (1992), and Sun and Sterling (1994). For nonisothermal conditions, suggested references include Abousleiman et al. (1996); Bai et al. (1995c, 1996) Li and Li (1992), Bai and Roegiers (1994b), and Nguyen and Selvadurai (1994).

1.1.3 Modeling Methodology

Modeling a real physical phenomenon involves solving a set of initial and boundary value problems. Existing analytical and numerical methodologies used in related subjects are briefly reviewed.

1.1.3.1 Analytical Methods

In general, analytical methods are the first choice in solving problems with simplified geometries and boundary and initial conditions, due primarily to their convenience and simplicity. Although analytical methods are typically restricted to solving geometrically simple problems containing few free parameters, they are still a dominant approach. Relatively popular analytical methods for subsurface hydrology are those using function transformations (e.g., Laplace, Fourier, and Hankel transforms).

Because it enables solution within Laplace space, where the time dimension degenerates to a single parameter, Laplace transforms are the most popular method to solve transient flow and transport problems (van Genuchten and Alves 1982; Javandel et al. 1984). It is also effective in tackling a limited suite of deformation problems such as viscoelasticity (Flügge 1967), and poro-viscoelasticity (Abousleiman et al. 1993). Using the combination of Laplace transforms and convolutional integrals, Ogata and Banks (1961) showed the difficulties in obtaining closed form solutions to a classical one-dimensional transport equation. Employing Laplace transforms, Tang et al. (1981) provided a closed form solution for a pseudo two-dimensional transport scenario of transport in orthogonal fractures by solving two decoupled transport equations. In the pioneering work on fluid flow by Warren and Root (1963), and on solute transport by Coats and Smith (1964), Laplace transforms showed significant restrictions in achieving closed form solutions. After Stehfest (1970) developed a straightforward numerical inversion procedure for Laplace transforms, the method has been increasingly utilized in semi-analytical approaches. Following this development, a series of inversion methods have been developed, including that by Crump (1976). Davies and Martin (1979) presented a comparison of several numerical inversion techniques. The double Laplace transform has been used for solving pseudo three-dimensional transport problems (Johns and Roberts 1991), and true two-dimensional transport problems of dual-porosity media (Bai, et al. 1999a). Hybrid methods related to Laplace-Fourier transforms can be found in Piessens and Huysmans (1984), Johns and Roberts (1991), and Ichikawa (1982). With the purpose of reducing numerical instability in time discretization, Laplace transform techniques have also been used in conjunction with numerical methods, such as the finite element method, for flow in homogeneous media (Sudicky 1989), and in fractured porous media (Sudicky and McLaren 1992).

Fourier transforms are a classic method to solve the initial and boundary value problems for physical domains of regular shape (e.g., for a 2-D rectangular geometry). Although solution by Fourier transforms is similar to the method of separation of variables, the former method is more general than the latter, since many functional unknowns are not theoretically separable. Unlike Laplace transforms, where integration is required in the closed form solution, the solution by Fourier transform is in the form of a discrete summation, which significantly simplifies the analytical procedure, avoiding numerical inversion. The method of Fourier transforms has been considered as an effective approach in defining classic solutions, such as those for heat transfer by Carslaw and Jaeger (1959), and thermoelasticity by Nowacki (1962). Despite limitations of the problem complexity, Fourier transforms can be used to solve coupled system of equations, and more complex coupled processes, such as dual-porosity poroelastic cases (Wilson and Aifantis 1982), and porothermoelastic consolidation (Bai and Abousleiman 1997).

Hankel transforms are often referred to as Bessel functions or Bessel transforms. This method has been dominant for solving problems within polar or cylindrical coordinates. For this reason, it has been predominantly applied to problems of reservoir engineering and hydrogeology, related to the performance of, or production from, pumped wells. Similar to Fourier transforms, the solution of Hankel transforms entails a discrete summation. Unlike Fourier transforms, however, the solution of Hankel transforms requires the solution of an additional eigenvalue problem (Wilson and Aifantis 1982; Bai et al. 1994b). The efficiency of Hankel transforms can be improved by adopting an algorithm that requires a minimum calculation of Bessel functions (Bai et al. 1994a). Even though Hankel transformation is more complicated than Fourier transformation, the solutions can be placed in a systematic format, circumventing the difficulties of comprehending the solution procedure.

Alternative analytical methods may include, but are not limited to, complex variables (O'Neil 1983), integral techniques (Kanwal 1971), and differential operators (Bai, et al. 1997a).

1.1.3.2 Numerical Methods

With the advent of increasingly powerful computers in the late 1950s, numerical methods have developed and risen markedly in application. If the purpose of an analytical method is to conceptualize a physical phenomenon, the aim of a numerical method is to replicate this phenomenon for the most realistic conditions imaginable. Three principal numerical methods are available for continuum mechanics: the boundary element method, the finite difference method, and the finite element method.

The boundary element method (BEM) is an efficient method of representing linear problems in infinite or semi-infinite media since the discretization is applied only along the bounding contour (Banerjee and Butterfield 1981). Boundary element methods efficiently reduce computational time by downgrading the dimensional space of the solution by one order, without changing the geometric configuration. The methods developed by Crouch and Starfield (1983), i.e., the displacement discontinuity method and the fictitious stress method, were popular due to their simplicity. Theoretical development of the BEM was summarized, for example, by Banerjee (1976), with practical applications discussed by Brebbia (1980). To enable the BEM to treat heterogeneities such as fractures, the hybrid models, combining the boundary element method with the finite element method, were also popular (Elsworth 1987). Further development of the BEM has been hindered by its inability to straightforwardly accommodate material nonlinearities and heterogeneities. Although nonlinear and heterogeneous behavior may be accommodated, interior meshing is required, dispelling some of the most attractive attributes of the method, i.e., the boundary-only meshing and solution requirement.

The finite difference method (FDM) is a direct discretization technique that breaks the continuous differential equations into amenable segments. This method is very popular due to its simplicity and diversity, and has not lost its popularity since inception. Many books document the method, including those by Richtmyer and Morton (1967), Smith (1978), Mitchell and Griffiths (1980), and Lapidus and Pinder (1982). With the rapid advance of computer technology, the FDM has grown in popularity, especially in its application to subsurface hydrology. Among numerous FDM software packages, the MODFLOW (3-D), algorithm is widely used (McDonald and Harbaugh 1988). The complexities of multiphase flow behavior are readily amenable to FDM techniques, as the formation and integration of the governing equations has a relatively low computational overhead. This low overhead is one of the reasons that many finite element methods use finite differences to pass through time (Huyakorn and Pinder 1983).

The finite element method (FEM) enables the true complex geometry of the physical system to be readily represented through the meshing of a mosaic of arbitrarily shaped elements. The FEM is perhaps the most powerful numerical means among all competing methodologies due to its ability to: (a) treat complex nonlinearities and material heterogeneities, (b) accommodate flexible geometric configurations, and (c) reach desired computational accuracy with the current computer technology. In fact, the FEM is so powerful, it has become the dominant numerical tool (Zienkiewicz 1983). The development of finite element methods has made them effective numerical tools capable of taking advantage of the typical sparseness of the system matrices, and of accommodating broad application to engineering problems (Akin 1982), to problems of solid mechanics (Smith and Griffiths 1988), and to subsurface flow (Huyakorn and Pinder 1983; Istok 1989). Finite element methods may also be combined with semi-analytical approaches to give time-continuous representation (Sudicky 1989) and approximation of infinite elements (Simoni and Schrefler 1987). Attempts have also been made to minimize the overhead in numerical integration and increase the order of interpolation using so-called collocation strategy (Lapidus and Pinder 1982), and to maintain consistent functional relationships representing poroelastic behavior using non-conforming elements (Zienkiewicz 1983), or higher order elements (Bai et al. 1999b). FEM should continue its dominance into the foreseeable future.

1.2 CONCEPTUAL PRELIMINARIES

The following provides some basic concepts, defines certain popular terminologies, and underlines necessary assumptions.

1.2.1 Concepts and Assumptions

1.2.1.1 Concepts

In a general sense, porous media comprise an assemblage of particles of various sizes and shapes, which form a "solid skeleton" incorporating a fluid-filled void structure. The porous medium can be envisaged as a mixture of solid grains and fluids. In view of the state of saturation, the porous rock may be considered as comprising:

- single-phase when in a dry condition
- two-phase when saturated
- multi-phase when partially saturated by each individual phase.

While the behavior of a porous medium in the first state has been described using the theory of elasticity, study in the second state has received less attention, although it has become increasingly important in the past few decades from knowledge of behavior using the theory of poroelasticity (Biot 1941). The basic difference between geologic media in the last two states can be attributed to the difference between linear and nonlinear formulations and treatments. Although the constitutive relations for nonlinear behavior are more general and less restrictive, difficulties of obtaining convergent and unique solutions make this approach more problematic than using linear approximations. As a result, applications of linear poroelastic theory dominate the current research on coupled flow-deformation systems.

The response of porous media involving multiple saturating fluids is relevant to the study of unsaturated soils and petroleum reservoirs, and is of increasing practical interest. Approaches to this difficult problem have principally involved application of general mixture theory (Crochet and Naghdi 1966). The mixture may be viewed as a superposition of a number of single-continua, each following its own motion. In addition, at any time, each position in the mixture is occupied simultaneously by several different constituents, each possessing particular characteristics. The theory of mixtures was originally developed as a thermodynamic framework to describe thermomechanical behaviors of materials consisting of more than one constituent (Atkin and Craine 1976). The theory was extended to fluid flow in porous media which was viewed as a composite-substance (Crochet and Naghdi 1966). To rationalize the behaviors of a multiple substance such as a fractured porous medium, Aifantis (1977, 1980) proposed a multi-porosity theory based on the theory of mixtures, declaring that any medium that exhibited finite discontinuities in the porosity field may be considered to possess a multi-porosity behavior.

It is important to understand conventions in the use of the term "phase." The term "multi-phase," in this work, represents a mixture of multiple fluids having different fluid characteristics, in addition to being immiscible. This differs from an alternative usage where "phase" may be defined as "state;" e.g., fluid and solid phases or states are equivalent. It is technically permissible to label a poroelastic medium as a "two-phase" medium. From a more general standpoint, a porous medium embedded with natural fractures can also be considered a "two-phase" medium due to the significantly different physical properties of the porous matrix and fractures. Although the terminology can be clarified through careful definition, the complexities involving multiple, as well as parallel, coupled phenomena and processes in flow-deformation systems must be fully comprehended. For convenience, in this work, the term "phase" is not reserved to merely represent multiple fluid components.

All of the systems described in this work contain "composite" materials; as a result, they may be treated as discontinua or as continua, depending on appropriate length and time scales. For appropriate length and time scales, behavior may be accommodated as a continuum, utilizing the appropriate tenets of mixture theory. In other words, the constituents behave collectively as a continuum. This "continuum" assumption is independent of the selected systems or coordinates, and is acceptable at both local and global levels.

1.2.1.2 Assumption

The following assumptions are implied in the formulations, descriptions, and analyses presented in this book:

- The concept of a representative elementary volume (REV)(Bear 1972) is applied. In terms of scales, the REV is substantially smaller than the studied domain, but significantly larger than the microscopic pore scale. Constitutive relations are established at the microscopic level.
- All fluids are Newtonian. Flow is laminar and linear where Darcy's flow velocity is applicable.
- Both rock deformation and fluid compressibility are assumed to be sufficiently small to maintain linear constitutive relationships (e.g., Hooke's law and Darcy's law), linear momentum and mass conservation (e.g., neglecting higher order nonlinear terms), and the validity of superposition. Fick's law is valid in diffusive transport; Fourier's law is valid in thermal transfer.
- Fractures and porous matrix blocks are treated as two distinct media. Any activities such as interactive flow and transport between these two media are viewed as internal.
- When discussion is confined to the classic dual-porosity approach, fluid flow in fractures and in the porous matrix blocks are considered as "separate" events, linked only by the leakage terms characterizing the interporosity flow as a result of the pressure difference between the two media. This mathematically "separate" but physically "overlapping" system represents the classic dual-porosity conceptualization first proposed by Barenblatt et al. (1960).
- As a result of this "separate" rule, flow parameters are assessed on an individual basis. For example, fracture porosity and matrix porosity are defined as the void fracture volume and void pore volume versus the total volume, respectively. In each flow equation, the porosity represents only a fraction of the void volume instead of the total void volume. With respect to corresponding volumes, matrix porosity should be greater than fracture porosity.
- Porous matrix is in general isotropic. However, anisotropic flow properties are permitted (e.g., allowing an anisotropic permeability tensor). Because flow in fractures is more significant than interstitial flow in the matrix pores, fracture permeability is typically larger than matrix permeability.

- There are in general two types of interporosity flow between the fractures and the porous matrix, i.e., transient flow and quasi-steady flow. The latter approach is adopted in the present descriptions.

- The impact of fluid pressures on solid equilibrium is incorporated in a lumped fashion, envisioned as separate seepage forces acting on the solid grains. Volumetric strain rate changes affect flux variations for both fractures and porous matrix blocks.

- In the finite element method, one element may comprise numerous fractures and matrix blocks of predetermined fracture and matrix characteristics that may differ from those of other elements. In addition to other degrees of freedom such as displacements, the fracture and matrix pressures, averaged within the element for fractures and for porous matrix blocks, are both designated at each nodal point. This "two-pressures at one-point" scenario seems unnatural, but is a typical, and eminently acceptable result of the averaging process.

1.2.2 Fundamental Formulations

Fundamental governing equations are examined from their mathematical relationships and physical implications.

1.2.2.1 Theoretical Aspects

Based on the mechanics of viscous flow, the Navier-Stokes equations are a set of equations accommodating conservation of mass and momentum, respectively, for incompressible fluid as

$$\nabla \cdot \mathbf{v} = 0 \tag{1.1}$$

$$\frac{\partial \mathbf{v}}{\partial t} + (\mathbf{v} \cdot \vec{\nabla})\mathbf{v} = -\frac{1}{\rho}\vec{\nabla} p + \frac{\mu}{\rho}\nabla^2 \mathbf{v} \tag{1.2}$$

where $\mathbf{v}$ and ρ are the general velocity and density, p is the pressure, μ is dynamic viscosity, and ∇ is the *del* operator, $\partial/\partial x_i$; these quantities may be applicable for both fluid and solid. t is the time. The derivation of Eqs. (1.1) and (1.2) will be presented in Chapter 3.

It should be noted that the Navier-Stokes equations are not restricted to any particular configuration of a flow system, even though they are typically applied in characterizing plane flow or pipe flow (Poiseuille flow)(Bear 1972). In addition, since the velocity term is associated with both solid and fluid, it is the derivatives of the displacements that are related to the stresses and strains. Furthermore, no linearity is implied in the Navier-Stokes equations. As a result, the Navier-Stokes system should be considered as a general framework.

Using the nondimensionalization technique, Eq. (1.2) can be transformed to:

$$\frac{\partial \mathbf{v}^*}{\partial t} + (\mathbf{v}^* \cdot \vec{\nabla})\mathbf{v}^* = -\vec{\nabla} p^* + \frac{1}{Re}\nabla^2 \mathbf{v}^* \tag{1.3}$$

where the superscript $*$ indicates the mapped variable, which is frequently omitted for simplicity; Re is the well-known Reynolds number, defined as

$$Re = \frac{v_a l_a \rho}{\mu} \tag{1.4}$$

where v_a and l_a are the average (or typical) flow velocity and flow geometry (e.g., length for a 1-D case).

The implication of the Reynolds number is that it defines the rate and intensity of the viscous flow. If $Re \ll 1$, a slow flow is implied, which is frequently termed "Stokes flow." If $Re \gg 1$, Eq. (1.3) reduces to an Euler equation where the viscous term disappears. This viscous term can be recovered by introducing thin viscous boundary layers (Bear 1972). It has been observed that an increase in Reynolds number may lead to unstable flow (i.e., turbulent flow), which may be related to the chaotic behavior of the Navier-Stokes equations.

1.2.2.2 Practical Aspects

By imagining that viscous materials (typically fluids) flow through porous media as movement through a bundle of uniform cylindrical pipes, the Navier-Stokes equations may be applied to flow in idealized subsurface domains. If the radius of the pipe, r_p, is sufficiently small that the Reynolds number is maintained within the limit of laminar flow, the fluid flux in each pipe can be expressed by Poiseuille flow as

$$\mathbf{v}_p = \frac{r_p^4 \pi}{8\mu} |\vec{\nabla} p| \tag{1.5}$$

where $\vec{\nabla} p$ is the pressure gradient along the pipe.

With certain simplifying assumptions, Eq. (1.5) resembles the expression for the well-known Darcy's velocity, i.e.,

$$\mathbf{v}_d = \frac{k}{\mu} |\nabla p| \tag{1.6}$$

where k is the permeability.

It should be mentioned that the validity of Darcy's velocity is based purely on experimental evidence, not on theoretical derivations. If k and μ in Eq. (1.6) are linear parameters, Darcy's velocity is apparently linear. Substituting $\mathbf{v}_d$ in Eq. (1.6) into Eq. (1.1) of the Navier-Stokes equations, a Laplace equation results.

On average, a porous medium may be characterized by the porosity (i.e., ratio between pore volume to total volume). Utilizing the concept of porosity is the most significant feature that distinguishes the flow in porous media from the flow defined in classical fluid mechanics (e.g., flow in a pipe). Porosity is a physical parameter, similar to permeability in Darcy's law. It may be established based on physical intuition, rather than from theoretical derivation. Incorporating variable density and porosity, conservation of mass for subsurface hydrology can be modified from the Navier-Stokes equation (1.1) for a compressible fluid as

$$\nabla \cdot \left(\frac{k}{\mu} \rho n \vec{\nabla} p \right) = \frac{\partial(\rho n)}{\partial t} \tag{1.7}$$

where n is the porosity.

1.2.3 Definition of Heterogeneity and Anisotropy

By definition, homogeneity and heterogeneity are associated with spatial distributions and corresponding variations, while isotropy and anisotropy are related to directional distributions and corresponding changes of a coupled system, respectively. In terms of these spatial and directional characteristics, the following four groups of porous media may be identified:

- Type 1: homogeneous and isotropic
- Type 2: homogeneous and anisotropic
- Type 3: heterogeneous and isotropic
- Type 4: heterogeneous and anisotropic.

In existing models, the first three types of porous media have been represented by the following counterpart models:

- Model 1: single-porosity poroelasticity
- Model 2: anisotropic single-porosity poroelasticity
- Model 3: dual-porosity poroelasticity.

To the knowledge of the authors, a theoretical model for the fourth type of porous media, i.e., anisotropic dual-porosity poroelasticity, is not currently available.

The heterogeneity and anisotropy mentioned so far is general. With reference to various length scales, different heterogeneous and anisotropic characteristics of the fractured porous medium may co-exist. The following three types of domains and related scales are considered:

- Scale 1: macroscopic scale reservoir
- Scale 2: fractured blocks on the mesoscopic scale
- Scale 3: single fracture or porous block on the microscopic scale.

In modeling, all scales and dimensionalities should be considered, but with reference to different aspects. The governing equations, along with boundary and initial conditions, are defined macroscopically. The modeling parameters (e.g., fracture spacing, matrix permeability, etc.) are characterized using the REV rationale. The constitutive laws (e.g., Darcy's law, effective stress law, etc.) are established at the microscopic scale without referring to any specific domains, since a single fracture or pore may be too tiny to be spatially significant.

The features of the heterogeneities and anisotropies at all three scales can be further divided between those for matrix blocks and those for fractures. As a result, it becomes a complex issue to distinguish different heterogeneities and anisotropies at various scales for different media.

In view of heterogeneities, for example, micromechanical characteristics for the solid matrix may show anisotropic behavior at the grain-scale, but at the same time may be envisioned as globally isotropic. By the same token, macroscopic heterogeneity may be exhibited in fractured porous media where the fracture network may show random patterns or arbitrary densities in the simulated domain. Conversely, an individual fracture may be microscopically homogeneous due to uniform infilling or constant roughness. Similarly, the porous matrix may be macroscopically or mesoscopically homogeneous in the reservoir domain or in the REV, but microscopically heterogeneous as a result of variations in grain sizes or void structures.

With respect to the anisotropy, for example, natural fractures typically introduce macroscopic or mesoscopic anisotropies regardless of whether the fractures are uniformly distributed with equal apertures or not, because the fractures themselves impose directional variations of material properties in the reservoir or in the REV domain. This is the primary reason a fractured medium is generally envisioned as heterogeneous (Bai et al. 1993), even though the latter implies much broader types of materials. Conversely, anisotropic characteristics within a single fracture may be microscopically different from those either on the macroscopic or on the mesoscopic scales, depending on the structure and material properties inside the fracture. Similar concepts may be applicable for the porous matrix. Even though the microscopic anisotropy at the grain scale may contribute to the general macroscopic or mesoscopic anisotropies at the reservoir or the REV scales, the nature of these anisotropies can be substantially different from each other. Significant differences may be demonstrated between macro- and meso-scales, with the latter cases frequently being considered as quantities averaged from the former domain.

Despite these complexities and differences, it is certain that if all the fractures and matrix blocks are microscopically homogeneous and isotropic, the domain of interest is homogeneous and isotropic. The reverse may not be true; i.e., the domain may look macroscopically or mesoscopically homogeneous and isotropic, but the fractures and the matrix blocks themselves can be microscopically heterogeneous and anisotropic.

1.2.4 Definition of Coupled Process

"Coupled processes" are defined as events that occur simultaneously and provide feedback to other parallel processes. A variety of coupled processes are present in the subsurface. The focus of this treatise is on fluid flow through deformable porous media, where the fluid potential may impose additional force to affect the equilibrium of an elastic system, with a reciprocal consequence that expansive or contractile body strains may exert internal influences over the variations of fluid flux. This is a typical poromechanical phenomenon. If the porous medium is also naturally fractured, fluid flow is not only affected by the rock deformation within the matrix blocks and fractures, but also by inter-porosity mass exchange between the matrix pores (primary porosity) and the fractures (secondary porosity). These multiple coupled processes are referred to as dual-porosity poroelastic behavior of fractured porous media.

Coupled processes may be more general than this flow-deformation system. A fully conservative system will maintain momentum, mass, and energy balances throughout its temporal and spatial evolution. In this aspect, comprehensive coupling is achievable for a poroelastic system. Theoretically, other processes may be coupled with this poroelastic response, such as (a) chemical reactions, (b) viscous and plastic behavior of rock constituents, (c) heterogeneities such as fractures, (d) nonlinearities, and (e) multi-phase fluids. In reality, however, meaningful and practical analyses are confined to the processes associated with a certain number of parallel processes.

A poroelastic approach is a preferred method to accurately characterize coupled fluid flow processes within deformable porous media. Consolidation phenomena are well understood where the porous skeleton is subjected to external loads and consolidates progressively as pore pressures dissipate. Compared with uncoupled fluid flow and solid deformation, the poroelastic effects show strong interaction between flow and deformation. Slower flow induces smaller displacement, or larger deformation leads to higher pressure perturbation. This mechanically affected flow may exhibit some unusual characteristics, different from a classical diffusive flow, such as an initial increase of the pore pressure over the original value, known as the Mandel effect (Mandel 1953). The effective stress law, originally proposed by Terzaghi (1923), provides a framework for the poroelastic concept. Since the total stress may sometimes be considered as invariant (e.g., a constant external load), the superposition of the effective stress and pore pressure results in a self-adaptive environment, i.e., an increase of the effective stress incurs a corresponding reduction of pore pressure, and vice versa. This retroactive response is the essence of a coupled poroelastic model, where total stresses may change with both spatial location and time.

Once the virtue of the poroelastic approach is retained, justification for a dual-porosity poroelastic model can be made by simply verifying that dual-porosity is a good prototype. In this prototype, fluid flow in naturally fractured porous formations can be described by the flows in two interacting media with distinctly different spatial scales and properties, commonly linked by a transfer function. For poroelastic media embedded with natural fractures, using the concept of an equivalent porous medium by merely increasing the equivalent permeability is not sufficient to replicate the observed response. In fact, it is the internal flow between the more conductive fractures and the matrix blocks with larger storativity that yields a pressure drawdown with a variable slope, closely replicating field observations. This interporosity flow, depicted at the local level, is perhaps far more important than any other mechanisms in characterizing flow in heterogeneous media.

Rock masses may be described as aggregates of blocks bounded by interconnected or isolated fractures, adding secondary porosity to the original porous materials. In general, fluid flow within the matrix blocks is different from the flow in a fracture network. In the former, the matrix blocks contain pore space with similar dimensions in lengths and widths, in addition to highly tortuous flow patterns. In the latter, the fractures provide more continuous apertures with lengths far in excess of their widths. If several closely spaced sets of joints are present, they may form a directly interconnected flow system. The combination of radial intergranular flow in the matrix blocks with linear flow in fractures represents a typical flow pattern within naturally fractured reservoirs.

1.3 NOTATION PRELIMINARIES

This section provides basic knowledge of tensor notation and the sign conventions used in mechanics and geosciences.

1.3.1 Tensor

Tensor notation is adopted throughout this book as a compact method of expressing complex mathematical expressions. A tensor is a quantity that describes a physical state or a physical phenomenon and is invariant, i.e, it remains unchanged when the frame of reference within which the quantity is defined is changed.

1.3.1.1 Tensor Notation

Scalar

In Cartesian coordinates, if the value of a quantity at a point in space can be described by a single number, the quantity is a scalar or a tensor of zero order.

Vector

A vector is a first order tensor. Consider a vector $\vec{x}$ whose components in a system of axes OX_i $(i = 1, 2, 3)$ are x_i. In a new system of rectangular axes $OX_i^{'}$, the components of $\vec{x}$ are given by $x_i^{'}$ with

$$\begin{cases} x_1^{'} = l_{11}x_1 + l_{12}x_2 + l_{13}x_3 \\ x_2^{'} = l_{21}x_1 + l_{22}x_2 + l_{23}x_3 \\ x_3^{'} = l_{31}x_1 + l_{32}x_2 + l_{33}x_3 \end{cases} \tag{1.8}$$

or

$$x_i^{'} = l_{ij}x_i(i, j = 1, 2, 3) \tag{1.9}$$

where l_{ij} are the components of the tensor and are the direction cosines of the axes of the new system with respect to the old one. Notice that summation signs, $\sum$, are neglected.

Second Order Tensor

A second order tensor $(\tilde{w} = w_{ij})$ is constructed by the tensor product $(\otimes)$ of two three-dimensional vectors. For instance,

$$\tilde{w} = u_i^{'} \otimes v_j^{'} = (l_{ik}u_k) \otimes (l_{jm}v_m) = l_{ik}l_{jm}\ u_k \otimes v_m \tag{1.10}$$

or, in rectangular Cartesian coordinates,

$$\tilde{w} = w_{ij}\ e_k \otimes e_m(k, m = 1, 2, 3) \tag{1.11}$$

where $e_k \otimes e_m$ is called a dyad. It is more convenient to write this as $\bar{\delta}_k\bar{\delta}_m$. Therefore, a second order tensor is the sum of nine dyadic components $\bar{\delta}_k\bar{\delta}_m$ which is each multiplied by its respective element w_{ij}. In three dimensions, for example,

$$e_2 \otimes e_3 = \bar{\delta}_2\bar{\delta}_3 = \begin{bmatrix} 0 & 0 & 0 \\ 0 & 0 & 1 \\ 0 & 0 & 0 \end{bmatrix}. \tag{1.12}$$

Higher Order Tensors

The general form in 3-D space can be expressed as

$$\tilde{t} = t_{i_1 i_2 \ldots\ldots i_n}\ e_{i_1} \otimes e_{i_2} \otimes \ldots\ldots \otimes e_{i_n} = t_{i_1 i_2 \ldots\ldots i_n}\ \bar{\delta}_{i_1}\bar{\delta}_{i_2}\ldots\ldots\bar{\delta}_{i_n} \tag{1.13}$$

For convenience the $e_{i_1} \otimes e_{i_2} \otimes \ldots\ldots \otimes e_{i_n}$ is omitted. So $\tilde{t} = t_{i_1 i_2 \ldots\ldots i_n}$.

In the following, all physical quantities overbarred with an *arrow* are *vectors*. For example $\vec{v}$ is the velocity vector. All quantities without an overbarred arrow, with only one lower case subscript, are vector components; e.g., v_i is a component of $\vec{v}$. "$\nabla\cdot$" denotes the divergence operator, and the gradient operator is defined as "$\vec{\nabla}$".

Quantities overbarred with a *hat* are tensors. Specifically, lower-case letters are used for second order tensors and upper-case letters are used for fourth order tensors (e.g., $\tilde{C}$ is a fourth order tensor whereas $\tilde{c}$ is a second order tensor). Any lower or upper case letter without hat or arrow but with two or four lower-case subscripts, is the element of the tensor corresponding to the same letter with a hat (e.g., M_{ijkl} and ε_{ij} are the components of $\tilde{M}$ and $\tilde{\varepsilon}$, respectively).

1.3.1.2 Tensor Operation

Following directly from the theory of elasticity, all tensors are of *even* order; the components always result in a *square matrix.* For example, the fourth order compliance tensor has 81 components (9×9 matrix).

Multiplication

Multiplication of tensors always results in tensors of higher order. The final order is the sum of the orders of the tensors being multiplied. Tensor multiplication is denoted by the symbol $\otimes$. For example, the multiplication of a second order tensor $\tilde{\chi}$ and a fourth order tensor $\tilde{\tau}$ gives a sixth order tensor

$$\tilde{\theta} = \tilde{\tau} \otimes \tilde{\chi} = \tau_{ijkl}\chi_{mn}\ \bar{\delta}_i\bar{\delta}_j\bar{\delta}_k\bar{\delta}_l \otimes \bar{\delta}_m\bar{\delta}_n \tag{1.14}$$

Multiplication "$\otimes$" is sometimes referred to as the outer product.

Contraction

Contraction of a tensor sets two indices as equal. This automatically reduces the order of the tensor by 2. Therefore it is only applicable to tensors of order greater than 1. Repeated contraction can reduce a higher order tensor to a lower order tensor or even a scalar, if the order is even.

The product of two tensors followed by double contraction is known as the *double inner product.* It is denoted by ":"; e.g.,

$$\tilde{\sigma} = \tilde{M}{:}\tilde{\varepsilon} \tag{1.15}$$

where $\tilde{\sigma}$ and $\tilde{\varepsilon}$ are stress and strain tensors, and $\tilde{M}$ is the stiffness tensor.

In Eq. (1.14), contracting twice means putting indices $j = l$ and $i = k$. This results in

$$\tilde{\theta}_{ij} = \tau_{ijkl}\chi_{kl}\ \bar{\delta}_i\bar{\delta}_j\bar{\delta}_k\bar{\delta}_l \tag{1.16}$$

Multiplication of τ_{ijkl} and χ_{kl} can be carried out as the multiplication of two matrices. τ_{ijkl} and χ_{kl} are represented by (9×9) and (9×1) matrices, respectively.

In fluid flow, determination of the Darcy velocity ($\vec{v}$) by multiplication of the permeability tensor (second order) and pressure gradient, $\vec{J}$, (which is a vector, or a first order tensor) is carried out by the *inner contraction*. The order of the product is three, and is reduced by two to generate the velocity vector (first order tensor). This product is denoted by "$\bullet$"; i.e.,

$$\vec{v} = -\frac{1}{\mu}\tilde{k}\bullet\vec{J} \tag{1.17}$$

where $\tilde{k}$ is the permeability tensor, μ is the dynamic viscosity of the fluid, and $\vec{J}$ is the vector of pressure gradients.

Summation Conventions

The present analysis assumes that the Greek and Latin subscripts have the values of 1, 2 and 1, 2, 3, respectively. A subscripted comma denotes differentiation, and summation is implied over the repeated Latin subscripts.

1.3.2 Sign Convention

The main conventions are related to the directions of forces and tensor-matrix notations.

1.3.2.1 Sign Conventions in Solid Mechanics and the Geosciences

The normal assumption in solid mechanics is that tension is positive and compression is negative. In geological applications, however, the compressive state of rock masses makes it more convenient to define compressive stresses and strains as positive. The convention of extension as positive is followed in this text.

For clarity, a brief comparison between the notations of solid mechanics and the geosciences is exhibited in the following.

Constitutive Law

In solid mechanics, the effective stress law (e.g., Biot 1941) is expressed as

$$\Delta\tilde{\sigma}^t = \Delta\tilde{\sigma}^e - \tilde{\alpha}\Delta p \tag{1.18}$$

where $\tilde{\sigma}^t$ and $\tilde{\sigma}^e$ are the total and the effective stress tensors, respectively; $\tilde{\alpha}$ is Biot's coefficient; and p is the fluid pressure.

In geomechanics, the effective stress law is usually written as

$$\Delta\tilde{\sigma}^t = \Delta\tilde{\sigma}^e + \tilde{\alpha}\Delta p \tag{1.19}$$

where compression is positive.

Equation of Solid Equilibrium

In solid mechanics, with the omission of gravitational and inertial forces, conservation of momentum defines the governing equation of solid equilibrium that may be written as

$$\vec{\nabla}\cdot(\tilde{\sigma}^e - \tilde{\alpha}p) = 0 \tag{1.20}$$

Equilibrium is defined in terms of total stresses, as are boundary and initial conditions. Deformation and failure are defined in terms of effective stresses, requiring that pore pressure distributions are evaluated. Conversely, in the geological sign convention, the equilibrium equation is defined as

$$\vec{\nabla}\cdot(\tilde{\sigma}^e + \tilde{\alpha}p) = 0. \tag{1.21}$$

Equation of Fluid Flow

Conservation of mass is typically applied as a secondary constraint to define the evolution of the effective stress and pore fluid pressure field. Where coupled with fluid sources that result from the contraction of porosity and the compressibility of the fluid, the fluid continuity equation can be expressed as

$$\vec{\nabla}\cdot(\frac{\tilde{k}}{\mu}\cdot\vec{\nabla}p) = c^*\frac{\partial p}{\partial t} + \tilde{\alpha}:\frac{\partial\tilde{\varepsilon}_v}{\partial t} \tag{1.22}$$

where c^* is the lumped compressibility and $\tilde{\varepsilon}_v$ is the volumetric strain tensor. In geomechanics, this equation is changed to

$$\vec{\nabla}\cdot(\frac{\tilde{k}}{\mu}\cdot\vec{\nabla}p) = c^*\frac{\partial p}{\partial t} - \tilde{\alpha}:\frac{\partial\tilde{\varepsilon}_v}{\partial t}. \tag{1.23}$$

1.3.2.2 Tensor-Matrix Notation

By definition, a tensor is a spatial parameter that defines directional properties, in contrast to a matrix, which has no similar requirement. For convenience, and its inherent compactness, tensor notation is primarily used in the following mathematical development. Matrix notation is used in the description of numerical operations. Where tensors can be alternatively expressed as a matrix, bold face symbols denote a tensor (i.e., a tensor in a vector notation), or a matrix.

Chapter 2

DEFORMATION

2.1 INTRODUCTION

Equilibrium requires that the sum of forces acting on a body is null. These include internal and external body forces and the action of external forces, defined through Newton's second law of motion, as

$$F = ma \tag{2.1}$$

where F represents forces acting on the body, inclusive of external loads, induced stresses, and gravitational or other static loads, and ma is the inertial component. This component comprises mass, m, and acceleration, a. Where the noninertial quasi-static system is considered, the right-hand side of Eq. (2.1) is simply zero.

Deformation of a body results in strains, defined as changes in the relative position of points in the undeformed and deformed states, relative to a normalizing length and the original separation of the two points. Strains may be partitioned relative to their causal mechanisms. Total strain may be the summation of the strains generated from different sources, which may be mathematically expressed as

$$\varepsilon = \varepsilon_e + \varepsilon_p + \varepsilon_t + \varepsilon_o \tag{2.2}$$

where the subscripts e, p, t, and o denote the sources of elastic, fluid, thermal, and other strains, respectively.

Whenever certain deformation processes become dominant, the other mechanisms are typically neglected. If the potential role of all contributing sources is ill-defined, inclusion of all reasonable sources is warranted. Both fundamental and coupled processes of deformation are investigated in the following.

2.2 MATHEMATICAL FORMULATION

Constructing the equations defining the solid equilibrium of the system using appropriate constitutive laws and physical relations, as well as compatibility conditions if necessary, is the primary task of the mathematical formulation.

2.2.1 Homogeneous Media

Although the basic concepts applied in structural mechanics to the manufacturing and construction industries and to the geosciences are similar, important differences remain. Where the system is fabricated, knowledge of material properties and structures are considerably better defined than in geologic applications where the data uncertainties are relatively large. This results in considerable differences in the application of modeling, requiring that geologic applications put greater emphasis on sensitivity analyses and on the selected use of judgment in interpreting results. In some instances, the role of uncertainty may be directly included in the analysis. To emphasize these differences, Table 2.1 contrasts the incremental differences.

Table 2.1. Comparison of the different issues involved in the modeling of fabricated and geologic systems.

Issue	Fabricated System	Geologic System
Governing theory	Elasticity	Elasticity
Applications	Manufacturing and fabrication	Subsurface engineering
Materials	Fabricated	Natural
Material behavior	Well constrained - deterministic	Poorly constrained - stochastic
Heterogeneity	Less prevalent	Ubiquitous
Solution precision	Relatively precise	Scoping and sensitivity studies

2.2.1.1 Solution Procedure Using the Theory of Elasticity

For a non-rotational quasi-static system, the simple equation of solid equilibrium can be expressed from Eq. (2.1) as

$$\sigma_{ij,j} = 0 \tag{2.3}$$

where σ_{ij} is the stress, which is a tensor. Eq. (2.3) represents three equations. Note that differentiation of the stress tensor is implied by the presence of a comma in the subscripted terms.

Eq. (2.3) can be transformed to the displacement based formulation as

$$Gu_{i,jj} + (\lambda + G)u_{k,ki} = 0 \tag{2.4}$$

where u is the displacement, λ is the *Lamé* constant, and G is the shear modulus, expressed as

$$G = \frac{E}{2(1+\nu)} \tag{2.5}$$

where E is the elastic modulus and ν is Poisson ratio. Eq. (2.4) represents three equations in the three-dimensional case.

The constitutive law, which relates stress components to strain components, may be defined as

$$\sigma_{ij} = 2G\varepsilon_{ij} + \lambda\varepsilon_{kk}\delta_{ij} \tag{2.6}$$

where ε_{ij} represents the individual components of the strain tensor, with the summation involved in ε_{kk} representing volumetric strain. The six equations represented in Eq. (2.6) may be inverted to yield

$$\varepsilon_{ij} = \frac{1+\nu}{E}\sigma_{ij} - \frac{\nu}{E}\sigma_{kk}\delta_{ij} \tag{2.7}$$

where σ_{kk} is the total stress. Eq. (2.7) also represents six equations.

Strains and displacements are correlated through the six relations identified as

$$\varepsilon_{ij} = \frac{1}{2}(u_{i,j} + u_{j,i}). \tag{2.8}$$

Eqs. (2.6), (2.7), and (2.8) link Eqs. (2.3) and (2.4) together. If the primary unknowns are stresses rather than displacements, then additional compatibility conditions must be satisfied. These are defined as

$$\nabla^2[(1+\nu)\sigma_{ij} - \nu\sigma_{kk}\delta_{ij}] + \frac{\partial^2\sigma_{kk}}{\partial x_i \partial x_j} = 0. \tag{2.9}$$

This now represents the complete suite of requirements necessary to forge a solution. Any solution that satisfies the equilibrium conditions, compatibility relations, and constitutive laws, and also meets the prescribed initial and boundary conditions, is an admissible solution.

2.2.1.2 Coupled Process

The previously defined elastic solution procedure can be transformed to solve coupled problems, such as those involving interstitial fluids (poromechanics) or thermal stresses. For poromechanical problems, for example, the total stress tensor is decomposed into the components representing intergranular (effective) stresses and fluid pressures. Consequently, substituting total stress σ in Eq. (2.3) with the effective stress law, yields the required relationship. It should be noted that σ is different from σ_{kk}, where the former is related to the Terzaghi or Biot effective stress laws (Terzaghi 1923; Biot 1941), and the latter is merely the summation of normal stresses. As a result, the coupled equilibrium equation becomes

$$\sigma_{ij,j} - \alpha p_{,i} = 0. \tag{2.10}$$

Where extension is positive, it is important to note that the difference between Eqs. (2.3) and (2.10) is that σ_{ij} is the total stress in the former, but the effective stress in the latter. The superscript in the latter is often omitted for simplicity.

Substituting the strain-stress relationship (2.7) and the strain-displacement relationship (2.8) into Eq. (2.10), the displacement based poroelastic formulation of the equilibrium equations results as

$$Gu_{i,jj} + (\lambda + G)u_{k,ki} - \alpha p_{,i} = 0. \tag{2.11}$$

An average fluid pressure can be defined for poromechanical problems involving multiple fluid phases. For a system with a wetting and a non-wetting phase, the average of the two may be defined similar to Bishop and Blight (1963) as

$$p = S_w p_w + S_n p_n \tag{2.12}$$

where S is the saturation, and the subscripts w and n represent wetting and non-wetting fluid phases, respectively.

2.2.2 Heterogeneous Media

Continuum or discontinuum methods may be applied in the analysis of heterogeneous media, including that of porous media intersected by fractures. In this work, only continuum approaches will be discussed, with these requiring the use of some homogenization technique. Of relevance is whether the mechanical properties are lumped together, and whether the fluid pressures are similarly homogenized. In the latter, the methods differ in the mode of fluid coupling, where the effective stress laws are either combined into a single relation, or separate effective stress laws are provided for matrix and for fractures.

2.2.2.1 Single Effective Stress Law

Using a single effective stress law, initially proposed by Wilson and Aifantis (1982), is the most popular method in the current literature. In actuality, this method does not consider the effect of fracture deformation in general, but incorporates the impact of fluid pressure on the change of the matrix deformation instead.

Using tensor notation, the effective stress law for a dual-porosity poroelastic medium may be written as (Wilson and Aifantis 1982)

$$\tilde{\sigma} = \tilde{\sigma}^e - (\tilde{\alpha}_1 p_1 + \tilde{\alpha}_2 p_2) \tag{2.13}$$

where subscripts 1 and 2, represent the matrix and fractures, respectively; $\tilde{\sigma}$ and $\tilde{\sigma}^e$ are the total and effective stress tensors; p is the fluid pressure, $\tilde{\alpha}$ is the pressure ratio factor tensor, compatible with Biot's coefficient (Biot 1941).

Substituting this single effective stress law into the equilibrium equation (2.10), and converting stresses (omitting the superscript e when necessary) to strains through the stress-strain relation (2.7), the equation of equilibrium for the dual-porosity poroelastic medium may be written as

$$\nabla \cdot (\tilde{D} : \tilde{\varepsilon} - \tilde{\alpha}_1 p_1 - \tilde{\alpha}_2 p_2) = 0 \tag{2.14}$$

where $\tilde{D}$ is the elastic stiffness tensor and $\tilde{\varepsilon}$ is the strain tensor.

For a linear poroelastic system and for isotropic media, with a further substitution of the strain-displacement relation (2.8), Eq. (2.14) can be simplified as

$$G \frac{\partial^2 u_i}{\partial x_j \partial x_j} + (\lambda + G) \frac{\partial^2 u_j}{\partial x_i \partial x_j} - \alpha_1 \frac{\partial p_1}{\partial x_i} - \alpha_2 \frac{\partial p_2}{\partial x_i} = 0. \tag{2.15}$$

Another explanation of Eq. (2.15) is that fluid pressures from both matrix and fractures are considered as additional seepage forces acting additively to define the equilibrium of the system.

2.2.2.2 Double Effective Stress Law

The advantage of using a single effective stress law, shown in Eq. (2.15), is that it provides a simplified expression for the lumped effects of matrix and fracture deformations. This is especially useful in defining the effects of deformation on fluid flow since the lumped effects are naturally incorporated, as will be discussed in Chapter 3. However, it should be noted that the deformations of the matrix blocks and the fractures are not distinguished in Eq. (2.15). In fact, as Lewis and Ghafouri (1997) acknowledged, fracture deformation is omitted in Eq. (2.15). In other words, the elastic parameters (e.g., λ and G) and associated deformations are related to the matrix only (Bai 1999). Although the elastic coefficients λ and G may also represent the lumped properties of the fractures, the unique modes of deformation of the fractures are not incorporated, where fracture deformations typically dominate the overall response. This physical difficulty may be circumvented by Elsworth and Bai (1992) through an alternative formulation where the deformations between the matrix and the fractures are separately identified, through the use of component parameters. Specifically, different from the assumption in Eq. (2.13), separate effective stress laws for the two media are assumed as follows:

$$\tilde{\sigma}_1^e = \tilde{\sigma}_1 + \tilde{\alpha}_1 p_1 \tag{2.16}$$

$$\tilde{\sigma}_2^e = \tilde{\sigma}_2 + \tilde{\alpha}_2 p_2. \tag{2.17}$$

The basic principles of subsequent derivations are to maintain stress equilibrium at the local level; i.e.

$$\tilde{\sigma} = \tilde{\sigma}_1 = \tilde{\sigma}_2, \tag{2.18}$$

while strains are lumped from those of the two media as

$$\tilde{\varepsilon} = \tilde{\varepsilon}_1 + \tilde{\varepsilon}_2. \tag{2.19}$$

Unlike in the majority of the dual-porosity poroelastic models, in which fracture deformation is omitted, these principles appear to be physically sensible since they offer individually defined elastic parameters for the matrix and the fractures by retaining separate deformations. Separate constitutive relationships can be defined as

$$\tilde{\sigma}_1^e = \tilde{D}_1 : \tilde{\varepsilon}_1 \tag{2.20}$$

$$\tilde{\sigma}_2^e = \tilde{D}_2 : \tilde{\varepsilon}_2 \tag{2.21}$$

where the elastic stiffness tensor $\tilde{D}$ can be related to its inverse, i.e., compliance tensor $\tilde{C}$; or, $\tilde{D} = \tilde{C}^{-1}$. As a result,

$$\tilde{\varepsilon}_1 = \tilde{C}_1 : \tilde{\sigma}_1^e \tag{2.22}$$

$$\tilde{\varepsilon}_2 = \tilde{C}_2 : \tilde{\sigma}_2^e. \tag{2.23}$$

Substituting Eqs. (2.16) and (2.17) into (2.22) and (2.23) results in

$$\tilde{\varepsilon}_1 = \tilde{C}_1 : (\tilde{\sigma}_1 + \tilde{\alpha}_1 p_1) \tag{2.24}$$

$$\tilde{\varepsilon}_2 = \tilde{C}_2 : (\tilde{\sigma}_2 + \tilde{\alpha}_2 p_2). \tag{2.25}$$

From Eqs. (2.18) and (2.19), the total strain can be expressed as

$$\tilde{\varepsilon} = (\tilde{C}_1 + \tilde{C}_2) : \tilde{\sigma} + (\tilde{C}_1 : \tilde{\alpha}_1 p_1 + \tilde{C}_2 : \tilde{\alpha}_2 p_2), \tag{2.26}$$

and a modified effective stress law can be written from Eq. (2.26) as

$$\tilde{\sigma} = \tilde{D}_{12} : \tilde{\varepsilon} - \tilde{D}_{12} : \tilde{C}_1 : \tilde{\alpha}_1 p_1 - \tilde{D}_{12} : \tilde{C}_2 : \tilde{\alpha}_2 p_2 \tag{2.27}$$

where

$$\tilde{D}_{12} = (\tilde{C}_1 + \tilde{C}_2)^{-1} \tag{2.28}$$

and $\tilde{D}_{12}$ is the elastic stiffness tensor evaluated from the lumping of deformations in both matrix and fractures.

Based on the modified effective stress law in Eq. (2.27), the governing equilibrium equation can be given as (Elsworth and Bai 1992)

$$\vec{\nabla}\cdot(\tilde{D}_{12}:\tilde{\varepsilon} - \tilde{D}_{12}:\tilde{C}_1:\tilde{\alpha}_1 p_1 - \tilde{D}_{12}:\tilde{C}_2:\tilde{\alpha}_2 p_2) = 0. \tag{2.29}$$

Eq. (2.29) can be rewritten in a similar form to Eq. (2.14), i.e.

$$\vec{\nabla}\cdot(\tilde{D}^*:\tilde{\varepsilon} - \tilde{\alpha}_1^* p_1 - \tilde{\alpha}_2^* p_2) = 0 \tag{2.30}$$

where

$$\tilde{D}^* = \tilde{D}_{12} \tag{2.31}$$

$$\tilde{\alpha}_1^* = \tilde{D}_{12}:\tilde{C}_1:\tilde{\alpha}_1 \tag{2.32}$$

$$\tilde{\alpha}_2^* = \tilde{D}_{12}:\tilde{C}_2:\tilde{\alpha}_2. \tag{2.33}$$

Comparing Eq. (2.29) with Eq. (2.14), it is apparent that only when the fracture compliance tensor $\tilde{C}_2$ vanishes in $\tilde{D}_{12}$ associated with the matrix [i.e., the term in Eq. (2.32)], or the matrix compliance tensor $\tilde{C}_1$ disappears in $\tilde{D}_{12}$ related to the fracture [i.e., the term in Eq. (2.33)], along with the elimination of $\tilde{C}_2$ in Eq. (2.31), will Eq. (2.29) become equivalent to Eq. (2.14). Clearly, therefore, the majority of the dual porosity poroelasticity models either neglect the parameters characterizing the fracture deformation or replace these parameters with those suitable only for matrix deformation. The advantage of using Eq. (2.30) instead of Eq. (2.14) is now clear in that the deformations of both matrix and fractures are individually incorporated in the former.

2.2.2.3 A Conversion Method

Because the use of a single effective stress law has the advantage of characterizing the effect of deformation on fluid flow, while the use of a double effective stress law enables respective deformations of the matrix and fractures to be rigorously defined, it is beneficial to derive a correlation between the two methods. The constitutive relations for solid deformation where a single effective stress law is applied may be defined in most general form as

$$\tilde{\varepsilon} = \tilde{C}:\tilde{\sigma} \tag{2.34}$$

where the compliance tensor $\tilde{C}$ can be written for a three-dimensional geometry as

$$\tilde{C} = \begin{bmatrix} C_{11} & C_{12} & C_{13} & 0 & 0 & 0 \\ C_{21} & C_{22} & C_{23} & 0 & 0 & 0 \\ C_{31} & C_{32} & C_{33} & 0 & 0 & 0 \\ 0 & 0 & 0 & C_{44} & 0 & 0 \\ 0 & 0 & 0 & 0 & C_{55} & 0 \\ 0 & 0 & 0 & 0 & 0 & C_{66} \end{bmatrix} \tag{2.35}$$

where

$$C_{11} = C_{22} = C_{33} = \frac{1}{E} = \psi_1 \tag{2.36}$$

$$C_{12} = C_{21} = C_{13} = C_{31} = C_{23} = C_{32} = -\frac{\nu}{E} = \psi_2 \tag{2.37}$$

$$C_{44} = C_{55} = C_{66} = \frac{2(1+\nu)}{E} = \psi_3 \tag{2.38}$$

and the corresponding two independent parameters are Young's modulus, E, and Poisson's ratio, ν.

Inversion of Eq. (2.34) results in

$$\tilde{\sigma} = \tilde{D}{:}\tilde{\varepsilon} \tag{2.39}$$

where the stiffness tensor $\tilde{D}$ can be given as

$$\tilde{D} = \tilde{C}^{-1} = \begin{bmatrix} D_{11} & D_{12} & D_{13} & 0 & 0 & 0 \\ D_{21} & D_{22} & D_{23} & 0 & 0 & 0 \\ D_{31} & D_{32} & D_{33} & 0 & 0 & 0 \\ 0 & 0 & 0 & D_{44} & 0 & 0 \\ 0 & 0 & 0 & 0 & D_{55} & 0 \\ 0 & 0 & 0 & 0 & 0 & D_{66} \end{bmatrix} \tag{2.40}$$

where

$$D_{11} = D_{22} = D_{33} = \frac{\psi_1}{\psi_1^2 - \psi_2^2} = \frac{E(1-\nu)}{(1-2\nu)(1+\nu)} = \lambda + 2G \tag{2.41}$$

$$D_{12} = D_{21} = D_{13} = D_{31} = D_{23} = D_{32} = -\frac{\psi_2}{\psi_1^2 - \psi_2^2} = \frac{E\nu}{(1-2\nu)(1+\nu)} \tag{2.42}$$

$$D_{44} = D_{55} = D_{66} = \frac{1}{\psi_3} = \frac{E}{2(1+\nu)} = G. \tag{2.43}$$

The constitutive relations defining solid deformation, in isolation, where a double effective stress law is incorporated, may be described as

$$\tilde{\varepsilon} = (\tilde{C}_1 + \tilde{C}_2)\tilde{\sigma} \tag{2.44}$$

where the matrix compliance tensor $\tilde{C}_1$ is identical to that defined in Eq.

(2.35); while the fracture compliance $\tilde{C}_2$ may be defined as

$$\tilde{C}_2 = \begin{bmatrix} \frac{1}{K_n s^*} & 0 & 0 & 0 & 0 & 0 \\ 0 & \frac{1}{K_n s^*} & 0 & 0 & 0 & 0 \\ 0 & 0 & \frac{1}{K_n s^*} & 0 & 0 & 0 \\ 0 & 0 & 0 & \frac{1}{K_{sh} s^*} & 0 & 0 \\ 0 & 0 & 0 & 0 & \frac{1}{K_{sh} s^*} & 0 \\ 0 & 0 & 0 & 0 & 0 & \frac{1}{K_{sh} s^*} \end{bmatrix} \tag{2.45}$$

where s^* is the fracture spacing, and K_n and K_{sh} are the fracture normal and shear stiffnesses, respectively.

The lumped compliance matrix $\tilde{C}_{12}$ can be written as

$$\tilde{C}_{12} = \tilde{C}_1 + \tilde{C}_2 \begin{bmatrix} C'_{11} & C'_{12} & C'_{13} & 0 & 0 & 0 \\ C'_{21} & C'_{22} & C'_{23} & 0 & 0 & 0 \\ C'_{31} & C'_{32} & C'_{33} & 0 & 0 & 0 \\ 0 & 0 & 0 & C'_{44} & 0 & 0 \\ 0 & 0 & 0 & 0 & C'_{55} & 0 \\ 0 & 0 & 0 & 0 & 0 & C'_{66} \end{bmatrix} \tag{2.46}$$

where $\tilde{C}_1$ is equal to $\tilde{C}$ in Eq. (2.35), and

$$C'_{11} = C'_{22} = C'_{33} = \frac{1}{E} + \frac{1}{K_n s^*} = \phi_1 \tag{2.47}$$

$$C'_{12} = C'_{21} = C'_{13} = C'_{31} = C'_{23} = C'_{32} = -\frac{\nu}{E} = \psi_2 = \phi_2 \tag{2.48}$$

$$C'_{44} = C'_{55} = C'_{66} = \frac{2(1+\nu)}{E} + \frac{1}{K_{sh} s^*} = \phi_3. \tag{2.49}$$

The lumped stiffness tensor $\tilde{D}_{12}$, as described in Eq. (2.28), can be expressed as

$$\tilde{D}_{12} = \tilde{C}_{12}^{-1} = \begin{bmatrix} D'_{11} & D'_{12} & D'_{13} & 0 & 0 & 0 \\ D'_{21} & D'_{22} & D'_{23} & 0 & 0 & 0 \\ D'_{31} & D'_{32} & D'_{33} & 0 & 0 & 0 \\ 0 & 0 & 0 & D'_{44} & 0 & 0 \\ 0 & 0 & 0 & 0 & D'_{55} & 0 \\ 0 & 0 & 0 & 0 & 0 & D'_{66} \end{bmatrix} \tag{2.50}$$

where

$$D'_{11} = D'_{22} = D'_{33} = \frac{\phi_1}{\phi_1^2 - \phi_2^2} \tag{2.51}$$

$$D'_{12} = D'_{21} = D'_{13} = D'_{31} = D'_{23} = D'_{32} = -\frac{\phi_2}{\phi_1^2 - \phi_2^2} \tag{2.52}$$

$$D'_{44} = D'_{55} = D'_{66} = \frac{1}{\phi_3}. \tag{2.53}$$

The elastic stiffness tensor $\tilde{D}$ and compliance tensor $\tilde{C}$ in the method of the single effective stress law do not contain the characteristic parameters for the fractures, such as the fracture spacing s^*, and normal and shear stiffnesses K_n and K_{sh}, which are shown in Eq. (2.45). As a result, using the conversion relations defined in this section is equivalent to applying the parameters $\tilde{D}_{12}$, $\tilde{C}_1$, and $\tilde{C}_2$ in lieu of $\tilde{D}$ and $\tilde{C}$ for those models using the single effective stress law, to incorporate approximated effects of the deformation of fractures.

Where multiple fluid phases are considered, the appropriate fluid pressures are the average quantities, defined relative to the proportional saturations of each fluid, in both matrix and fractures; i.e.,

$$p_1 = S_{1w}p_{1w} + S_{1n}p_{1n} \tag{2.54}$$

$$p_2 = S_{2w}p_{2w} + S_{2n}p_{2n} \tag{2.55}$$

where S is the saturation, subscripts $1w$ and $2w$ represent wetting fluid phases, and $1n$ and $2n$ represent non-wetting fluid phases of the matrix and the fracture, respectively.

2.3 PARAMETRIC STUDY

With the general underpinnings of the relations defined, focus can move to specifics of the effective stress laws, the parametric relations that are relevant to coupled processes, and the incorporation of anisotropic properties.

2.3.1 Effective Stress Law

Effective stresses control deformation and failure in fluid-infiltrated media. The widely accepted form of the relation partitions intergranular and fluid loads, as conditioned by the ratio of grain and skeletal moduli. The parameters controlling behavior are identified in the following, with the presentation focusing on both macroscopic and microscopic scales, and for homogeneous (single porosity) and heterogeneous (multiple porosity) media.

2.3.1.1 Homogeneous Media

Macroscopic Scale

The law of effective stresses is defined at the macroscopic scale, including a representative volume of solid material and aggregate intergranular forces into an overall magnitude. Effective stresses govern the magnitude of induced strains through the modulating influence of the elastic constants. Effective stresses that modulate deformations and strains are defined relative to total stresses that are required to define equilibrium through knowledge of pore fluid pressures.

In general, Terzaghi's law of effective stresses may be defined as

$$\sigma_{ij}^e = \sigma_{ij} + p\delta_{ij}, \tag{2.56}$$

where σ_{ij}^e is the effective stress tensor and σ_{ij} is the total stress tensor. This relationship was defined empirically, and is only exact where the solid grains of a porous medium are incompressible. Where grain compressibility is finite, a stress ratio term must be added.

A general effective stress law can be expressed as (Biot 1941):

$$\sigma_{ij}^e = \sigma_{ij} + \alpha p\delta_{ij}, \tag{2.57}$$

where α is a factor describing the form of fluid-solid coupling.

A historical development of the effective stress laws was provided by Lade and Boer (1997). Terzaghi first suggested $\alpha = 1$ (Terzaghi 1923, 1943). Others advocating this choice include Hubbert and Rubey (1959, 1960), Skempton (1960) and many others (Handin 1958; Robinson 1959; Handin et al. 1963; Murrell 1963; Walsh 1965; Brace and Byerlee 1965; Brace and Martin 1968; Garg and Nur 1973). This assumption has been extensively used for unconsolidated materials such as sand, where its use appears appropriate.

Hoffman (1928) and Fillunger (1930) recognized that $\alpha = 1$ was not sufficient to characterize different porous materials. They suggested that $\alpha = n$ where n is the porosity. This postulation was supported later by Terzaghi (1945), Lubinski (1954) and Biot (1955).

Other suggested forms of α are:

- (Schiffman 1970)

$$n \ll \alpha \ll 1 \tag{2.58}$$

- (Suklje 1969)

$$\alpha = 1 - (1-n)\frac{K}{K_s} \tag{2.59}$$

where n is the porosity and K and K_s are the elastic moduli of the porous medium and solid grains, respectively.

- (Skempton and Bishop 1954)

$$\alpha = 1 - g_c \tag{2.60}$$

where g_c is the effective grain contact area per unit area in a traversing plane. This form was also supported by Bishop (1955) and by Skempton (1960).

- (Biot 1941)

$$\alpha = 1 - \frac{K}{K_s} \tag{2.61}$$

which has been the dominant form of α, and supported by many researchers (Gassmann 1951; Biot and Willis 1957; Geertsma 1957; Skempton 1960; Serafim 1964; Nur and Byerlee 1971; Bishop 1973). Nur and Byerlee's (1971) theoretical conceptualization demonstrated that this form of α appears to be the most suitable representation of the pore pressure coupling for poroelastic applications, especially for those in rock and rock-like materials. As a result, this form is adopted in this work.

Microscopic Scale

Conversely, the law of effective stresses may be defined at the microscopic scale, where deformations are defined at a local level. In particular, its relation to volume or porosity changes, which is indirectly associated with flow, may be defined.

Based on the conceptual loading configuration proposed by Nur and Byerlee (1971), the total loading on the porous medium may be decomposed into two parts: the hydrostatic loading (I) of pore pressures on the solid matrix and the deviatoric loading (II) by the effective stress on the dry porous medium. This loading configuration is schematically depicted in Figure 2-1. The strains due to the first and second loads can be expressed, respectively, as

$$\varepsilon_{ij}^{I} = -\frac{p}{3K_s}\delta_{ij} \tag{2.62}$$

where K_s is the bulk modulus of solid grains, and

$$\varepsilon_{ij}^{II} = \frac{1+\nu}{E}(\sigma_{ij} + p\delta_{ij}) - \frac{\nu}{E}(\sigma_{kk} + 3p)\delta_{ij} \tag{2.63}$$

where E and ν are the elastic modulus and Poisson ratio of the porous medium, and σ_{kk} is the total stress.

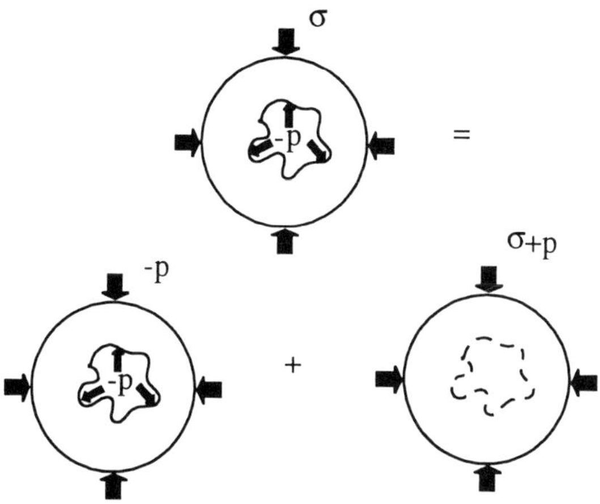

Figure 2-1. Schematic loading for single-porosity media.

From these two equations, the total strain can be written in a general form as

$$\varepsilon_{ij} = \varepsilon_{ij}^{I} + \varepsilon_{ij}^{II} = \frac{1+\nu}{E}(\sigma_{ij}^{e}) - \frac{\nu}{E}(\sigma_{kk}^{e})\delta_{ij} \tag{2.64}$$

where

$$\sigma_{ij}^{e} = \sigma_{ij} + \alpha p \delta_{ij} \tag{2.65}$$

$$\sigma_{kk}^{e} = \sigma_{kk} + 3\alpha p \delta_{ij} \tag{2.66}$$

$$K = \frac{E}{3(1-2\nu)} \tag{2.67}$$

where K is the bulk modulus of the porous medium and α is defined in Eq. (2.61).
The total strain can be expressed as

$$\varepsilon_{kk} = \frac{\Delta V}{V} = \frac{1}{K}(\frac{\sigma_{kk}}{3} + p) - \frac{p}{K_s}. \tag{2.68}$$

The total volume of the porous medium, V, can be defined as

$$V = V_s + V_p \tag{2.69}$$

where V_s and V_p are the solid and pore volumes, respectively. Since the pore volume V_p is a function of both stress and pore pressure, the total derivatives can be expressed as

$$\frac{dV_p}{V_p} = \frac{1}{V_p}\frac{\partial V_p}{\partial \bar{\sigma}}d\bar{\sigma} + \frac{1}{V_p}\frac{\partial V_p}{\partial p}dp \tag{2.70}$$

where $\bar{\sigma}$ is the mean stress, defined as

$$\bar{\sigma} = \frac{\sigma_{kk}}{3}. \tag{2.71}$$

For this particular loading case, the applied mean stress is equal to the applied pore pressure; i.e.,

$$d\bar{\sigma} = -dp. \tag{2.72}$$

Also, the relative pore volume change is due to the compressibility of the solid matrix (loading I); i.e.,

$$\frac{dV_p}{V_p} = -\frac{dp}{K_s}. \tag{2.73}$$

Considering Eqs. (2.72) and (2.73), Eq. (2.70) can be rewritten in the form of a partial derivative, since the formats of all the functions are now determined, as

$$\frac{\partial V_p}{V_p} = -\frac{\partial p}{K_s} + \frac{1}{V_p}\frac{\partial V_p}{\partial \bar{\sigma}}(\partial \bar{\sigma} + \partial p). \tag{2.74}$$

Using Betty's reciprocal theorem (Charlez 1991), the following relation is obtained as

$$\partial V_p \cdot \partial p = \partial V \cdot \partial \bar{\sigma}. \tag{2.75}$$

Or, rearranging Eq. (2.75), it follows that

$$\frac{\partial p}{\partial V} = \frac{\partial \bar{\sigma}}{\partial V_p}. \tag{2.76}$$

From Eq. (2.68), it is known that

$$\frac{\partial V}{V} = \frac{\sigma_{kk}}{3K} - \left(\frac{1}{K_s} - \frac{1}{K}\right)p. \tag{2.77}$$

Differentiating Eq. (2.77) with respect to p yields

$$\frac{1}{V}\frac{\partial V}{\partial p} = -\left(\frac{1}{K_s} - \frac{1}{K}\right). \tag{2.78}$$

Or, rearranging Eq. (2.78), one has

$$\frac{\partial p}{\partial V} = \frac{K}{V\alpha} \tag{2.79}$$

where α is defined in Eq. (2.61). Combining Eqs. (2.76) and (2.79), gives

$$\frac{\partial V_p}{\partial \bar{\sigma}} = \frac{V\alpha}{K}. \tag{2.80}$$

Substituting Eq. (2.80) into Eq. (2.74), noting the definition of porosity (i.e., $n = V_p/V$), yields

$$\frac{\partial V_p}{V_p} = -\frac{\partial p}{K_s} + \frac{\alpha}{nK}(\partial \bar{\sigma} + \partial p). \tag{2.81}$$

By definition,

$$\frac{\partial V_s}{V_s} = \frac{\partial(V - V_p)}{(V - V_p)} = \frac{1}{1-n}\left(\frac{\partial V}{V} - n\frac{\partial V_p}{V_p}\right). \tag{2.82}$$

Substituting Eqs. (2.77) and (2.81) into Eq. (2.82), gives

$$\frac{\partial V_s}{V_s} = \frac{(1-\alpha)}{(1-n)K}(\frac{\partial \sigma_{kk}}{3} + \partial p) - \frac{\partial p}{K_s}. \tag{2.83}$$

Substituting the following relationship,

$$\frac{1}{\rho_s}\frac{\partial \rho_s}{\partial t} = -\frac{1}{V_s}\frac{\partial V_s}{\partial t} \tag{2.84}$$

into Eq. (2.83), yields

$$\frac{\partial \rho_s}{\rho_s} = \frac{1}{(1-n)}\left[\frac{(\alpha - n)}{K_s}p - (1-\alpha)\varepsilon_{kk}\right]. \tag{2.85}$$

The time derivative of the relation in Eq. (2.85) is used in Chapter 3 describing flow behavior.

2.3.1.2 Heterogeneous Media

The theoretical underpinnings developed in the following are limited to dual-porosity poroelasticity.

Macroscopic Scale

As indicated previously, the application of a single effective stress law to dual-porosity poroelastic media, as shown in Eq. (2.13), may result in difficulty distinguishing between the component deformations. Consequently, the discussion in this section examines the effective stress laws at the macroscopic scale and exposes any conflicts in interpretation embedded in some existing models.

By extending the analytical treatment first expounded for single-porosity poroelastic media (Nur and Byerlee 1971), Wilson and Aifantis (1982) were among the first to provide an effective stress law for dual-porosity poroelastic models. The relations may be recovered from Eq. (2.13) developed for isotropic media, as

$$\sigma_{ij}^e = \sigma_{ij} + \sum_{i=1}^{2} \alpha_i p_i \delta_{ij} \tag{2.86}$$

where the pressure ratio factors α_i were expressed as

$$\alpha_1 = \frac{K^*}{K}\left(1 - \frac{K}{K_s}\right) \tag{2.87}$$

$$\alpha_2 = 1 - \frac{K^*}{K} \tag{2.88}$$

and K^*, K, and K_s are assumed as the bulk moduli of fractured rock, rock without fractures (i.e., porous matrix), and solid grains, respectively. The nomenclature used here is slightly different from that used by Wilson and Aifantis (1982) and others, as it has been modified here for consistency.

Tuncay and Corapcioglu (1995) offered a similar formulation using the concept of a volume average. The only difference in their formulation is the definition of α_i:

$$\alpha_1 = \frac{K^*}{K}\left(1 - \beta_f - \frac{K}{K_s}\right), \tag{2.89}$$

$$\alpha_2 = 1 - (1 - \beta_f)\frac{K^*}{K} \tag{2.90}$$

where β_f is the volume fraction of fractures. If β_f vanishes, Tuncay and Corapcioglu's (1995) formulation is identical to that of Wilson and Aifantis (1982).

Even though the preceding has the advantage that only a single effective stress law is needed to define dual-porosity response, the drawback is that fracture and matrix compressions are linked. As a result, the respective parameters are less independent, and the concept becomes intrinsically complex. It is difficult to define α_1 and α_2 according to their individual characteristics. Similar to the definition of $\tilde{C}_2$ in Eq. (2.45), it is expected that α_2 is related to the fracture stiffnesses K_n, K_{sh}, and spacing s^*, rather than related to the properties of the porous matrix, such as elastic modulus E and Poisson ratio ν.

While seeking the conditions that enable conversion between the parameters defining single-porosity and the dual-porosity poroelastic behavior, Wilson and Aifantis (1982) noted that Eq. (2.86) collapses to Biot's coefficient if the pore pressures are identical between the matrix and fractures ($p_1 = p_2 = p$), i.e.,

$$\sigma_{ij}^e = \sigma_{ij} + (\alpha_1 + \alpha_2)p\delta_{ij} = \sigma_{ij} + \alpha p \delta_{ij} \tag{2.91}$$

where

$$\alpha = 1 - \frac{K^*}{K_s}. \tag{2.92}$$

Even though the pressures are identical in the fractures and the matrix, the medium itself should still reflect the two-phase characteristics. In other words, the dual-porosity medium merges to the single-porosity medium only when the fractures become infinitely stiff. In fact, K^* in Eq. (2.92) does not collapse to the bulk modulus of the porous medium, K, providing evidence that fractures have an effect on deformation. As a result, equal pressures in the matrix and the fractures should not be considered as a precondition for the equivalence between the single-porosity and dual-porosity poroelastic models.

The specific form of α_1 and α_2 in Eqs. (2.87) and (2.88) remains to be confirmed. Because the matrix pressure ratio factor α_1 is related to the bulk modulus of the fractured rock K^*, while the fracture pressure ratio factor α_2 is associated with the bulk modulus of the porous medium K, individual experimental determination of these factors appears impossible. Furthermore, confusion regarding the exact form of α_i is exacerbated for multi-porosity poroelastic media (e.g., triple-porosity media) due to the complexity of the system.

As an alternative approach to using a single effective stress law, the following conceptualization, extended from Nur and Byrlee's theoretical model (1971), is used to overcome the principal ambiguities in Eqs. (2.87) and (2.88). For ease of illustration, stress is assumed to be a scalar quantity applied uniformly (hydrostatic) in loading.

Two-stage loading is used in Nur and Byrlee's (1971) model for homogeneous porous media. The first stage relates the pore pressure loading to the compressibility of solid grains, while the second stage imposes total stresses in determining the compressibility of the porous medium skeleton. The effective variation in volume can be expressed as

$$\frac{\Delta V}{V} = -\frac{p}{K_s} + \frac{\sigma + p}{K} \tag{2.93}$$

where V is the total volume, K and K_s are the bulk moduli of the porous medium and solid grains, respectively, and σ is the total stress.

Incorporating Eq. (2.93), the effective stress σ^e is described as

$$\sigma^e = K\frac{\Delta V}{V} = \sigma + \alpha p \tag{2.94}$$

where α is defined in Eq. (2.61).

For fractured porous media (dual-porosity), consider the following steps:

- In the first stage of loading, the pore pressures are summed from both matrix and fractures using the principle of superposition

- In the second stage of loading, the two fluid pressures are subtracted from the total stress

- Finally, the effective volume change for a dual-porosity medium may be defined as

$$\frac{\Delta V}{V} = -\frac{p_1}{K_s} - \frac{p_2}{K_{fr}} + \frac{\sigma + p_1 + p_2}{K^\#} \tag{2.95}$$

  where p_1 and p_2 are the pore pressures in matrix and fractures, and $K^\#$, K_s, and K_{fr} are the bulk moduli of the fractured porous medium, solid grains, and fractured medium, respectively.

$K^\#$ in Eq. (2.95) is different from K^* in Eqs. (2.87) and (2.88). The former is for the dual-porosity medium, while the latter is for the equivalent single-porosity medium, i.e., a fractured medium containing impermeable matrix blocks. In addition, bulk modulus of the fractured medium K_{fr} can be characterized by the fracture normal and shear stiffnesses K_n and K_{sh}, along with the fracture spacing s^*.

As a result, the effective stress for the dual-porosity medium σ^e can be deduced as

$$\sigma^e = K^\# \frac{\Delta V}{V} = \sigma + \alpha_1 p_1 + \alpha_2 p_2 \tag{2.96}$$

where

$$\alpha_1 = 1 - \frac{K^\#}{K_s} \tag{2.97}$$

$$\alpha_2 = 1 - \frac{K^\#}{K_{fr}}. \tag{2.98}$$

Even though $K^\#$ represents the bulk modulus averaged from the matrix and fractures, Eqs. (2.97) and (2.98) provide similar forms but distinguish the individual properties of pressure ratio factors for the matrix and fractures. For this reason, the formulation in Eqs. (2.96), (2.97), and (2.98) can be easily extended to the multi-porosity case.

If the pore pressures in the matrix and fractures are identical (i.e., $p_1 = p_2 = p$), the effective stress in Eq. (2.96) reduces to

$$\sigma^e = \sigma + \left[2 - K^\# \left(\frac{1}{K_s} + \frac{1}{K_{fr}}\right)\right] p. \tag{2.99}$$

In contrast to other formulations, e.g., Eq. (2.86) when $p_1 = p_2 = p$, it may be noted that Eq. (2.99) does not convert to a similar form for α as in the single-porosity poroelastic formulation (Biot 1941). This is physically correct because the dual-porosity medium still exists, although the pressures are equalized between the two media. However, if the fractures vanish, which is equivalent to letting $K_{fr} = K^{\#} = K$, or setting $\alpha_2 = 0$ in Eq. (2.96), then Eq. (2.99) reduces to

$$\sigma^e = \sigma + \left(1 - \frac{K}{K_s}\right) p \tag{2.100}$$

which is identical to the form of Biot's coefficient in the single-porosity poroelastic model (Biot 1941).

Microscopic Scale

The form of the pore pressure coefficients may also be determined for dual-porosity media at the local scale. The total loading on the fractured porous medium can be decomposed into three parts: the loading by the matrix pore pressure on the solid matrix (I), by the fracture pore pressure on the fractured medium (II), and the deviatoric loading (III) by the modified effective stress on the dry fractured porous medium. The loading is shown schematically in Figure 2-2. The strains due to the three-part loadings can be expressed, respectively, as

$$\varepsilon_{ij}^{I} = -\frac{p_1}{3K_s}\delta_{ij} \tag{2.101}$$

where K_s is the bulk modulus of solid grains, and

$$\varepsilon_{ij}^{II} = -\frac{p_2}{3K^*}\delta_{ij} \tag{2.102}$$

where K^* is the bulk modulus of fractured medium, and

$$\varepsilon_{ij}^{III} = \frac{1+\nu^{\#}}{E^{\#}}(\sigma_{ij} + p_1\delta_{ij} + p_2\delta_{ij}) - \frac{\nu^{\#}}{E^{\#}}(\sigma_{kk} + 3p_1 + 3p_2)\delta_{ij} \tag{2.103}$$

where $E^{\#}$ and $\nu^{\#}$ are the elastic modulus and Poisson ratio of the fractured porous medium. From these three equations, the total strain can be written in a general form as

$$\varepsilon_{ij} = \varepsilon_{ij}^{I} + \varepsilon_{ij}^{II} + \varepsilon_{ij}^{III} = \frac{1+\nu^{\#}}{E^{\#}}(\sigma_{ij}^{e}) - \frac{\nu^{\#}}{E^{\#}}(\sigma_{kk}^{e})\delta_{ij} \tag{2.104}$$

where the effective stress and total stresses are

$$\sigma_{ij}^{e} = \sigma_{ij} + \alpha_1 p_1 \delta_{ij} + \alpha_2 p_2 \delta_{ij} \tag{2.105}$$

$$\sigma_{kk}^{e} = \sigma_{kk} + 3\alpha_1 p_1 \delta_{ij} + 3\alpha_2 p_2 \delta_{ij} \tag{2.106}$$

and pressure ratio factors (Biot's coefficient in a dual-porosity medium) are

$$\alpha_1 = 1 - \frac{K^{\#}}{K_s} \tag{2.107}$$

$$\alpha_2 = 1 - \frac{K^{\#}}{K_{fr}} \tag{2.108}$$

where $K^{\#}$ is the bulk modulus of the fractured porous medium, which can be further defined as

$$K^{\#} = \frac{E^{\#}}{3(1-2\nu^{\#})}. \tag{2.109}$$

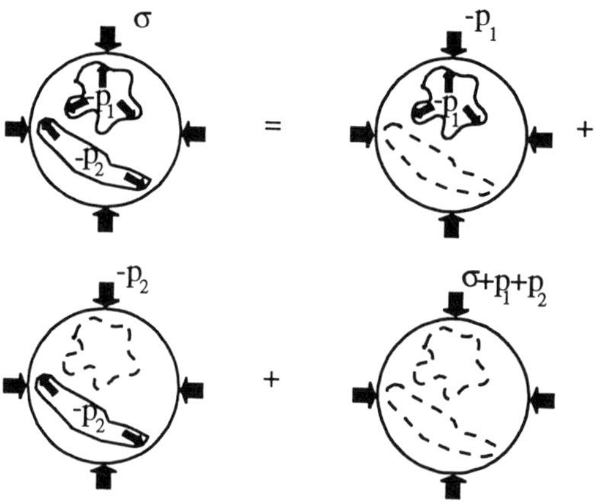

Figure 2-2. Schematic loading for dual-porosity media.

The counterpart of $K^{\#}$ for the porous medium is K; both are macroscopic parameters. In contrast, the counterpart of K_{fr} for the porous medium is K_s; both are microscopic parameters. The expressions for the latter parameters are generally defined by physical intuition rather than by theoretical definition, such as for the former parameters. Specifically, for porous media, K is defined in Eq. (2.67), while K_s is generally given as an input parameter. For fractured porous media, $K^{\#}$ is defined in Eq. (2.109), while K_{fr} is generally given as an input parameter, which can be described as

$$K_{fr} = K_n s^* \tag{2.110}$$

where K_n is the fracture normal stiffness and s^* is the fracture spacing. In this postulation, the shear rigidity of the fractures is omitted.

The total strain, which is related to the ratio of volume change, can be expressed as

$$\varepsilon_{kk} = \frac{\Delta V}{V} = \frac{1}{K\#}(\frac{\sigma_{kk}}{3} + p_1 + p_2) - \frac{p_1}{K_s} - \frac{p_2}{K_{fr}}. \tag{2.111}$$

The total volume of the porous medium, V, can be defined as

$$V = V_s + V_{p1} + V_{p2} \tag{2.112}$$

where V_s, V_{p1} and V_{p2} are the solid volume, pore volume of the matrix and pore volume of the fractures, respectively. Dividing through by total volume, V in Eq. (2.112), and by definition

$$1 = \frac{V_s}{V} + \frac{V_{p1}}{V} + \frac{V_{p2}}{V} = n_s + n_1 + n_2. \tag{2.113}$$

Since the pore volume V_p is a function of both stress and pore pressures, the total derivatives lead to

$$\frac{dV_p}{V_p} = \frac{1}{V_p}\frac{\partial V_p}{\partial \bar{\sigma}} d\bar{\sigma} + \frac{1}{V_p}\frac{\partial V_p}{\partial p_1} dp_1 + \frac{1}{V_p}\frac{\partial V_p}{\partial p_2} dp_2 \tag{2.114}$$

where $\bar{\sigma}$ is the mean stress, defined in Eq. (2.71). For this particular loading case, the applied mean stress is equal to the applied pore pressures; i.e.,

$$d\bar{\sigma} = -(dp_1 + dp_2). \tag{2.115}$$

Also, the relative pore volume change is due to the compressibilities of the solid matrix (loading I) and fractured medium (loading II); i.e.,

$$\frac{dV_p}{V_p} = -\frac{dp_1}{K_s} - \frac{dp_2}{K_{fr}}. \tag{2.116}$$

Considering Eqs. (2.115) and (2.116), as well as transformation of total derivatives to partial derivatives since the formats of all functions are now known, Eq. (2.114) can be rewritten as

$$\frac{\partial V_p}{V_p} = -\frac{\partial p_1}{K_s} - \frac{\partial p_2}{K_{fr}} + \frac{1}{V_p}\frac{\partial V_p}{\partial \bar{\sigma}}(\partial \bar{\sigma} + \partial p_1 + \partial p_2). \tag{2.117}$$

Using a modified form of Betty's reciprocal theorem, the following relation is obtained:

$$\partial V_p \; \partial(p_1 + p_2) = \partial V \; \partial \bar{\sigma}. \tag{2.118}$$

Rearranging Eq. (2.118) yields

$$\frac{\partial p_1}{\partial V} + \frac{\partial p_2}{\partial V} = \frac{\partial \bar{\sigma}}{\partial V_p}. \tag{2.119}$$

From Eq. (2.111), it is known that

$$\frac{\partial V}{V} = \frac{\sigma_{kk}}{3K^{\#}} - \left(\frac{1}{K_s} - \frac{1}{K^{\#}}\right) p_1 - \left(\frac{1}{K_{fr}} - \frac{1}{K^{\#}}\right) p_2. \tag{2.120}$$

Taking partial derivatives of Eq. (2.120) with respect to p_1 and p_2, yields

$$\begin{cases} \dfrac{1}{V}\dfrac{\partial V}{\partial p_1} = -\left(\dfrac{1}{K_s} - \dfrac{1}{K^{\#}}\right) \\ \dfrac{1}{V}\dfrac{\partial V}{\partial p_2} = -\left(\dfrac{1}{K_{fr}} - \dfrac{1}{K^{\#}}\right) \end{cases} \tag{2.121}$$

and Eq. (2.121) can be rewritten as

$$\begin{cases} \dfrac{\partial p_1}{\partial V} = -\left[V\left(\dfrac{1}{K_s} - \dfrac{1}{K^{\#}}\right)\right]^{-1} \\ \dfrac{\partial p_2}{\partial V} = -\left[V\left(\dfrac{1}{K_{fr}} - \dfrac{1}{K^{\#}}\right)\right]^{-1} \end{cases} \tag{2.122}$$

From Eq. (2.122), one has

$$\frac{\partial p_1}{\partial V} + \frac{\partial p_2}{\partial V} = \frac{K^{\#}}{V}\left(\frac{1}{\alpha_1} + \frac{1}{\alpha_2}\right) \tag{2.123}$$

where α_1 and α_2 are defined in Eqs. (2.107) and (2.108). Combining Eq. (2.119) with Eq. (2.123), gives

$$\frac{\partial V_p}{\partial \bar{\sigma}} = \frac{V}{K^{\#}}\left(\frac{1}{\alpha_1} + \frac{1}{\alpha_2}\right)^{-1}. \tag{2.124}$$

Substituting Eq. (2.124) into Eq. (2.117), noting the definition of porosity in Eq. (2.113), yields

$$\frac{\partial V_p}{V_p} = -\frac{\partial p_1}{K_s} - \frac{\partial p_2}{K_{fr}} + \frac{1}{(n_1+n_2)K^{\#}}\left(\frac{1}{\alpha_1} + \frac{1}{\alpha_2}\right)^{-1}(\partial\bar{\sigma} + \partial p_1 + \partial p_2). \tag{2.125}$$

In addition,

$$\frac{\partial V_s}{V_s} = \frac{\partial(V - V_p)}{(V - V_p)} = \frac{1}{1 - n_1 - n_2}\left[\frac{\partial V}{V} - (n_1 + n_2)\frac{\partial V_p}{V_p}\right] \tag{2.126}$$

and substituting Eqs. (2.120) and (2.125) into Eq. (2.126), gives:

$$\frac{\partial V_s}{V_s} = \frac{1}{(1-n_1-n_2)K\#}\left[1-\left(\frac{1}{\alpha_1}+\frac{1}{\alpha_2}\right)^{-1}\right](\frac{\partial\sigma_{kk}}{3}+\partial p_1+\partial p_2)-\frac{\partial p_1}{K_s}-\frac{\partial p_2}{K_{fr}}. \tag{2.127}$$

Substituting the relationship in Eq. (2.84) into Eq. (2.127), and defining the following relationship (Meng 1998):

$$\alpha_3 = \left(\frac{1}{\alpha_1}+\frac{1}{\alpha_2}\right)^{-1} = \frac{\alpha_1\alpha_2}{\alpha_1+\alpha_2} \tag{2.128}$$

results in the following relation:

$$\frac{\partial\rho_s}{\rho_s} = \frac{1}{(1-n_1-n_2)}\left[\frac{\alpha_3-n_1-n_2}{K_s}p_1+\frac{\alpha_3-n_1-n_2}{K_{fr}}p_2-(1-\alpha_3)\varepsilon_{kk}\right]. \tag{2.129}$$

The time derivative of the relation in Eq. (2.129) is also used in Chapter 3, which describes flow.

2.3.2 Parametric Relations in Coupled Processes

Behavior may be described separately for homogeneous and heterogeneous media.

2.3.2.1 Homogeneous Media

For isotropic linear elasticity, the two parameters of elastic modulus E and Poisson ratio ν uniquely define mechanical response. For isotropic linear poroelasticity, the basic parameters are extended to five. In addition to E and ν, the other three parameters are the intrinsic permeability κ ($\kappa = \frac{k}{\mu}$ where k is the permeability and μ is the fluid dynamic viscosity), the Biot (1941) coefficient α, and the Biot (1941) modulus M. Among these poroelastic parameters, however, the last two are not totally independent. In fact, α can be related to the bulk modulus K via Eqs. (2.61) and (2.67), where, besides E and ν, the bulk modulus of the solid grains is treated as an independent parameter. The Biot modulus, M, is viewed as the inverse of the storage coefficient (Biot and Willis 1957; Green and Wang 1990). Depending on the assumptions made in defining the specific storage of the porous medium, M is frequently written as

$$\frac{1}{M} = \frac{\alpha-n}{K_s}+\frac{n}{K_f} \tag{2.130}$$

where porosity n and fluid bulk modulus K_f may be considered as other independent parameters. Therefore, the following seven are the minimum set of independent parameters in describing linear poroelastic behavior: E, ν, k, μ, K_s, n and K_f.

Because a poroelastic medium can be envisioned as a "two-phase" medium containing interacting solid and fluid components, it is important to be able to identify the responses of the solid and fluid subjected to the external loads, both separately and collectively. This identification delineates the responses that enable parametric determination, especially using laboratory experimental techniques. Theoretically, with an analogy to the definition of solid strain $\Delta\varepsilon$, defining change in solid volume relative to a reference volume, fluid content ζ is defined as a fluid "strain" that records the fluid volume change due to mass transport. The incremental form of the generated strain energy W in its normal and reversible forms may be described as (Detournay and Cheng 1993)

$$dW = \sigma_{ij} d\varepsilon_{ij} + p d\zeta = \varepsilon_{ij} d\sigma_{ij} + \zeta dp \tag{2.131}$$

The following relationship is readily recovered from the above equation:

$$\frac{\partial \varepsilon_{ij}}{\partial p} = \frac{\partial \zeta}{\partial \sigma_{ij}}. \tag{2.132}$$

The coupled "solid strain"-stress and "fluid strain"-pressure constitutive relationships can be defined as (Ghaboussi and Wilson 1973)

$$\varepsilon_{ij} = \frac{1+\nu}{E}\sigma_{ij} - \frac{\nu}{E}\sigma_{kk}\delta_{ij} + \alpha p \delta_{ij}, \tag{2.133}$$

$$\zeta = \alpha \varepsilon_{kk} + \frac{p}{M} \tag{2.134}$$

where M is the Biot modulus defined in Eq. (2.130).

The fluid content ζ may be used as a primary unknown or dependent variable, for instance, as used to evaluate surface subsidence as a result of fluid flow activated by a regional earthquake (Segall 1985). However, direct measurement is difficult, and it is more frequently used as an intermediate step to define pressure variation. Nevertheless, its introduction has some theoretical and practical implications. Defining ζ enables the poroelastic formulation to be expressed in a consistent manner for both solid deformation and fluid flow. In other words, ζ can be related to strain ε in an analogous manner that pore pressure p is associated with stress σ, as extensively developed in the previous equations. Pragmatically, the introduction of ζ leads directly to the concept of drained and undrained poroelastic responses, eliciting the design of laboratory experiments for parameter determination. For example, Skempton (1954) provided a mechanism to determine the Skempton pore pressure coefficient B based on an undrained test in which fluid leakage from the tested samples was fully restricted (e.g., $\zeta = 0$). This undrained test enables the derivation of the essential poroelastic parameter α (Biot coefficient) through the following expression (Detournay and Cheng 1993):

$$B = \frac{\alpha K_f}{[\alpha - n(1-\alpha)]K_f + nK} \tag{2.135}$$

where K_f and K are the bulk moduli of fluid and solid skeleton, respectively, and n is the porosity. Alternatively (Detournay and Cheng 1993), this may be defined as

$$B = \frac{3(\nu_u - \nu)}{\alpha(1-2\nu)(1+\nu_u)} \tag{2.136}$$

where ν_u and ν are the undrained and drained Poisson ratios, respectively.

The introduction of additional parameters, such as undrained Poisson ratio ν_u and undrained bulk modulus K_u, yields a physically based interpretation of the poroelastic parameters. The determination of poroelastic parameters becomes more flexible with the addition of these extra "drained" and "undrained" parameters. Alternatively, confusion may result as to which parameters (i.e., drained or undrained) should be used in the calculation of transient processes since these parameters only reflect the two extreme cases (i.e., $t = 0$ and $t \to \infty$) and are time-independent. In most applications of poroelastic theory, the undrained concept has been restricted to simplified or limiting cases (e.g., fast loading) with a decoupled poroelastic formulation, or cases involving experimental determination of material parameters. Consequently, the drained parameters are dominant choices even for the transient analysis of linear poroelasticity.

In Biot's theory, the volumetric strain can be written as

$$\varepsilon_{ii} = \frac{1}{K}\left[\frac{1}{3}(trace(\sigma_{ij}) + \frac{K}{H}p\right], \tag{2.137}$$

where K is the effective modulus of the skeleton, and H is a constant (Biot 1941), defined as

$$H = \frac{1}{K} - \frac{1}{K_s}. \tag{2.138}$$

For isotropic stress conditions, one has

$$\varepsilon_{ii} = \frac{1}{K}\sigma^e \tag{2.139}$$

where effective stress σ^e may be redefined as

$$\sigma^e = \frac{1}{3}\sigma_{kk} + \alpha p \tag{2.140}$$

where α is the Biot coefficient defined in Eq. (2.61) and σ_{kk} is the total hydrostatic stress. This equation reduces to Terzaghi's theory only when α is unity.

Although α may approach unity for compressible rocks (Kranz et al. 1979; Walsh 1981), the magnitude may be determined from both the degree of rock fracturing and solid bulk modulus. Geertsma (1957) and Skempton (1960) proposed, on experimental grounds, Eq. (2.61). Nur and Byerlee (1971) verified that relationship theoretically. It is only when the effective compressibility of the dry aggregate is much greater than the intrinsic compressibility of the solid grains ($K \ll K_s$) that Terzaghi's effective stress relationship, with $\alpha = 1$, is valid.

In the Biot approach (1941), the lumped compressibility c^* is defined as

$$c^* = \frac{1}{R} - \frac{\alpha}{H} \tag{2.141}$$

where R is an additional Biot constant. Biot's component parameters (H, R etc.) have not been widely adopted in practice, however, due to their lack of direct physical interpretation.

Biot's field equation for stress-based analysis, which has a different format than that of the displacement-based method of Eq. (2.11), was expressed by Cleary (1977) as

$$Gu_{i,kk} + \frac{G}{1-2\nu}u_{k,ki} - \frac{3(\nu_u - \nu)}{B(1+\nu_u)(1-2\nu)}p_{,i} = 0 \tag{2.142}$$

where ν_u and ν are the undrained and drained Poisson ratios, respectively; B is Skempton's constant (Skempton 1954) which is defined as the change in pore pressure per unit change in confining pressure under undrained conditions.

For the stress-based method, the governing equation in the fluid phase may be written as

$$\frac{k}{\mu}p_{,kk} = \frac{9(\nu_u - \nu)}{B^2E(1+\nu_u)}\dot{p} + \frac{3(\nu_u - \nu)}{B(1+\nu_u)(1-2\nu)}\dot{\varepsilon}_{kk}. \tag{2.143}$$

With the same unknown distribution of fluid pressure, present in both the displacement-based method (refer to Chapter 3 on flow) and the stress-based method, the following relationships are readily obtained

$$\alpha = \frac{3(\nu_u - \nu)}{B(1+\nu_u)(1-2\nu)} \tag{2.144}$$

$$c^* = \frac{9(\nu_u - \nu)}{B^2E(1+\nu_u)}. \tag{2.145}$$

From Eq. (2.61), the Biot coefficient α may be modified as

$$\alpha = \frac{E}{3(1-2\nu)}(\frac{1}{K} - \frac{1}{K_s}). \tag{2.146}$$

Comparing Eq. (2.144) with Eq. (2.146), the Skempton constant may be expressed as

$$B = \frac{9(\nu_u - \nu)}{E(1+\nu_u)} \left(\frac{1}{K} - \frac{1}{K_s}\right)^{-1}. \tag{2.147}$$

For the case $K \ll K_s$, Eq. (2.147) collapses to Eq. (2.144) with $\alpha = 1$, consistent with the definitions. The coefficient B can be derived from Eq. (2.145) as

$$B = \sqrt{\frac{9(\nu_u - \nu)}{c^* E(1+\nu_u)}}. \tag{2.148}$$

Equating Eq. (2.147) to Eq. (2.148), undrained and drained Poisson ratios ν_u, ν can be described by

$$\nu_u = \frac{\nu + \psi^*}{1 - \psi^*}, \tag{2.149}$$

and

$$\nu = \nu_u - \psi^*(1+\nu_u), \tag{2.150}$$

where

$$\psi^* = \frac{E}{9c^*}\left(\frac{1}{K} - \frac{1}{K_s}\right)^2. \tag{2.151}$$

It is understood from Eq. (2.151) that $\psi^* \geq 0$; therefore, $\nu_u \geq \nu$. If $\psi^* = 0$ and $\nu_u = \nu$, then $\alpha = 0$, and fluid flow is fully decoupled from the solid deformation.

Rice and Cleary (1976) pointed out that

$$\nu \leq \nu_u \leq \frac{1}{2}. \tag{2.152}$$

Referring to Eqs. (2.151) and (2.152), the following relation results:

$$\frac{1-2\nu}{3} \geq \psi^* \geq 0. \tag{2.153}$$

Substituting Eq. (2.149) into Eq. (2.148), Skempton's constant B reduces to

$$B = \frac{1}{c^*}\left(\frac{1}{K} - \frac{1}{K_s}\right). \tag{2.154}$$

Using the expression for c^* proposed by Rice and Cleary (1976), B in Eq. (2.154) can be rewritten as

$$B = \left[\frac{1}{K} - \frac{1}{K_s}\right]\left[\frac{n}{K_f} - \frac{1+n}{K_s} + \frac{1}{K}\right]^{-1}. \tag{2.155}$$

When $K \ll K_s$, which is often the case for soil, B in Eq. (2.155) reduces to

$$B = \left[1 + \frac{nK}{K_f}\right]^{-1}. \tag{2.156}$$

This is the identical form of B to that proposed by Skempton (1954).

Experimental determination of the parameters for the single-porosity model is by no means trivial. Careful preparation, setup, and execution are needed to control and measure the interactive behavior of pore pressure, external loads, and induced stresses, as well as displacements. In view of the theoretical basis for the parametric relationships, interested readers are referred to Biot and Willis (1957) and Geertsma (1957) for further details.

2.3.2.2 Heterogeneous Media

A similar analysis may be extended to dual-porosity media, representative of fractured porous media. Inclusion of additional physical representations in the mathematical formulation requires that extra phenomenological parameters be determined. In contrast to single-porosity poroelasticity, the number of required parameters are almost doubled for the dual-porosity poroelastic formulation, especially in the representation of fluid flow (refer to Chapter 3 on flow), since the parametric definitions are made separately for the two distinct media. For the latter formulation, the number of required parameters is extended to eight, including elastic modulus E, Poisson ratio ν, intrinsic permeabilities κ_1 and κ_2, pressure ratio factors α_1 and α_2 (Bai et al. 1993), and coupling moduli M_1 and M_2, where the subscripts 1 and 2 represent matrix and fractures, respectively. Similar to the single-porosity formulation, these parameters are not completely independent. For example, analogous to α in Eq. (2.61), which is α_1 for the present case, α_2 can be expressed as

$$\alpha_2 = 1 - \frac{K_2}{K_n s^*} \tag{2.157}$$

where K_2 is the bulk modulus of the fractured medium, K_n is the fracture normal stiffness, and s^* is the fracture spacing.

In a similar manner as for M in Eq. (2.130) (M_1 for the present case), M_2 can be written as

$$\frac{1}{M_2} = \frac{n_2}{K_f} + \frac{\alpha_2 - n_2}{K_n s^*} \tag{2.158}$$

where n_2 is the fracture porosity.

As a result, the following 12 are the minimum number of independent parameters for the dual-porosity system: E, ν, μ, K_f, K_s, k_1, n_1, k_2, n_2, K_2, K_n, and s^*. This number can be reduced with various simplifying assumptions. For example, K_2 may be excluded if the difference between the bulk modulus of the fractured medium and that of the porous medium is neglected.

The main characteristic of the dual-porosity model is to distinguish between the fracture and intergranular flows. In view of deformation, the difference between the single-porosity poroelastic and dual-porosity poroelastic model is minimized if a single effective stress law is applied. If the double effective stress law is used, the parameters are separate, but are frequently combined to form lumped representation of deformation effects.

The matrix pressure ratio factor α_1 is described by Eq. (2.61). Determination of the fracture pressure ratio factor α_2 appears to be more difficult because of its nonlinear dependence on the stress history. Robin (1973) suggested

$$\alpha_2 = 1 - v_p \beta_s \frac{\partial v_p}{\partial p} \tag{2.159}$$

where v_p is the pore volume, β_s is the rock compressibility in the fractured medium, and $\frac{\partial v_p}{\partial p}$ is the rate of change in pore volume with applied hydrostatic pressure for a natural fracture without pore fluid.

Based on experimental results, Walsh (1981) noted that α_2 varies between 0.5 and 1. Specifically reported were, $\alpha_2 = 0.9$ for fractures with polished surfaces and $\alpha_2 = 0.56$ for a tension fracture. Kranz et al. (1979) expressed their results in a similar manner, proposing that α_2 should be less than 1 for jointed rock, and that α_2 should approach unity for whole rock. They pointed out that the stress dependence of α_2 is a function of both surface roughness and ambient pressure.

For dual-porosity media, parametric determination through experimental testing becomes substantially more difficult than that for single-porosity media, both as a result of the increased number of parameters and the difficulty in separating the influences between the fracture and the matrix. The experimental determination of dual-porosity poroelastic parameters has been the subject of limited discussion (Wilson and Aifantis 1982, Berryman and Wang 1995). The primary approaches in either field tests or laboratory tests attempt to isolate the tested rocks between two distinct groups, i.e., intact and fractured rocks. Attempts to directly test fractured poroelastic rocks is difficult, at best, since the parameters describing behavior relate to the individual components. The deconvolution of matrix and fracture response from that of the aggregate specimen is not trivial. Laboratory experiments are feasible, since the testing environment can be properly controlled. Field experiments are difficult because the response of the two distinct components (i.e., matrix and fractures) within the total system are usually lumped, and are difficult to distinguish where only restricted control may be applied over the test conditions.

Where testing of the component parts is feasible, results can be compared between the two different single and dual porosity media in accommodating different length and time scales (Wilson and Aifantis 1982). Noticeable differences between the spatial and temporal responses of the intact and fractured media result when subjected to external loading. For example, the matrix blocks typically possess significant porosity and resulting storage potential, and the fractures exhibit the converse, i.e., low storage but high conductivity. Flow response is most rapid in the fractures with substantially slower response times in the porous blocks. These concepts are useful to interpret the responses of porous or fractured media while designing the field flow tests, such as drawdown tests.

With reference to the length and time scales, it is also important to recognize the coupling between elastic deformation and fluid pressure dissipation within the respective media. For early time response, sometimes referred to as undrained response, fracture pressures immediately following the initial loading are typically higher than matrix pressures, due to the significantly larger fracture compliance. Fluid pressures within the fractures, however, typically dissipate rapidly following this initial build-up. Distinguishing the impacts of rock deformation from fluid flow in a coupled system can be accomplished by performing separate tests either with or without mechanical loading (Bai and Meng 1997).

2.3.3 Anisotropic Properties

Material constitutive relations may incorporate anisotropic properties, viewed both at the macroscopic and microscopic scales.

2.3.3.1 Material Anisotropy

For general anisotropic linear elastic behavior, the stress ($\tilde{\sigma}$) and strain ($\tilde{\varepsilon}$) relation may be expressed as

$$\tilde{\sigma} = \tilde{\Psi}:\tilde{\varepsilon} \tag{2.160}$$

where $\tilde{\Psi}$ is an elastic modulus tensor that can be expanded in Cartesian coordinates and expressed as a multiplication of matrices, i.e, $\sigma_{ij} = \Psi_{ijkl}\varepsilon_{kl}$ $(i, j, k, l = 1, 2, 3)$. Ψ has four indices as its subscript; as a result, it is a fourth order tensor. Ψ_{ijkl} may be substituted by Ψ_{ij} for notational convenience. However, the contraction is not carried out on these tensors in the present analysis. Therefore, the appearance of these four indices cannot be omitted. Furthermore $\sigma_{ij} = \Psi_{ijkl}\varepsilon_{kl}(i, j, k, l = 1, 2, 3)$ is coherent with the Einstein summation convention, which can be expressed as

$$\sigma_{ij} = (\sigma_{11},\ \ \sigma_{12},\ \ \sigma_{13},\ \ \sigma_{21},\ \ \sigma_{22},\ \ \sigma_{23},\ \ \sigma_{31},\ \ \sigma_{32},\ \ \sigma_{33})^T \tag{2.161}$$

$$
\Psi_{ijkl} = \begin{bmatrix}
\Psi_{1111} & \Psi_{1112} & \Psi_{1113} & \Psi_{1121} & \Psi_{1122} & \Psi_{1123} & \Psi_{1131} & \Psi_{1132} & \Psi_{1133} \\
\Psi_{1211} & \cdot & \cdot & \cdot & \cdot & \cdot & \cdot & \cdot & \cdot \\
\Psi_{1311} & \cdot & \Psi_{1313} & \cdot & \cdot & \cdot & \cdot & \cdot & \cdot \\
\cdot & \cdot & \cdot & \cdot & \cdot & \cdot & \cdot & \cdot & \cdot \\
\cdot & \cdot & \cdot & \cdot & \Psi_{2222} & \cdot & \cdot & \cdot & \cdot \\
\Psi_{2311} & \cdot & \cdot & \cdot & \cdot & \cdot & \cdot & \cdot & \Psi_{2333} \\
\cdot & \cdot & \cdot & \cdot & \cdot & \cdot & \Psi_{3131} & \cdot & \cdot \\
\cdot & \cdot & \cdot & \cdot & \cdot & \cdot & \cdot & \cdot & \cdot \\
\Psi_{3311} & \cdot & \cdot & \cdot & \Psi_{3322} & \cdot & \cdot & \cdot & \Psi_{3333}
\end{bmatrix} \tag{2.162}
$$

$$
\varepsilon_{ij} = (\varepsilon_{11}, \quad \varepsilon_{12}, \quad \varepsilon_{13}, \quad \varepsilon_{21}, \quad \varepsilon_{22}, \quad \varepsilon_{23}, \quad \varepsilon_{31}, \quad \varepsilon_{32}, \quad \varepsilon_{33})^T. \tag{2.163}
$$

Due to the symmetry of the strain tensor, i.e,

$$
\varepsilon_{kl} = \varepsilon_{lk} \tag{2.164}
$$

results in

$$
\sigma_{ij} = \Psi_{ijkl}\varepsilon_{kl} = \Psi_{ijlk}\varepsilon_{lk} \tag{2.165}
$$

hence,

$$
\Psi_{ijkl} = \Psi_{ijlk}. \tag{2.166}
$$

Therefore, Eq. (2.162) reduces to

$$
\begin{Bmatrix} \sigma_{11} \\ \sigma_{12} \\ \sigma_{13} \\ \sigma_{21} \\ \sigma_{22} \\ \sigma_{23} \\ \sigma_{31} \\ \sigma_{32} \\ \sigma_{33} \end{Bmatrix} = \begin{bmatrix}
\Psi_{1111} & \Psi_{1112} & \Psi_{1113} & \Psi_{1122} & \Psi_{1123} & \Psi_{1133} \\
\Psi_{1211} & \cdot & \cdot & \cdot & \cdot & \cdot \\
\Psi_{1311} & \cdot & \Psi_{1313} & \cdot & \cdot & \cdot \\
\cdot & \cdot & \cdot & \cdot & \cdot & \cdot \\
\cdot & \cdot & \cdot & \Psi_{2222} & \cdot & \cdot \\
\Psi_{2311} & \cdot & \cdot & \cdot & \cdot & \Psi_{2333} \\
\cdot & \cdot & \cdot & \cdot & \cdot & \cdot \\
\cdot & \cdot & \cdot & \cdot & \cdot & \cdot \\
\Psi_{3311} & \cdot & \cdot & \Psi_{3322} & \cdot & \Psi_{3333}
\end{bmatrix} \begin{Bmatrix} \varepsilon_{11} \\ 2\varepsilon_{12} \\ 2\varepsilon_{13} \\ \varepsilon_{22} \\ 2\varepsilon_{23} \\ \varepsilon_{33} \end{Bmatrix} \tag{2.167}
$$

Furthermore, because of the symmetry of the stress tensor

$$
\begin{cases} \sigma_{ij} = \Psi_{ijkl}\varepsilon_{kl} \\ \sigma_{ji} = \Psi_{jikl}\varepsilon_{kl} \end{cases} \tag{2.168}
$$

hence,

$$
\Psi_{ijkl} = \Psi_{jikl}. \tag{2.169}
$$

Due to linear dependency, i.e., $\sigma_{ij} = \sigma_{ji}$ $(i \neq j)$, Eq. (2.167) further reduces to

$$\begin{Bmatrix} \sigma_{11} \\ \sigma_{12} \\ \sigma_{13} \\ \sigma_{22} \\ \sigma_{23} \\ \sigma_{33} \end{Bmatrix} = \begin{bmatrix} \Psi_{1111} & \Psi_{1112} & \Psi_{1113} & \Psi_{1122} & \Psi_{1123} & \Psi_{1133} \\ \Psi_{1211} & \cdot & \cdot & \cdot & \cdot & \cdot \\ \Psi_{1311} & \cdot & \Psi_{1313} & \cdot & \cdot & \cdot \\ \Psi_{2211} & \cdot & \cdot & \Psi_{2222} & \cdot & \cdot \\ \Psi_{2311} & \cdot & \cdot & \cdot & \cdot & \Psi_{2333} \\ \Psi_{3311} & \cdot & \cdot & \Psi_{3322} & \cdot & \Psi_{3333} \end{bmatrix} \begin{Bmatrix} \varepsilon_{11} \\ 2\varepsilon_{12} \\ 2\varepsilon_{13} \\ \varepsilon_{22} \\ 2\varepsilon_{23} \\ \varepsilon_{33} \end{Bmatrix} \tag{2.170}$$

The 9×9 matrix of $\tilde{\Psi}$ in Eq. (2.162) has been reduced to a 6×6 matrix of $\tilde{\Psi}$ in Eq. (2.170). For the latter matrix $\tilde{\Psi}$, it is shown that it is also symmetric.

Elastic potential energy in a linear elastic system is given by

$$U = \frac{1}{2}\tilde{\varepsilon}{:}\tilde{\Psi}{:}\tilde{\varepsilon} = \frac{1}{2}\Psi_{ijkl}\varepsilon_{ij}\varepsilon_{kl}. \tag{2.171}$$

Taking derivatives with respect to the respective strain tensor yields

$$\frac{dU}{d\varepsilon_{ij}} = \frac{1}{2}\Psi_{ijkl}\varepsilon_{kl} = \frac{1}{2}\sigma_{ij} \tag{2.172}$$

and

$$\frac{dU}{d\varepsilon_{kl}} = \frac{1}{2}\Psi_{ijkl}\varepsilon_{ij} = \frac{1}{2}\sigma_{kl}. \tag{2.173}$$

It is permissible to interchange kl and ij in Eq. (2.173), yielding

$$\frac{dU}{d\varepsilon_{ij}} = \frac{1}{2}\Psi_{klij}\varepsilon_{ij} = \frac{1}{2}\sigma_{ij}. \tag{2.174}$$

Equating Eqs. (2.172) and (2.174), results in

$$\Psi_{ijkl} = \Psi_{klij}. \tag{2.175}$$

Hence, the symmetry of the $\tilde{\Psi}$ matrix in Eq. (2.170) is apparent. The proofs hold for all fourth order elastic modulus tensors, and a similar derivation can be developed for fourth order compliances. The number of independent coefficients in Eq. (2.170) is 21 for complete anisotropic media. This number of independent coefficients is decreased as symmetry is increased, as summarized in Table 2.2.

Table 2.2 Independent parameters for the elastic modulus tensor.

Case of study	Number of independent parameters
anisotropy	21
monocliny	13
orthotropy	9
axisymmetry	5
isotropy	2

In addition to the elastic modulus tensor $\tilde{M}$, other elastic tensors include:

- Compliance tensor $\tilde{C}$

  This tensor is the inverse of the elastic modulus tensor [Eq. (2.160)] and can be rewritten as

$$\tilde{\varepsilon} = \tilde{C}:\tilde{\sigma} \tag{2.176}$$

  where the compliance tensor is

$$\tilde{C} = C_{ijkl} = \Psi_{ijkl}^{-1} = \begin{bmatrix} C_{1111} & C_{1122} & C_{1133} & C_{1112} & C_{1113} & C_{1123} \\ C_{2211} & C_{2222} & C_{2233} & C_{2212} & C_{2213} & C_{2223} \\ C_{3311} & C_{3322} & C_{3333} & C_{3312} & C_{3313} & C_{3323} \\ C_{1211} & C_{1222} & C_{1233} & C_{1212} & C_{1213} & C_{1223} \\ C_{1311} & C_{1322} & C_{1333} & C_{1312} & C_{1313} & C_{1323} \\ C_{2311} & C_{2322} & C_{2333} & C_{2312} & C_{2313} & C_{2323} \end{bmatrix} \tag{2.177}$$

  This tensor or matrix is symmetric.

- Modified Skempton tensor

  This tensor accounts for anisotropic stress changes due to pore pressure changes under undrained conditions. It is defined as

$$\tilde{B} = B_{ij} = \begin{bmatrix} B_{11} & B_{12} & B_{13} \\ B_{12} & B_{22} & B_{23} \\ B_{13} & B_{32} & B_{33} \end{bmatrix} \tag{2.178}$$

  and acts on the components of normal stress.

- Modified Biot's coefficient tensor

  This tensor is a coupling term to link the influence of stress with that of pore pressure. It is given as

$$\tilde{\alpha} = \alpha_{ij} = \begin{bmatrix} \alpha_{11} & \alpha_{12} & \alpha_{13} \\ \alpha_{12} & \alpha_{22} & \alpha_{23} \\ \alpha_{13} & \alpha_{23} & \alpha_{33} \end{bmatrix} \tag{2.179}$$

  and α_{ij} in Eq. (2.179) are symmetric, with all component coefficients assumed to be obtained under drained condition.

2.3.3.2 Material Constitutive Relations

These relations are discussed separately for homogeneous and heterogeneous media.

Homogeneous Media

The relation between the elastic modulus tensor $\tilde{\Psi}$ and the compliance tensor $\tilde{C}$ may be described by

$$\Psi_{ijkl}C_{klmn} = \frac{1}{2}(\delta_{im}\delta_{jn} + \delta_{in}\delta_{jm}). \tag{2.180}$$

δ_{ij} in Eq. (2.180) is not a unit second order tensor $\tilde{\delta}$, but the Kronecker symbol, where $\delta_{ij} = 1$ for $i = j$ and $\delta_{ij} = 0$ for $i \neq j$.

Pore pressure p and strain $\tilde{\varepsilon}$ may be related by the following equation:

$$p = M(\zeta - \tilde{\alpha}{:}\tilde{\varepsilon}) \tag{2.181}$$

where ζ is the fluid content, $\tilde{\alpha}$ is the Biot coefficient tensor, and M is the Biot modulus (Detournay and Cheng 1993), which is a scalar, signifying the storage capacity of the porous medium.

Macroscopic Scale: The total strain tensor can be written as

$$\tilde{\varepsilon} = \tilde{C}{:}\tilde{\sigma}^e \tag{2.182}$$

where $\tilde{\sigma}^e$ is the effective stress, which can be expressed as

$$\tilde{\sigma}^e = \tilde{\sigma} + \tilde{\alpha}p \tag{2.183}$$

where $\tilde{\sigma}$ is the total stress.

Substituting Eq. (2.183) into Eq. (2.182), results in

$$\tilde{\varepsilon} = \tilde{C}{:}(\tilde{\sigma} + \tilde{\alpha}p). \tag{2.184}$$

It is known that

$$\Psi_{ijkl}C_{klmn} = \frac{1}{2}(\delta_{im}\delta_{jn} + \delta_{in}\delta_{jm}). \tag{2.185}$$

Using the relation in Eq. (2.185), Eq. (2.184) can be rewritten as

$$\tilde{\sigma} = \tilde{\Psi}{:}\tilde{\varepsilon} - \tilde{\alpha}p. \tag{2.186}$$

The pore pressure may be related to the change in water content and solid strain by Eq. (2.181) (Rice and Cleary 1976).

For the undrained condition, $\zeta = 0$, then, from Eq. (2.181)

$$p = M\tilde{\alpha}{:}\tilde{\varepsilon}. \tag{2.187}$$

Substituting Eq. (2.187) into Eq. (2.186), yields

$$\tilde{\sigma} = \tilde{\Psi}^u{:}\tilde{\varepsilon} \tag{2.188}$$

where $\tilde{\Psi}^u$ is the undrained elastic modulus, and is given as

$$\tilde{\Psi}^u = \tilde{\Psi} + M\tilde{\alpha} \otimes \tilde{\alpha}. \tag{2.189}$$

where $\otimes$ represents tensor multiplication.

Inversion of Eq. (2.188), leads to

$$\tilde{\varepsilon} = \tilde{C}^u{:}\tilde{\sigma} \tag{2.190}$$

where

$$\tilde{C}^u = (\tilde{\Psi} + M\tilde{\alpha} \otimes \tilde{\alpha})^{-1}. \tag{2.191}$$

If the solid deformation is negligible, then $\tilde{\varepsilon} = 0$, Eq. (2.181) reduces to

$$p = M\zeta. \tag{2.192}$$

Under undrained conditions, substituting Eq. (2.190) into Eq. (2.181) with $\tilde{\zeta} = 0$, yields

$$p = -M\tilde{\alpha}{:}\tilde{C}^u{:}\tilde{\sigma}. \tag{2.193}$$

Eq. (2.193) can be rewritten as

$$p = -\tilde{B}{:}\tilde{\sigma} \tag{2.194}$$

where $\tilde{B}$ is the modified Skempton coefficient tensor, which is described by

$$\tilde{B} = M\tilde{\alpha}{:}\tilde{C}^u. \tag{2.195}$$

Microscopic Scale: The bulk parameters derived in the previous section may also be related to micromechanical parameters. Development of a correlation between strains and porosity allows deformation to be directly related to the fluid flow without explicitly resorting to the effective stress law.

The effective strain can be expressed as

$$\tilde{\varepsilon} = \tilde{C}{:}\tilde{\sigma}^e - \tilde{\eta}p \tag{2.196}$$

where $\tilde{\eta}$ is a tensor related to solid grain compressibility, which may be expressed as

$$\tilde{\eta} = \frac{\tilde{K}}{K_s} \tag{2.197}$$

where $\tilde{K}$ is the elastic bulk modulus tensor, and K_s is the bulk modulus of the solid grains (a scalar).

Substituting a general effective stress law similar to that of Terzaghi (1943), i.e.,

$$\tilde{\sigma}^e = \tilde{\sigma} + \tilde{\delta} p \tag{2.198}$$

into Eq. (2.196), yields

$$\tilde{\varepsilon} = \tilde{C}:\tilde{\sigma} + (\tilde{C}:\tilde{\delta} - \tilde{\eta})p. \tag{2.199}$$

Considering the solid strain only, then

$$\tilde{\varepsilon}^s = \tilde{C}^s:\tilde{\sigma}^s + (\tilde{C}^s:\tilde{\delta} - \tilde{\eta})^s p \tag{2.200}$$

where $\tilde{C}^s$ and $\tilde{\sigma}^s$ are the compliance and stress of the solid constituent and $\tilde{\eta}$ is related to the solid grains directly.

Separating the total strain into solid and fluid components results in

$$\tilde{\varepsilon} = (1-n)\tilde{\varepsilon}^s + n\tilde{\varepsilon}^p \tag{2.201}$$

where n is the porosity, $\tilde{\varepsilon}^s$ is the strain of the solid constituent, and $\tilde{\varepsilon}^p$ is the strain in the fluid.

The total stress, distributed proportionally to the porosity, can be given as

$$\tilde{\sigma} = (1-n)\tilde{\sigma}^s - n\tilde{\delta} p \tag{2.202}$$

enabling $\tilde{\sigma}^s$ to be defined directly as

$$\tilde{\sigma}^s = \frac{\tilde{\sigma}}{1-n} + \frac{n}{1-n}\tilde{\delta} p. \tag{2.203}$$

If the total stress in Eq. (2.202) is equivalent to the load ($-\tilde{\delta}P$), and the load is also equal to the pressure p, Eq. (2.202) can be rewritten as

$$-\tilde{\delta}P = (1-n)\tilde{\sigma}^s - n\tilde{\delta}P. \tag{2.204}$$

Rearranging, Eq. (2.204) reduces to

$$\tilde{\sigma}^s = -\tilde{\delta}P. \tag{2.205}$$

Substituting p with P in Eq. (2.200), and inserting the relation of Eq. (2.205), Eq. (2.200) reduces to

$$\tilde{\varepsilon}^s = -\tilde{\eta}^s P. \tag{2.206}$$

Correspondingly, it is clear that $\tilde{\eta}^s$ is related directly to the properties of the solid grains.

The fluid deformation can be conceptualized as the following two parts:

$$\tilde{\varepsilon}^p = \tilde{\varepsilon}^i + \tilde{\varepsilon}^e \tag{2.207}$$

where $\tilde{\varepsilon}^i$ denotes fluid internal expansion and contraction, and $\tilde{\varepsilon}^e$ represents fluid external expulsion and injection. With this expression, Eq. (2.201) can be rewritten as

$$\tilde{\varepsilon} = (1-n)\tilde{\varepsilon}^s + n(\tilde{\varepsilon}^i + \tilde{\varepsilon}^e). \tag{2.208}$$

$\tilde{\varepsilon}^e$ is mainly responsible for the variation of fluid volume; i.e.,

$$\zeta = n\tilde{\varepsilon}^e{:}\tilde{\delta}. \tag{2.209}$$

Substituting the expression for $\tilde{\varepsilon}^e$ from Eq. (2.208) into Eq. (2.209), yields

$$\zeta = [\tilde{\varepsilon} - (1-n)\tilde{\varepsilon}^s - n\tilde{\varepsilon}^i]{:}\tilde{\delta}. \tag{2.210}$$

It is known that

$$\tilde{\varepsilon}^i = -c_f p\tilde{\delta} \tag{2.211}$$

where c_f is the fluid compressibility.

Substituting $\tilde{\varepsilon}$ from Eq. (2.199), $\tilde{\varepsilon}^s$ from Eq. (2.200), $\tilde{\sigma}^s$ from Eq. (2.203), and $\tilde{\varepsilon}^i$ from Eq. (2.211) into Eq. (2.210), yields

$$\zeta = \tilde{\delta}{:}(\tilde{C} - \tilde{C}^s){:}\tilde{\sigma} + [(\tilde{C} - \tilde{C}^s){:}\tilde{\delta} + (\tilde{\eta}^s - \tilde{\eta}) + n(c_f\tilde{\delta} - \tilde{\eta})]{:}\tilde{\delta}p. \tag{2.212}$$

Assuming $\zeta = 0$ in Eq. (2.212), the modified Skempton coefficient tensor, $\tilde{B}$, is defined as

$$p = -\tilde{B}{:}\tilde{\sigma} \tag{2.213}$$

where

$$\tilde{B} = \frac{\tilde{\delta}{:}(\tilde{C} - \tilde{C}^s)}{[\tilde{\delta}{:}(\tilde{C} - \tilde{C}^s) + (\tilde{\eta}^s - \tilde{\eta}) + n(c_f\tilde{\delta} - \tilde{\eta})]{:}\tilde{\delta}}. \tag{2.214}$$

Heterogeneous Media

The double effective stress law for the dual-porosity poroelastic media can be expressed as (Elsworth and Bai 1992):

$$\begin{cases} \tilde{\sigma}_1 = \tilde{\sigma}_1^e - \tilde{\alpha}_1 p_1 \\ \tilde{\sigma}_2 = \tilde{\sigma}_2^e - \tilde{\alpha}_2 p_2 \end{cases} \tag{2.215}$$

where subscripts 1 and 2 represent matrix and fractures, respectively.

The effective stress and strain relations can be described by

$$\begin{cases} \tilde{\sigma}_1^e = \tilde{\Psi}_1 : \tilde{\varepsilon}_1 \\ \tilde{\sigma}_2^e = \tilde{\Psi}_2 : \tilde{\varepsilon}_2. \end{cases} \tag{2.216}$$

Alternatively, the inverse strain-effective stress relations may be defined as

$$\begin{cases} \tilde{\varepsilon}_1 = \tilde{C}_1 : \tilde{\sigma}_1^e \\ \tilde{\varepsilon}_2 = \tilde{C}_2 : \tilde{\sigma}_2^e. \end{cases} \tag{2.217}$$

As for single-porosity media, the pressure-strain relation for the dual-porosity medium can be given as

$$\begin{cases} p_1 = M_1(\zeta_1 - \tilde{\alpha}_1 : \tilde{\varepsilon}_1) \\ p_2 = M_2(\zeta_2 - \tilde{\alpha}_2 : \tilde{\varepsilon}_2) \end{cases} \tag{2.218}$$

where M is the storage term.

Interestingly, Berryman and Wang (1995) illustrated that a change in fluid content in one solid component (e.g., matrix) is coupled to changes in an adjacent component (e.g., fractures); i.e.,

$$\begin{cases} p_1 = a_{11}\zeta_1 + a_{12}\varepsilon + a_{13}\zeta_2 \\ p_2 = a_{21}\zeta_1 + a_{22}\varepsilon + a_{23}\zeta_2 \end{cases} \tag{2.219}$$

where ε is the total strain and a_{ij} $(i = 1, 2; j = 1, 2, 3)$ are phenomenological constants determined experimentally. Unlike other dual-porosity poroelastic models, Eq. (2.219) shows that a_{13} and a_{21} are not equal to zero.

The total strain tensor, representing components accumulated from matrix and fractures, is given as

$$\tilde{\varepsilon} = \tilde{\varepsilon}_1 + \tilde{\varepsilon}_2. \tag{2.220}$$

Substituting Eq. (2.217) into Eq. (2.220) yields

$$\tilde{\varepsilon} = \tilde{C}_1 : \tilde{\sigma}_1^e + \tilde{C}_2 : \tilde{\sigma}_2^e. \tag{2.221}$$

Substituting Eq. (2.215) into the above equation leads to

$$\tilde{\varepsilon} = \tilde{C}_1 : (\tilde{\sigma}_1 + \tilde{\alpha}_1 p_1) + \tilde{C}_2 : (\tilde{\sigma}_2 + \tilde{\alpha}_2 p_2). \tag{2.222}$$

Local equilibrium requires that changes in total stress within adjacent phases must remain in equilibrium (Elsworth and Bai 1992); i.e.,

$$\tilde{\sigma} = \tilde{\sigma}_1 = \tilde{\sigma}_2. \tag{2.223}$$

Using this relation, Eq. (2.222) can be modified and expressed with total stress as

$$\tilde{\sigma} = \tilde{\Psi}_{12} : (\tilde{\varepsilon} - \tilde{C}_1 : \tilde{\alpha}_1 p_1 - \tilde{C}_2 : \tilde{\alpha}_2 p_2) \tag{2.224}$$

where

$$\tilde{\Psi}_{12}\colon = \tilde{C}^{-1} = (\tilde{C}_1 + \tilde{C}_2)^{-1}. \tag{2.225}$$

Under the undrained condition, $\zeta_1 = \zeta_2 = 0$, and from Eq. (2.218), one has

$$\begin{cases} p_1 = -M_1\tilde{\alpha}_1 : \tilde{\varepsilon}_1 \\ p_2 = -M_2\tilde{\alpha}_2 : \tilde{\varepsilon}_2 \end{cases} \tag{2.226}$$

Substituting Eqs. (2.225) and (2.226) into Eq. (2.224), one has

$$\tilde{\sigma} = \tilde{\Psi}_{12} : (\tilde{\varepsilon} - M_1\tilde{\alpha}_1^* \otimes \tilde{\alpha}_1 : \tilde{\varepsilon}_1 - M_2\tilde{\alpha}_2^* \otimes \tilde{\alpha}_2 : \tilde{\varepsilon}_2) \tag{2.227}$$

where $\otimes$ represents tensor multiplication, and

$$\begin{cases} \tilde{\alpha}_1^* = \tilde{\Psi}_{12} : \tilde{C}_1 : \tilde{\alpha}_1 \\ \tilde{\alpha}_2^* = \tilde{\Psi}_{12} : \tilde{C}_2 : \tilde{\alpha}_2. \end{cases} \tag{2.228}$$

Substituting Eq. (2.220) into Eq. (2.227) and rearranging, Eq. (2.227) reduces to

$$\tilde{\sigma} = \tilde{\Psi}_1^u : \tilde{\varepsilon}_1 + \tilde{\Psi}_2^u : \tilde{\varepsilon}_2 \tag{2.229}$$

where $\tilde{\Psi}_1^u$ and $\tilde{\Psi}_2^u$ are undrained elastic modulus tensors, which are given as

$$\begin{cases} \tilde{\Psi}_1^u = \tilde{\Psi}_{12} + M_1\tilde{\alpha}_1^* \otimes \tilde{\alpha}_1 \\ \tilde{\Psi}_2^u = \tilde{\Psi}_{12} + M_2\tilde{\alpha}_2^* \otimes \tilde{\alpha}_2. \end{cases} \tag{2.230}$$

In the absence of solid deformation, Eq. (2.218) reduces to

$$\begin{cases} p_1 = M_1\zeta_1 \\ p_2 = M_2\zeta_2. \end{cases} \tag{2.231}$$

If interchangeably $\tilde{\varepsilon}_1 = 0$ and $\tilde{\varepsilon}_2 = 0$, Eq. (2.229) reduces to

$$\begin{cases} \tilde{\sigma}_1 = \tilde{\Psi}_1^u : \tilde{\varepsilon}_1 \\ \tilde{\sigma}_2 = \tilde{\Psi}_2^u : \tilde{\varepsilon}_2 \end{cases} \tag{2.232}$$

which can be compared to Eq. (2.216). Similarly, from Eq. (2.232)

$$\begin{cases} \tilde{\varepsilon}_1 = \tilde{C}_1^u : \tilde{\sigma}_1 \\ \tilde{\varepsilon}_2 = \tilde{C}_2^u : \tilde{\sigma}_2 \end{cases} \tag{2.233}$$

which can be compared to Eq. (2.217). It is understood that

$$\begin{cases} \tilde{C}_1^u = (\tilde{\Psi}_1^u)^{-1} \\ \tilde{C}_2^u = (\tilde{\Psi}_2^u)^{-1}. \end{cases} \tag{2.234}$$

For $\zeta_1 = \zeta_2 = 0$, substituting Eq. (2.233) into Eq. (2.218) and rearranging, results in

$$\begin{cases} p_1 = -\tilde{B}_1 : \tilde{\sigma}_1 \\ p_2 = -\tilde{B}_2 : \tilde{\sigma}_2 \end{cases} \tag{2.235}$$

where $\tilde{B}$ is the modified Skempton coefficient tensor, and is expressed as

$$\begin{cases} \tilde{B}_1 = M_1 \tilde{\alpha}_1 : \tilde{C}_1^u \\ \tilde{B}_2 = M_2 \tilde{\alpha}_2 : \tilde{C}_2^u \end{cases} \tag{2.236}$$

which can be compared to the $\tilde{B}$ for single-porosity media in Eq. (2.195).

Heterogeneous material constitutive relations may also be evaluated at the microscopic scale, through application of a similar procedure as described for the homogeneous media. This is not included here.

Chapter 3

FLOW

3.1 INTRODUCTION

The movement of fluids in porous media may be determined from the same principles applied to deformation. Conservation of mass is combined with the requirement that pressures are continuous. The constitutive relation is Darcy's law, linking pressure or head gradients with fluid fluxes. These three relations, conservation, continuity and a constitutive relation, together with initial and boundary conditions are sufficient to solve initial and boundary value problems. In solving for groundwater flows or the removal of fluids from petroleum reservoirs, a common assumption is that total stresses remain constant, enabling the system to be solved in uncoupled format. The release of fluids due to fluid compressibility and skeletal compaction is assumed to be controlled by the scalar magnitude of fluid pressure. This simplification is typically adequate, but in some circumstances, consideration of mechanical effects cannot be ignored. In general, for coupled analyses where fluid flow is related to other influences or mechanisms, the total rate changes of fluid flux due to various sources, may be symbolically expressed as

$$\dot{q} = \dot{q}_e + \dot{q}_f + \dot{q}_t + \dot{q}_o \tag{3.1}$$

where the subscripts e, p, t, o denote the sources of elastic, fluid, thermal, and other types of fluid fluxes, respectively. These sources may be internal, at local or microscopic levels, or externally applied at the boundary.

The basic flow processes, as well as the intrinsic coupled relationships, are investigated in the following.

3.2 MATHEMATICAL FORMULATION

The significant differences between behavior for homogeneous and heterogeneous media warrants separate discussion.

3.2.1 Homogeneous Media

Since the definition of "homogeneous media" is broad, the present analysis is restricted to the situation in which the properties of the medium are invariant with respect to space.

3.2.1.1 Single-Phase Fluid

The fluid flow equation can be obtained from the mass conservation relation, which in turn is a subset of the Navier-Stokes equations as described in the following.

Navier-Stokes Equations

The most general form of the flow equation can be defined in terms of the Navier-Stokes system. The fluid may be either a liquid or a gas, and relative motion may include both fluid and solid. It is assumed that the locally defined velocity vector field $\mathbf{v}$ is a twice continuously differentiable function of $\mathbf{x}$ and t. For isothermal conditions, conservation of mass and momentum, together with an equation of state, characterize the fluid motion.

The differential form of the equation of conservation of mass can be expressed as

$$\frac{\partial \rho}{\partial t} + \nabla \cdot (\rho \mathbf{v}) = 0 \tag{3.2}$$

where ρ is the general density (i.e., fluid or solid).

In accordance with Newton's second law, conservation of momentum for the fluid can be written as

$$\frac{\partial \rho \mathbf{v}}{\partial t} + \nabla \cdot (\rho \mathbf{v}\mathbf{v}) = \nabla \cdot \tilde{\sigma} \tag{3.3}$$

where $\tilde{\sigma}$ is the stress tensor, which, for a viscous fluid, can be simply written as

$$\nabla \cdot \tilde{\sigma} = -\vec{\nabla} p + \nabla \cdot \tilde{\tau} \tag{3.4}$$

where p is the fluid pressure, $\tilde{\tau}$ is the shear stress tensor, which, from the Navier-Stokes system, can be expressed as

$$\tilde{\tau} = \mu \nabla^2 \mathbf{v} \tag{3.5}$$

where μ is the fluid dynamic viscosity. In this definition, the fluid is assumed to be incompressible; i.e.,

$$\nabla \cdot \mathbf{v} = 0. \tag{3.6}$$

The equation of state relates ρ to p, which is discussed later. Eq. (3.3) can be rewritten as

$$\rho \left[\frac{\partial \mathbf{v}}{\partial t} + (\mathbf{v} \cdot \vec{\nabla}) \mathbf{v} \right] = \nabla \cdot \tilde{\sigma} \tag{3.7}$$

From Eqs. (3.4), (3.6), and (3.7), the Navier-Stokes equation can be obtained as follows:

$$\frac{\partial \mathbf{v}}{\partial t} + (\mathbf{v} \cdot \vec{\nabla})\mathbf{v} = -\frac{1}{\rho}\vec{\nabla}p + \frac{\mu}{\rho}\nabla^2 \mathbf{v}. \tag{3.8}$$

The left-hand side of the Navier-Stokes equation is the fluid acceleration, with its associated operator being defined as the total derivative

$$\frac{D}{Dt} = \frac{\partial}{\partial t} + \mathbf{v} \cdot \vec{\nabla}. \tag{3.9}$$

This derivative is also known as the material derivative, substantial derivative, or Lagrangian derivative, with the latter signifying a moving coordinate system. The total derivative given in Eq. (3.9) enables a more general definition for the flow and transport system than the traditional fixed Eulerian coordinate system. Here, Eq. (3.2) is the focus of the discussion on irrotational fluid flow.

Flow Through Porous Media

Eq. (3.2) may be suitable only for flow in an open environment, such as in channels and pipes. Where the open volume is filled with material of prescribed porosity, the usual approach is to combine the conservation equation with a constitutive relation. For this reason, the porosity term must be included in the flow formulation (Bear 1972). Because the porosity n is defined as the ratio of total void volume to the total volume of the porous medium, the total mass for the solid (M_s) and fluid (M_f) can be written as

$$M_s = \int_V (1-n)\rho_s dV \tag{3.10}$$

$$M_f = \int_V n\rho_f dV \tag{3.11}$$

where V is the control volume, and ρ_s and ρ_f are the densities of solid and fluid, respectively. The porosity n can be treated as a variable in a general case.

Due to the nature of interstitial flow, compressibilities of both fluid and solid grains in the porous medium must be considered; i.e., the flow equation becomes

$$\frac{(1-n)}{\rho_s}\frac{\partial \rho_s}{\partial t} + \frac{n}{\rho_f}\frac{\partial \rho_f}{\partial t} + \dot{v}^f_{i,i} = 0. \tag{3.12}$$

By definition of the relationship between density and pressure, the following equation can be defined:

$$\frac{1}{\rho_s}\frac{\partial \rho_s}{\partial t} = \frac{1}{K_s}\frac{\partial p}{\partial t} - \frac{K}{(1-n)K_s^2}\frac{\partial p}{\partial t} = \frac{\alpha - n}{(1-n)K_s}\frac{\partial p}{\partial t} \tag{3.13}$$

where α is the Biot coefficient (Biot 1941), which is defined in Eq. (2.61) of Chapter 2, and

$$\frac{n}{\rho_f}\frac{\partial \rho_f}{\partial t} = \frac{n}{K_f}\frac{\partial p}{\partial t} \tag{3.14}$$

where K_f is the bulk modulus of the fluid.

Assuming flow is in the laminar range where the Darcy's velocity is valid, this leads to

$$\dot{v}^f_{i,i} = -\frac{k}{\mu}\vec{\nabla} p \tag{3.15}$$

where k is the permeability.

From Eqs. (3.13) and (3.15), the flow equation (3.12) can be rewritten as

$$\frac{k}{\mu}\nabla^2 p = c^*\frac{\partial p}{\partial t} \tag{3.16}$$

where c^* is the lumped compressibility, i.e.,

$$c^* = \frac{\alpha - n}{K_s} + \frac{n}{K_f}. \tag{3.17}$$

Eq. (3.16) is well-known in the literature as the pressure diffusion equation, or simply as the flow equation for homogeneous media.

Coupled Processes

In the flow equation, if the effects of solid deformation on the rate variations of fluid mass cannot be omitted, the process is intrinsically coupled. Mass conservation of the solid can be expressed as (Bear 1972):

$$\frac{\partial (1-n)\rho_s}{\partial t} + \dot{u}^s_i \cdot \vec{\nabla}(1-n)\rho_s + (1-n)\rho_s \nabla \cdot \dot{u}^s_i = 0 \tag{3.18}$$

where ρ_s is the solid density, u^s_i is a vector of solid displacements, n is the porosity, and superscript "· indicates time derivative.

The substantial time derivative can be defined as

$$\frac{D\Omega}{Dt} = \frac{\partial \Omega}{\partial t} + \dot{u}^s_i \frac{\partial \Omega}{\partial x_i} \tag{3.19}$$

where Ω is a dummy variable and x_i is a vector of coordinates. Application of Eq. (3.19) enables the expanded form of Eq. (3.18) to be written in an abridged form as

$$\frac{(1-n)}{\rho_s}\frac{D\rho_s}{Dt} - \frac{Dn}{Dt} + (1-n)\dot{u}^s_{i,i} = 0. \tag{3.20}$$

In contrast, mass conservation restrictively applied to a mobile fluid but a relatively immobile solid can be written as (Bear 1972):

$$\frac{\partial n\rho_f}{\partial t} + \dot{u}_i^f \cdot \vec{\nabla} n\rho_f + n\rho_f \vec{\nabla} \cdot \dot{u}_i^f = 0 \tag{3.21}$$

where ρ_f is the fluid density, and u_i^f is a vector of fluid displacements. Defining relative flow velocity, $\dot{v}_i^f$, as

$$\dot{v}_i^f = n\dot{u}_i^f - n\dot{u}_i^s \tag{3.22}$$

Eq. (3.22) represents the relative velocity between fluid and solid.

Eq. (3.22) can be further modified as

$$\dot{u}_{i,i}^f = -\frac{\dot{v}_i^f}{n^2}\frac{\partial n}{\partial x_i} + \frac{\dot{v}_{i,i}^f}{n} + \dot{u}_{i,i}^s. \tag{3.23}$$

Substituting Eq. (3.23) into the expanded form of Eq. (3.21), and incorporating Eq. (3.19), Eq. (3.21) can be expressed in short form as

$$\frac{n}{\rho_f}\frac{D\rho_f}{Dt} + \frac{Dn}{Dt} + \frac{\dot{v}_i^f}{\rho_f}\frac{\partial \rho_f}{\partial x_i} + \dot{v}_{i,i}^f + n\dot{u}_{i,i}^s = 0. \tag{3.24}$$

Since the porous medium comprises both solid and fluid, the total mass conservation can be obtained through summing Eqs. (3.20) and (3.24); i.e.,

$$\frac{(1-n)}{\rho_s}\frac{D\rho_s}{Dt} + \frac{n}{\rho_f}\frac{D\rho_f}{Dt} + \frac{\dot{v}_i^f}{\rho_f}\frac{\partial \rho_f}{\partial x_i} + \dot{v}_{i,i}^f + \dot{u}_{i,i}^s = 0. \tag{3.25}$$

Eq. (3.25) can be considered as an expanded form of the previous Eq. (3.12).

To convert fluid density ρ_f to fluid pressure p in Eq. (3.25), one has

$$\frac{1}{\rho_f}\frac{D\rho_f}{Dt} = \frac{1}{K_f}\frac{Dp}{Dt} \tag{3.26}$$

where K_f is the fluid bulk modulus.

To relate solid density ρ_s to fluid pressure and solid displacement requires that the effective stress principle is invoked. Following the theoretical approach of Nur and Byerlee (1971), Charlez (1991) offered a framework for this derivation, which leads to the following expression:

$$\frac{1}{\rho_s}\frac{D\rho_s}{Dt} = \frac{1}{K_s}\frac{Dp}{Dt} - \frac{K}{(1-n)K_s}\left(\dot{u}_{i,i}^s + \frac{1}{K_s}\frac{Dp}{Dt}\right) \tag{3.27}$$

where K and K_s are the macroscopic bulk modulus of the porous medium and the microscopic bulk modulus of the solid grains, respectively.

Neglecting the convective term in the substantial time derivative, and after some manipulation, Eq. (3.27) can be reduced to

$$\frac{1}{\rho_s}\frac{\partial \rho_s}{\partial t} = \frac{1}{(1-n)}\left[\frac{(\alpha - n)}{K_s}\dot{p} - (1-\alpha)\dot{u}^s_{i,i}\right] \tag{3.28}$$

where α is Biot coefficient, defined in Eq. (2.61) of Chapter 2.

Without taking the derivative with respect to t and also noting $u^s_{i,i} = \varepsilon_{kk}$, the salient derivation of Eq. (3.28) can be found in Chapter 2, where Eq. (2.85) is equivalent to Eq. (3.28).

Eq. (3.27) can be rewritten as

$$\frac{1}{\rho_s}\frac{D\rho_s}{Dt} = \frac{(\alpha - n)}{(1-n)K_s}\frac{Dp}{Dt} - \frac{K}{(1-n)K_s}\dot{u}^s_{i,i}. \tag{3.29}$$

Substituting Eqs. (3.26) and (3.29) into Eq. (3.25), the single-porosity poroelastic flow equation can be given as

$$c^*\frac{Dp}{Dt} + \dot{v}^f_{i,i} + \alpha\dot{u}^s_{i,i} + \frac{\dot{v}^f_i}{\rho_f}\frac{\partial \rho_f}{\partial x_i} = 0 \tag{3.30}$$

where c^* is the lumped compressibility, which is defined in Eq. (3.17).

Assuming that the velocity of the solid, $\dot{u}^s_i$, is relatively slow in comparison with the fluid velocity, Eq. (3.19) may be reduced to

$$\frac{D\Omega}{Dt} \approx \frac{\partial \Omega}{\partial t}. \tag{3.31}$$

Also, it is assumed that the flow velocity, $\dot{v}^f_i$, satisfies Darcy's velocity, which is defined in Eq. (3.15).

In addition to these two assumptions, the convective term in Eq. (3.30), $\frac{\dot{v}^f_i}{\rho_f}\frac{\partial \rho_f}{\partial x_i}$, is further assumed to be negligible. As a result, Eq. (3.30) can be finally simplified as

$$\frac{k}{\mu}\nabla^2 p = c^*\frac{\partial p}{\partial t} + \alpha\frac{\partial \varepsilon^s_{kk}}{\partial t} \tag{3.32}$$

where $\varepsilon^s_{kk} = u^s_{i,i}$, which is the solid volumetric strain.

Eq. (3.32) is of identical form to the flow equation in Biot's theory of poroelasticity (Biot 1941). By admitting certain assumptions, the present derivation shows that Biot's poroelastic model is indeed supported on a rigorous theoretical basis.

If, assuming a new parameter that may be related to an undrained experiment, i.e., the change of fluid mass content per unit volume, ζ, which may be described as

$$\zeta = A^{\#}(\sigma_{kk} + \frac{3}{B}p) \tag{3.33}$$

where B is Skempton's constant (Skempton 1954), and

$$A^{\#} = \frac{3(\nu_u - \nu)}{BE(1 + \nu_u)} \tag{3.34}$$

where ν_u and ν are the undrained and drained Poisson's ratios, respectively, E is the elastic modulus. Then, mass conservation can be satisfied by combining Darcy's velocity, v_i, with the continuity requirement to form

$$v_{i,i} - \dot{\zeta} = 0 \tag{3.35}$$

then substituting into Eq. (3.33) to yield an equation similar to Eq. (3.32) as

$$\frac{k}{\mu}\vec{\nabla}^2 p = A^{\#}(\frac{3}{B}\dot{p} + \dot{\sigma}_{kk}) \tag{3.36}$$

where the superscripted dot denotes differentiation with respect to time.

Combining the compatibility condition given by Rice and Cleary (1976) and Eq. (3.36), upon consideration of Eq. (3.33), results in

$$c_d \nabla^2 \zeta = \dot{\zeta} \tag{3.37}$$

where c_d is the diffusivity coefficient (Rice and Cleary 1976). Interestingly, Eq. (3.37) carries a similar form to a normal diffusion equation, but the dependent variable is the fluid strain.

3.2.1.2 Two-Phase Fluid

In the previous analysis, the porous medium was assumed to contain only a single fluid (e.g., water, or air) where the void spaces are completely filled with that fluid. For a porous medium containing multiple fluids, the percentage of each occupying fluid, with respect to the total void volume (i.e., respective saturation), becomes an additional function to be determined. In addition, where the flowing fluid mixtures are immiscible, associated complications are further represented by the nonlinear relationships among the saturations, capillary pressures (i.e., the pressure that characterizes the difference between wetting and nonwetting phases), and relative permeabilities (ratio of individual phase permeability to the absolute total permeability) for the wetting and nonwetting phases.

Flow Through Porous Media

For a two-phase system, if one fluid has a greater tendency to imbibe onto the adjacent solid, it is defined as the wetting fluid phase, while the other fluid is considered the nonwetting fluid phase. For example, water is typically the wetting phase in oil-water mixtures, while oil is typically the wetting phase in an oil-gas mixture. The wettability is also determined by the rock properties. For example, some types of rock are water-wet, while other types of rock are oil-wet. In general, the nonwetting and wetting phase pressures (p_n and p_w) are related by the capillary pressure (p_c), such as

$$p_c = p_n - p_w \tag{3.38}$$

where the existence of the capillary pressure p_c is interpreted as the result of capillary or interfacial tension.

Specifically, this relationship for the oil-water mixture or oil-gas mixture yields

$$p_{cow} = p_o - p_w \tag{3.39}$$

$$p_{cog} = p_g - p_o \tag{3.40}$$

where p_{cow} and p_{cog} are the capillary pressures for oil-water and oil-gas systems; p_o, p_w and p_g are the fluid pressures for oil, water, and gas phases, respectively.

The aggregate pore pressure within a two-phase fluid occupying the void space of a porous medium can be written as

$$p = S_w p_w + S_n p_n \tag{3.41}$$

where S is the saturation of the appropriate phase. For a two-phase system,

$$S_w + S_n = 1 \tag{3.42}$$

describing the complete occupation of the pore space by the dual fluids.

Conservation of mass can be expressed for the wetting and nonwetting fluids as

$$\frac{\partial(1-n)\rho_s}{\partial t} + \frac{\partial n S_w \rho_w}{\partial t} + \frac{\partial n S_w \rho_w \dot{v}_{wi}}{\partial x_i} = 0 \tag{3.43}$$

$$\frac{\partial(1-n)\rho_s}{\partial t} + \frac{\partial n S_n \rho_n}{\partial t} + \frac{\partial n S_n \rho_n \dot{v}_{ni}}{\partial x_i} = 0 \tag{3.44}$$

where the fluid flow velocities are defined as

$$\dot{v}_{wi} = \frac{\dot{v}^*_{wi}}{n S_w} \tag{3.45}$$

$$\dot{v}_{ni} = \frac{\dot{v}^*_{ni}}{n S_n} \tag{3.46}$$

where $\dot{v}^*_{wi}$ and $\dot{v}^*_{ni}$ are the intrinsic flow velocities.

For a more general derivation, substituting Eqs. (3.45) and (3.46) into Eqs. (3.43) and (3.44), then applying partial differentiation of the variables, yields

$$(1-n)\frac{\partial \rho_s}{\partial t} - \rho_s \frac{\partial n}{\partial t} + nS_w \frac{\partial(\rho_w)}{\partial t} + n\rho_w \frac{\partial(S_w)}{\partial t} +$$

$$S_w \rho_w \frac{\partial(n)}{\partial t} + \dot{v}^*_{wi} \frac{\partial(\rho_w)}{\partial x_i} = 0 \tag{3.47}$$

$$(1-n)\frac{\partial \rho_s}{\partial t} - \rho_s \frac{\partial n}{\partial t} + nS_n \frac{\partial(\rho_n)}{\partial t} + n\rho_n \frac{\partial(S_n)}{\partial t} +$$

$$S_n \rho_n \frac{\partial(n)}{\partial t} + \dot{v}^*_{ni} \frac{\partial(\rho_n)}{\partial x_i} = 0. \tag{3.48}$$

Following a similar procedure of defining the compressibilities as for the single-phase fluid, one has

$$\frac{D\rho_s}{\rho_s Dt} = \frac{Dp}{K_s Dt} \tag{3.49}$$

$$\frac{D\rho_w}{\rho_w Dt} = \frac{Dp_w}{K_w Dt} \tag{3.50}$$

$$\frac{D\rho_n}{\rho_n Dt} = \frac{Dp_n}{K_{nw} Dt} \tag{3.51}$$

where K_s, K_w, and K_{nw} are the bulk moduli of the solid, wetting fluid, and nonwetting fluid phases, respectively.

With all terms defined, the governing equations of flow continuity can be described as

$$(A_1 + A_2)\dot{p}_w + A_2 \left(\frac{dp_c}{dS_w}\dot{S}_w - \frac{p_c}{S_n}\dot{S}_w \right) + \dot{v}^*_{wi,i} + n\dot{S}_w + \frac{\dot{v}^*_{wi}}{\rho_w}\frac{\partial(\rho_w)}{\partial x_i} = 0 \tag{3.52}$$

$$(A_2 + A_3)\dot{p}_w + A_3 \frac{dp_c}{dS_w}\dot{S}_w - A_2 \frac{p_c}{S_w}\dot{S}_w + \dot{v}^*_{ni,i} + n\dot{S}_n + \frac{\dot{v}^*_{ni}}{\rho_n}\frac{\partial(\rho_n)}{\partial x_i} = 0 \tag{3.53}$$

where

$$A_1 = S_w \left[\frac{S_w(1-n)}{K_s} + \frac{n}{K_w} \right] \tag{3.54}$$

$$A_2 = \frac{(1-n)S_w S_n}{K_s} \tag{3.55}$$

$$A_3 = S_n \left[\frac{S_n(1-n)}{K_s} + \frac{n}{K_{nw}} \right]. \tag{3.56}$$

In this derivation, the nonwetting pore pressure is eliminated because capillary pressure is substituted as a primary unknown. By doing so, only the explicit relationship between the saturation of the wetting phase and the capillary pressure should be required. All other correlations with the nonwetting phase, including the pore pressure, can be calculated through the relationships shown in Eqs. (3.38) and (3.42).

In the relations governing flow, Eqs. (3.52) and (3.53), the flow velocities may be assumed to follow Darcy' law, and can be expressed as

$$\dot{v}^*_{wi} = -\kappa_{wij}(p_{w,j} + \rho_w h^*_{,j}) \tag{3.57}$$

$$\dot{v}^*_{ni} = -\kappa_{nij}(p_{n,j} + \rho_n h^*_{,j}) \tag{3.58}$$

where h^*_j is the elevation relative to a prescribed datum and κ is the mobility, which can be evaluated by

$$\kappa_{wij} = \frac{k_{wij} k_{rw}}{\mu_w} \tag{3.59}$$

$$\kappa_{nij} = \frac{k_{nij} k_{rn}}{\mu_n} \tag{3.60}$$

where k_w and k_n are the absolute permeabilities, k_{rw} and k_{rn} are the relative permeabilities which are functions of saturation, and μ_w and μ_n are fluid dynamic viscosities.

In addition, due to the volume changes that result when the fluid is produced from a reservoir, the density ρ at the standard condition for each phase may have to be converted to the field condition through the following relation:

$$\rho = \frac{(\rho_i)_{std}}{B_i} \tag{3.61}$$

where $(\rho_i)_{std}$ is the density of each phase at standard conditions and B_i is the formation volume factor.

Coupled Processes

Because the porosity n is defined as the ratio of total void volume to the total volume of the porous medium, the total mass for the solid (M_s) and fluid (M_f) can be written as (Li et al. 1990):

$$M_s = \int_V (1-n)\rho_s dV \tag{3.62}$$

$$M_f = \int_V n\rho_f dV \tag{3.63}$$

where V is the control volume and ρ_s and ρ_f are the densities of solid and fluid, respectively. The porosity n can be treated as a variable for the present analysis.

Using the total derivative defined as before (Huyakorn and Pinder 1983), conservation of mass for the solid and the fluid phases can be expressed as

$$\frac{\partial(1-n)\rho_s}{\partial t}+\frac{\partial(1-n)\rho_s\dot{u}_{si}}{\partial x_i}+\frac{\partial nS_w\rho_w}{\partial t}+\frac{\partial nS_w\rho_w\dot{v}_{wi}}{\partial x_i}=0 \tag{3.64}$$

$$\frac{\partial(1-n)\rho_s}{\partial t}+\frac{\partial(1-n)\rho_s\dot{u}_{si}}{\partial x_i}+\frac{\partial nS_n\rho_n}{\partial t}+\frac{\partial nS_n\rho_n\dot{v}_{ni}}{\partial x_i}=0. \tag{3.65}$$

If the fluid flow velocities are considered relative with respect to the velocity of the moving solid as well as to the general coordinate system, they can be defined as

$$\dot{v}_{wi}=\frac{\dot{v}^*_{wi}}{nS_w}+\dot{u}_{si} \tag{3.66}$$

$$\dot{v}_{ni}=\frac{\dot{v}^*_{ni}}{nS_n}+\dot{u}_{si}. \tag{3.67}$$

Since in most cases $\dot{u}_s \ll \dot{v}_w$ and $\dot{u}_s \ll \dot{v}_n$, the previous two equations are frequently simplified as

$$\dot{v}_{wi}=\frac{\dot{v}^*_{wi}}{nS_w} \tag{3.68}$$

$$\dot{v}_{ni}=\frac{\dot{v}^*_{ni}}{nS_n}. \tag{3.69}$$

For a more general derivation, substituting Eqs. (3.66) and (3.67) into Eqs. (3.64) and (3.65), then applying partial differentiation of the variables, gives

$$\begin{aligned}(1-n)\frac{\partial\rho_s}{\partial t}-\rho_s\frac{\partial n}{\partial t}+(1-n)\rho_s\dot{u}_{si,i}+\dot{u}_{si}\frac{\partial(1-n)\rho_s}{\partial x_i}+\\ \frac{\partial(nS_w\rho_w)}{\partial t}+\frac{\partial(nS_w\rho_w)\dot{u}_{si}}{\partial x_i}+\frac{\partial(\rho_w)\dot{v}^*_{wi}}{\partial x_i}=0\end{aligned} \tag{3.70}$$

$$\begin{aligned}(1-n)\frac{\partial\rho_s}{\partial t}-\rho_s\frac{\partial n}{\partial t}+(1-n)\rho_s\dot{u}_{si,i}+\dot{u}_{si}\frac{\partial(1-n)\rho_s}{\partial x_i}+\\ \frac{\partial(nS_n\rho_n)}{\partial t}+\frac{\partial(nS_n\rho_n)\dot{u}_{si}}{\partial x_i}+\frac{\partial(\rho_n)\dot{v}^*_{ni}}{\partial x_i}=0.\end{aligned} \tag{3.71}$$

Eqs. (3.70) and (3.71) can be written in an abridged form using Eq. (3.9), as

$$\frac{(1-n)D\rho_s}{\rho_s Dt}+\frac{nD\rho_w}{\rho_w Dt}+\frac{nDS_w}{S_w Dt}+\dot{u}_{si,i}+\frac{\dot{v}^*_{wi,i}}{S_w}+\frac{\dot{v}^*_{wi}}{S_w\rho_w}\frac{\partial(\rho_w)}{\partial x_i}=0 \tag{3.72}$$

$$\frac{(1-n)D\rho_s}{\rho_s Dt}+\frac{nD\rho_n}{\rho_n Dt}+\frac{nDS_n}{S_n Dt}+\dot{u}_{si,i}+\frac{\dot{v}^*_{ni,i}}{S_n}+\frac{\dot{v}^*_{ni}}{S_n\rho_n}\frac{\partial(\rho_n)}{\partial x_i}=0 \tag{3.73}$$

From the density-mass-volume relationship, the density, ρ (either fluid or solid), can be generally defined as

$$\rho = \frac{m}{V} \tag{3.74}$$

where m is the mass and V is the volume. Assuming that both density and the control volume are functions of time, their differentiation with respect to time in Eq. (3.74) results in

$$\frac{1}{\rho}\frac{\partial \rho}{\partial t} = \frac{V}{m}\partial(\frac{m}{V})\frac{1}{\partial t} = -\frac{1}{V}\frac{\partial V}{\partial t}. \tag{3.75}$$

Or, more generally,

$$\frac{1}{\rho}\frac{D\rho}{Dt} = -\frac{1}{V}\frac{DV}{Dt}. \tag{3.76}$$

Under isothermal conditions, the compressibility can be defined as

$$c_p^* = \frac{1}{K} = \frac{1}{\rho}\frac{D\rho}{Dp} \tag{3.77}$$

where p is the pore pressure which should be evaluated from Eq. (3.41) and K is the bulk modulus of the porous medium.

From the equation of state, one has

$$\frac{D\rho}{\rho} = \frac{1}{\rho}\left(\frac{D\rho}{Dp}\right) Dp. \tag{3.78}$$

Substituting Eq. (3.77) into this equation, gives

$$\frac{D\rho}{\rho} = c_p^* Dp = \frac{Dp}{K}. \tag{3.79}$$

Using these expressions, the total derivatives with respect to the density ρ can be described by the relevant pore pressures, i.e.,

$$\frac{D\rho_s}{\rho_s Dt} = \frac{1}{1-n}\left[\frac{\alpha - n}{K_s}\frac{Dp}{Dt} - (1-\alpha)\dot{u}_{si,i}\right] \tag{3.80}$$

where K_s is the bulk modulus of the solid grain material.

The bulk moduli of the wetting fluid and nonwetting fluid phases, K_w and K_{nw}, are defined in Eqs. (3.50) and (3.51), respectively. Even though the wetting fluid tends to be preferentially absorbed to the solid, it is assumed that both fluids are in contact with the solid phase, as shown in Eq. (3.41).

In deriving Eq. (3.80), the relationship described in Eq. (2.85) of Chapter 2 is assumed, with the consideration of time derivative.

In the previous equations, $\dot{u}_{si,i}$ also represents the time derivative of the volumetric strain $\dot{\varepsilon}_{kk}$, identical to that for single-phase poroelasticity (Biot 1941) in Eq. (3.32). With this substitution, together with the substitution of Eqs. (3.80), (3.50), and (3.51) with Eqs. (3.72) and (3.73), the governing equations of flow continuity can be described as

$$(A_1+A_2)\dot{p}_w+A_2\left(\frac{dp_c}{dS_w}-\frac{p_c}{S_n}\right)\dot{S}_w+\alpha S_w\dot{\varepsilon}_{kk}+\dot{v}^*_{wi,i}+n\dot{S}_w+\frac{\dot{v}^*_{wi}}{\rho_w}\frac{\partial(\rho_w)}{\partial x_i}=0 \quad (3.81)$$

$$(A_2+A_3)\dot{p}_w+\left(A_3\frac{dp_c}{dS_w}-A_2\frac{p_c}{S_w}\right)\dot{S}_w+\alpha S_n\dot{\varepsilon}_{kk}+\dot{v}^*_{ni,i}+n\dot{S}_n+\frac{\dot{v}^*_{ni}}{\rho_n}\frac{\partial(\rho_n)}{\partial x_i}=0 \quad (3.82)$$

where A_1, A_2 and A_3 are given in Eqs. (3.54), (3.55), and (3.56).

In the governing flow equations (3.81) and (3.82), the flow velocities, $\dot{v}^*_{wi}$ and $\dot{v}^*_{ni}$, may be assumed to follow Darcy's law, as shown in Eqs. (3.57) and (3.58).

3.2.2 Heterogeneous Media

In comparison with homogeneous media, the definition of "heterogeneous media" is even broader. Here, this definition is confined to media that show dependency on location in space. In addition, primary focus of the discussion is limited to fractured porous media based on continuum conceptualizations.

3.2.2.1 Single-Phase Fluid

A typical representation of a fractured porous medium is as a dual-porosity system in which both fractures and matrix blocks possess fluid storage characterized by the respective porosities.

Barenblatt et al.'s Model

Natural fractures are a ubiquitous component of all lithified reservoirs and aquifers and comprise a major proportion of the total resource-bearing formations. Due to their importance and unique characteristics, study of the behavior of fractured formations has become increasingly more important. In the four decades since the proposal of a prototype by Barenblatt et al. (1960), the dual-porosity formulation has been considered as an adequate mathematical representation of naturally fractured formations. The dual-porosity conceptualization of a fractured formation is the simplest possible approach that represents the salient features of flow within complex fractured porous media. The nonlinear behavior of a fractured formation, as represented by the temporal pressure curves measured in the field, may be represented by a linear dual-porosity model utilizing a transfer function to represent the internal process of fluid exchange between fractures and matrix blocks. This internal flow is activated after initial pumping, and modifies the rate of pressure change.

Although the dual-porosity behavior of naturally fractured reservoirs is often masked by other physical factors, such as wellbore storage and skin effects (Moench 1984; Streltsova 1988), many field measurements (Crawford et al. 1976) are consistent with dual-porosity response. However, the matching of field data with mathematical models does not automatically prove the validity of the theory itself. As Chilingarian et al. (1992) pointed out, it is difficult to believe that a real reservoir (Figure 3-1) could behave in accordance with a model, which assumes that all block sizes are identical. This discrepancy may be manifested at the usual well measuring location because the flow at the point represents an average quantity yielded from matrix blocks and fractures. Therefore, it may not be a correct interpretation that the transient region representing the stabilization of the pressure curve, as shown in Figure 3-2, is a result of interporosity flow between matrix blocks and fractures. In addition to the crossflow between stratified layers and other exchange mechanisms, the closing of fractures due to fluid drainage may also lead to a temporary reduction in the pressure response at the well.

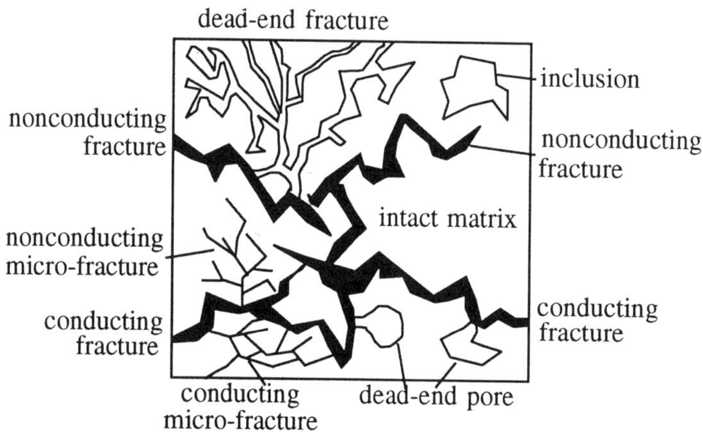

Figure 3-1. Visually inspected naturally fractured formation.

Dual-porosity response is usually described as (Barenblatt et al. 1960):

$$\frac{k_1}{\mu}\nabla^2 p_1 = c_1^* n_1 \frac{\partial p_1}{\partial t} + \Gamma(p_1 - p_2) \tag{3.83}$$

$$\frac{k_2}{\mu}\nabla^2 p_2 = c_2^* n_2 \frac{\partial p_2}{\partial t} - \Gamma(p_1 - p_2) \tag{3.84}$$

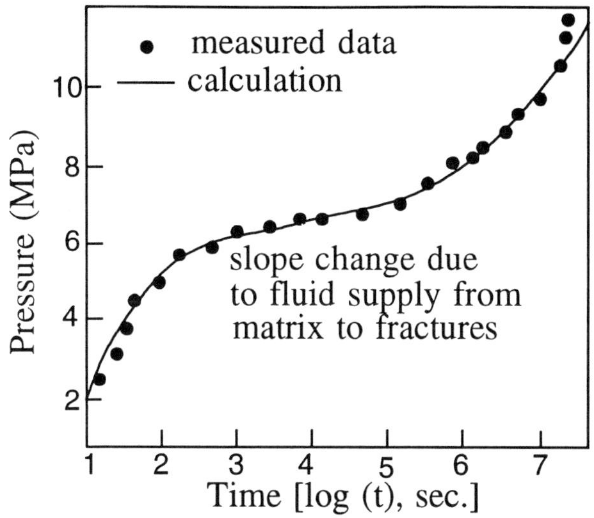

Figure 3-2. Pressure transient in a typical dual-porosity reservoir.

where subscripts 1 and 2 represent matrix and fractures, respectively; p is the fluid pressure, k is the permeability; μ is the dynamic viscosity; n is the porosity; c^* is the compressibility; t is the time; and Γ is the geometric leakage factor.

Based on the usual observation that the matrix permeability is substantially smaller than the fracture permeability, fluid flow in the matrix may have a minimal contribution to flow in the total system, and may therefore be neglected. This rationale leads to the following simplification of the matrix phase in Eq. (3.83) (Barenblatt et al. 1960; Warren and Root 1963):

$$c_1^* n_1 \frac{\partial p_1}{\partial t} = -\Gamma(p_1 - p_2). \tag{3.85}$$

The general physical description of dual-porosity behavior can be represented by the temporal pressure semi-log plot recorded from the pressure measurement at the well (Figure 3-2). At the initial stages of pumping, fluid flow occurs mainly within the fractures. After exhausting the storage, flow begins to occur primarily between the matrix and fractures showing the reduction of pressure gradient, which is termed "pressure stabilization." After pressure equilibrium between the matrix and fractures is reached, fluid flows continuously to the well through fractures as well as through matrix blocks.

In modeling, the strata encompassing the pumping well are assumed to comprise regularly spaced porous blocks intercepted by orthogonal fractures with identical apertures. In reality, the fracture network may not have a regular pattern. Visible vugs, stylolites, and cavities, interwoven with unconnected pores, may be a more typical composition of a fractured porous formation. Although using an average pressure appears more appropriate, for the present, matrix pressure and fracture pressure are retained as separate, albeit fictitious parameters, in this application of mixture theory.

Adding Eqs. (3.83) and (3.84) gives

$$\frac{k_1}{\mu}\nabla^2 p_1 + \frac{k_2}{\mu}\nabla^2 p_2 = c_1^* n_1 \frac{\partial p_1}{\partial t} + c_2^* n_2 \frac{\partial p_2}{\partial t}. \tag{3.86}$$

Similar to the behavior of composite materials, Eq. (3.86) is a lumped representation of mass conservation, where no dual-porosity profile (as shown in Figure 3-2) can be generated.

Subtracting Eq. (3.84) from (3.83) yields

$$\frac{k_1}{\mu}\nabla^2 p_1 - \frac{k_2}{\mu}\nabla^2 p_2 = c_1^* n_1 \frac{\partial p_1}{\partial t} - c_2^* n_2 \frac{\partial p_2}{\partial t} + 2\Gamma(p_1 - p_2). \tag{3.87}$$

A quantitative evaluation indicates that dual-porosity response will be manifested if, on the right side of Eq. (3.87), the effect of the third term is larger than the sum of the other two terms. When equilibrium is reached between the fractures and matrix (i.e., $p_1 = p_2 = p$), then from Eqs. (3.86) and (3.87),

$$\frac{k_1 + k_2}{\mu}\nabla^2 p = (c_1^* n_1 + c_2^* n_2)\frac{\partial p}{\partial t} \tag{3.88}$$

$$\frac{k_1 - k_2}{\mu}\nabla^2 p = (c_1^* n_1 - c_2^* n_2)\frac{\partial p}{\partial t}. \tag{3.89}$$

Eqs. (3.88) and (3.89) indicate that at a nonvanishing fluid pressure, the uniqueness of the solution warrants the complete closure of the fracture space (i.e., $n_2 = k_2 = 0$). As a result,

$$\frac{k_1}{\mu}\nabla^2 p = c_1^* n_1 \frac{\partial p}{\partial t}. \tag{3.90}$$

Eq. (3.90) represents single-porosity flow in a homogeneous reservoir. The appearance of pressure stabilization, shown in Figure 3-2 (for a closed reservoir with a finite external boundary), depends on the magnitude of the term "$\Gamma(p_1 - p_2)$" in Eq. (3.87), which acts as a broadly distributed sink or a source, preventing the fluid pressure from further decline or buildup.

Warren and Root's Model

Barrenblatt et al.'s dual-porosity formulation (1960), although conceptually comprehensive, is difficult for reservoir engineers to derive relevant analytical solutions. As a result, the simplified model proposed independently by Warren and Root (1963) is more popular among practitioners. In Warren and Root's model, the flow within the matrix blocks is assumed to be negligible, which leads to the uncoupling of fluid pressures between the matrix and fractures within the Laplace domain, as shown in the following:

$$0 = c_1^* n_1 \frac{\partial p_1}{\partial t} + \Gamma(p_1 - p_2) \tag{3.91}$$

$$\frac{k_2}{\mu}\nabla^2 p_2 = c_2^* n_2 \frac{\partial p_2}{\partial t} - \Gamma(p_1 - p_2). \tag{3.92}$$

Eq. (3.92) is identical to Eq. (3.84), as Eq. (3.91) is identical to Eq. (3.85). The relations may be normalized through the following dimensionless quantities:

$$\begin{cases} P_{Di} = \dfrac{2\pi k_2 h_r (p_0 - p_i)}{q\mu} \quad (i = 1, 2) \\ r_D = \dfrac{r}{r_w}, \quad \lambda^* = \dfrac{\Gamma \mu r_w^2}{k_2} \\ t_D = \dfrac{k_2 t}{\mu(n_1 c_1^* + n_2 c_2^*) r_w^2} \\ \omega = \dfrac{n_2 c_2^*}{n_1 c_1^* + n_2 c_2^*} \end{cases} \tag{3.93}$$

where h_r is the reservoir thickness, p_0 is the initial reservoir pressure, ω is the ratio of fracture storativity, q is the flow rate at the well, and r_w is the wellbore radius.

The geometric leakage factor Γ is considered as a function of the shape factor $\alpha^{\#}$ in Warren and Root's (1963) approach, expressed as

$$\Gamma = \alpha^{\#} \frac{k_1}{\mu} \tag{3.94}$$

where

$$\alpha^{\#} = \frac{4j^*(j^* + 2)}{\xi_w^2} \tag{3.95}$$

where j^* =1, 2, or 3 in terms of the number of orthogonal sets of fractures, ξ_w is the characteristic dimension of the matrix blocks (Figure 3-3), and can be written as

$$\xi_w = \frac{3a^{\#}b^{\#}c^{\#}}{a^{\#}b^{\#} + b^{\#}c^{\#} + c^{\#}a^{\#}}. \tag{3.96}$$

For a regularly spaced parallelepiped block-type matrix model, frequently referred to as the "sugar cube" model, one has

$$j^* = 3, \quad a^{\#} = b^{\#} = c^{\#} = s^* \tag{3.97}$$

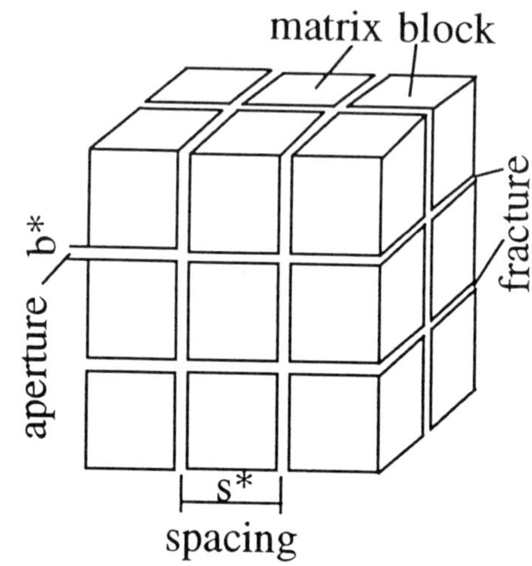

Figure 3-3. Warren and Root "sugar-cube" model.

where s^* is the average fracture spacing, so that

$$\alpha^* = \frac{60}{(s^*)^2}. \tag{3.98}$$

Substituting the dimensionless terms given in Eq. (3.93) into the governing Eqs. (3.91) and (3.92), suitably modified in cylindrical coordinates, results in

$$\lambda^*(P_{D2} - P_{D1}) = (1-\omega)\frac{\partial P_{D1}}{\partial t_D} \tag{3.99}$$

$$\frac{1}{r_D}\frac{\partial}{\partial r_D}(r_D\frac{\partial P_{D2}}{\partial r_D}) + \lambda^*(P_{D1} - P_{D2}) = \omega\frac{\partial P_{D2}}{\partial t_D}. \tag{3.100}$$

The boundary and initial conditions for an infinite reservoir with constant flow rate at the inner boundary and constant pressure at the outer boundary may be described as

$$\begin{cases} P_{D2}(\infty, t_D) = 0 \qquad (\textit{outer boundary}) \\ \\ \frac{\partial P_{D2}}{\partial r_D}|_{r_D=1} = -1 \qquad (\textit{inner boundary}) \\ \\ P_{D1}(r_D, t_D = 0) = P_{D2}(r_D, t_D = 0) = 0 \qquad (\textit{initial condition}) \end{cases} \tag{3.101}$$

Using the Laplace transform to convert the partial differential equations (3.91) and (3.92) to ordinary differential equations and solving these equations in the Laplace domain, gives

$$\lambda^*(\overline{P}_{D2} - \overline{P}_{D1}) = (1 - \omega)s\overline{P}_{D1} \tag{3.102}$$

$$\frac{d^2\overline{P}_{D2}}{dr_D^2} + \frac{1}{r_D}\frac{d\overline{P}_{D2}}{dr_D} + \lambda^*(\overline{P}_{D1} - \overline{P}_{D2}) = \omega s\overline{P}_{D2}. \tag{3.103}$$

Solving these two equations simultaneously yields

$$\frac{d^2\overline{P}_{D2}}{dr_D^2} + \frac{1}{r_D}\frac{d\overline{P}_{D2}}{dr_D} - sf(s)\overline{P}_{D2} = 0 \tag{3.104}$$

where s is a Laplace parameter, and

$$f(s) = \frac{\lambda^* + s\omega(1 - \omega)}{\lambda^* + s(1 - \omega)} \tag{3.105}$$

where $f(s)$ is also known as the system modifier. For single porosity, $f(s) = 1$.

It may be noted that Eq. (3.104) is the product of Eqs. (3.102) and (3.103), within the Laplace space. The elimination of the matrix pressure within the Laplace domain is due to the uncoupling of the pore pressures in the continuity equation for the matrix, such as

$$\overline{P}_{D1} = \frac{\lambda^*}{\lambda^* + s(1 - \omega)}\overline{P}_{D2}. \tag{3.106}$$

Boundary conditions in the Laplace domain can be written as

$$\begin{cases} \overline{P}_{D2}(\infty, s) = 0 & (\textit{outer boundary}) \\ \frac{\partial \overline{P}_{D2}}{\partial r_D}\Big|_{r_{eD}=1} = -\frac{1}{s} & (\textit{inner boundary}) \end{cases} \tag{3.107}$$

Eq. (3.104) is a Bessel equation with the following solution:

$$\overline{P}_{D2} = A^* K_0\left[r_D\sqrt{sf(s)}\right] + B^* I_0\left[r_D\sqrt{sf(s)}\right] \tag{3.108}$$

where K_0 and I_0 are the modified Bessel functions of zero order, and A^* and B^* are constants to be determined. Since $I_0(r_D, s)$ is unbounded, a meaningful solution of $\overline{P}_{D2}$ requires

$$B^* = 0. \tag{3.109}$$

From boundary condition (3.107), A^* is derived as

$$A^* = \frac{1}{s\sqrt{sf(s)}K_1(\sqrt{sf(s)})} \tag{3.110}$$

Therefore, one has

$$\overline{P}_{D2} = \frac{K_0(r_D\sqrt{sf(s)})}{s\sqrt{sf(s)}K_1(\sqrt{sf(s)})}. \tag{3.111}$$

where K_1 is also a modified Bessel function.

Inversion of $\overline{P}_{D2}$ can be performed using any convenient numerical inversion technique, such as the Stehfest (1970) method. Some analytical solutions can be obtained under certain special circumstances. For example, in the late time approximation, the Bessel functions can be simplified as $K_0(z) \approx \ln(z) - \ln 2 - 0.5772$ and $K_1(z) \approx \frac{1}{z}$. The fluid pressure in the fractures can be inverted from Eq. (3.111) into

$$P_{D2} = \frac{1}{2}\left\{\ln(t_D) + 0.80909 + E_i\left[-\frac{\lambda t_D}{\omega(1-\omega)}\right] - E_i\left(-\frac{\lambda t_D}{1-\omega}\right)\right\} \tag{3.112}$$

where

$$E_i(-x) = -\int_x^\infty \frac{e^{-u}}{u}du = 0.5772 + \ln(x) + \sum_{n=1}^{\infty}\frac{(-1)^n x^n}{n\, n!}. \tag{3.113}$$

Flow Through Fractured Porous Media

The models proposed by Barenblatt et al. (1960) and by Warren and Root (1963) were based on phenomenological conceptualizations that, even though physically sensible, lack a rigorous theoretical basis. This deficiency may be partially alleviated in the following model.

Extrapolating from Eq. (3.12) for homogeneous media, the flow equation for dual-porosity media can be written as

$$\frac{(1-n_1-n_2)}{\rho_s}\frac{\partial \rho_s}{\partial t} + \frac{n_1}{\rho_{1f}}\frac{\partial \rho_{1f}}{\partial t} + \dot{v}^f_{1i,i} = 0 \tag{3.114}$$

$$\frac{(1-n_1-n_2)}{\rho_s}\frac{\partial \rho_s}{\partial t} + \frac{n_2}{\rho_{2f}}\frac{\partial \rho_{2f}}{\partial t} + \dot{v}^f_{2i,i} = 0 \tag{3.115}$$

where $\frac{1}{\rho_s}\frac{\partial \rho_s}{\partial t}$ can be extrapolated from Eq. (3.13) by considering the lumped impact of matrix and fractures as

$$\frac{1}{\rho_s}\frac{\partial \rho_s}{\partial t} = \frac{1}{1-n_1-n_2}\left[\frac{(\alpha_1-n_1-n_2)}{K_s}\frac{\partial p_1}{\partial t} + \frac{(\alpha_2-n_1-n_2)}{K_{fr}}\frac{\partial p_2}{\partial t}\right] \tag{3.116}$$

where K_{fr} is the microscopic bulk modulus of solid fractured media, and α_1 and α_2 are defined in Eqs. (2.107) and (2.108) of Chapter 2.

As before, the fluid density-pressure relationship can be expressed as

$$\frac{1}{\rho_{1f}}\frac{\partial \rho_{1f}}{\partial t} = \frac{1}{K_f}\frac{\partial p_1}{\partial t} \tag{3.117}$$

$$\frac{1}{\rho_{2f}}\frac{\partial \rho_{2f}}{\partial t} = \frac{1}{K_f}\frac{\partial p_2}{\partial t}. \tag{3.118}$$

Substituting Eqs. (3.116), (3.117), and (3.118) into Eqs. (3.114) and (3.115) gives

$$c_{11}^{\#}\frac{\partial p_1}{\partial t} - c_{12}^{\#}\frac{\partial p_2}{\partial t} + \dot{v}_{1i,i}^{f} = 0 \tag{3.119}$$

$$c_{21}^{\#}\frac{\partial p_2}{\partial t} - c_{22}^{\#}\frac{\partial p_1}{\partial t} + \dot{v}_{2i,i}^{f} = 0 \tag{3.120}$$

where

$$\begin{cases} c_{11}^{\#} = \dfrac{\alpha_1 - n_1 - n_2}{K_s} + \dfrac{n_1}{K_f} \\[2ex] c_{12}^{\#} = -\dfrac{\alpha_2 - n_1 - n_2}{K_{fr}} \\[2ex] c_{21}^{\#} = \dfrac{\alpha_2 - n_1 - n_2}{K_{fr}} + \dfrac{n_2}{K_f} \\[2ex] c_{22}^{\#} = -\dfrac{\alpha_1 - n_1 - n_2}{K_s}. \end{cases} \tag{3.121}$$

The cross-influence terms in Eqs. (3.119) and (3.120) are only related to the respective pore pressures, which is consistent with a view of dual-porosity media (e.g., $c_{12}^{\#}$ is associated with p_2 while $c_{22}^{\#}$ is related to p_1 only).

Again, defining Darcy velocities for the dual-porosity medium as

$$\dot{v}_{1i}^{f} = -\frac{k_1}{\mu}\nabla p_1 \tag{3.122}$$

$$\dot{v}_{2i}^{f} = -\frac{k_2}{\mu}\nabla p_2 \tag{3.123}$$

and adding the quasi-steady interporosity flow terms (treated as internal flow sources/sinks), Eqs. (3.119) and (3.120) can be further simplified as

$$\frac{k_1}{\mu}\nabla^2 p_1 = c_{11}^{\#}\frac{\partial p_1}{\partial t} - c_{12}^{\#}\frac{\partial p_2}{\partial t} + \Gamma(p_1 - p_2) \tag{3.124}$$

$$\frac{k_2}{\mu}\nabla^2 p_2 = c_{21}^{\#}\frac{\partial p_2}{\partial t} - c_{22}^{\#}\frac{\partial p_1}{\partial t} - \Gamma(p_1 - p_2) \tag{3.125}$$

where Γ is the leakage coefficient (Barenblatt et al. 1960).

It can be readily verified that when the effects of fractures disappear (e.g., n_2, α_2, p_2, and Γ all vanish), Eqs. (3.124) and (3.125) collapse to the single-porosity representation of Eq. (3.16).

The second terms on the right-hand side of Eqs. (3.124) and (3.125) are cross-influence terms, which appear to be similar to, but in fact are different from, those suggested by Barenblatt et al. (1960) [refer to the definitions of $c_{12}^{\#}$ and $c_{22}^{\#}$ in Eq. (3.121)]. Noticeable differences between Eqs. (3.124), (3.125) and the majority of dual-porosity poroelastic models (e.g., Wilson and Aifantis 1982) can also be found in the definitions of compressibilities shown in Eq. (3.121). Most dual-porosity models do not have cross-influence terms because the total porosity is not treated as the summation of the individual porosities. Furthermore, the interporosity flow terms are necessary because they are the only terms linking the two separate flow equations for those dual-porosity formulations that do not include the cross-influence terms. Since the formulations shown in Eqs. (3.119) and (3.120) already include the cross-influence terms, the interporosity flow terms, which are purely phenomenological, can be neglected.

Coupled Processes

Substituting n with the summation of n_1 and n_2, the mass conservation of solid in a dual-porosity system becomes

$$\frac{(1-n_1-n_2)}{\rho_s}\frac{D\rho_s}{Dt} - \frac{Dn_1}{Dt} - \frac{Dn_2}{Dt} + (1-n_1-n_2)\dot{u}_{i,i}^s = 0. \tag{3.126}$$

Neglecting the convective terms, two separate flow equations result, one for each solid component, in prescribing mass conservation of the fluid. The conservation equation may be written as

$$\frac{n_1}{\rho_{1f}}\frac{D\rho_{1f}}{Dt} + \frac{Dn_1}{Dt} + \dot{v}_{1i,i}^f + n_1\dot{u}_{i,i}^s = 0 \tag{3.127}$$

$$\frac{n_2}{\rho_{2f}}\frac{D\rho_{2f}}{Dt} + \frac{Dn_2}{Dt} + \dot{v}_{2i,i}^f + n_2\dot{u}_{i,i}^s = 0. \tag{3.128}$$

In a typical dual-porosity poroelastic system, the influence of solid deformation on fluid flow is considered as a lumped effect. Based on this concept, Eq. (3.126) should be added to Eq. (3.127) and to Eq. (3.128), respectively, to yield

$$\frac{(1-n_1-n_2)}{\rho_s}\frac{D\rho_s}{Dt} + \frac{n_1}{\rho_{1f}}\frac{D\rho_{1f}}{Dt} - \frac{Dn_2}{Dt} + \dot{v}_{1i,i}^f + (1-n_2)\dot{u}_{i,i}^s = 0 \tag{3.129}$$

$$\frac{(1-n_1-n_2)}{\rho_s}\frac{D\rho_s}{Dt}+\frac{n_2}{\rho_{2f}}\frac{D\rho_{2f}}{Dt}-\frac{Dn_1}{Dt}+\dot{v}^f_{2i,i}+(1-n_1)\dot{u}^s_{i,i}=0. \tag{3.130}$$

Conversion of solid density in Eqs. (3.129) and (3.130), as a function of fluid pressure and solid displacement, requires the application of the appropriate effective stress laws. For dual-porosity poroelastic media, these may be written as

$$\frac{1}{\rho_s}\frac{D\rho_s}{Dt}=\frac{1}{1-n_1-n_2}\left[\frac{(\alpha_3-n_1-n_2)}{K_s}\frac{Dp_1}{Dt}+\right.$$

$$\left.\frac{(\alpha_3-n_1-n_2)}{K_{fr}}\frac{Dp_2}{Dt}-(1-\alpha_3)\dot{u}^s_{i,i}\right]. \tag{3.131}$$

Apart from being the derivative with respect to time t, Eq. (3.131) is analogous to Eq. (2.129) (Chapter 2 on deformation). Neglecting the convective term in the substantial time derivative, Eq. (3.131) can be simplified as

$$\frac{1}{\rho_s}\frac{\partial\rho_s}{\partial t}=\frac{1}{1-n_1-n_2}\left[\frac{(\alpha_3-n_1-n_2)}{K_s}\frac{\partial p_1}{\partial t}+\right.$$

$$\left.\frac{(\alpha_3-n_1-n_2)}{K_{fr}}\frac{\partial p_2}{\partial t}-(1-\alpha_3)\dot{u}^s_{i,i}\right]. \tag{3.132}$$

Without considering the time derivatives and recalling that $u^s_{i,i}=\varepsilon_{kk}$, Eq. (3.132) is now equivalent to Eq. (2.129) (Chapter 2 on deformation). As before, the fluid density-pressure relationship can be expressed as

$$\frac{1}{\rho_{1f}}\frac{D\rho_{1f}}{Dt}=\frac{1}{K_f}\frac{Dp_1}{Dt} \tag{3.133}$$

$$\frac{1}{\rho_{2f}}\frac{D\rho_{2f}}{Dt}=\frac{1}{K_f}\frac{Dp_2}{Dt}. \tag{3.134}$$

Substituting Eqs. (3.131), (3.133), and (3.134) into Eqs. (3.129) and (3.130), gives

$$c^*_{11}\frac{Dp_1}{Dt}-c^*_{12}\frac{Dp_2}{Dt}+\dot{v}^f_{1i,i}+(\alpha_3-n_2)\dot{u}^s_{i,i}-\frac{Dn_2}{Dt}=0 \tag{3.135}$$

$$c^*_{21}\frac{Dp_2}{Dt}-c^*_{22}\frac{Dp_1}{Dt}+\dot{v}^f_{2i,i}+(\alpha_3-n_1)\dot{u}^s_{i,i}-\frac{Dn_1}{Dt}=0 \tag{3.136}$$

where c_{11}^*, c_{12}^*, c_{21}^* and c_{22}^* are defined as

$$\begin{cases} c_{11}^* = \dfrac{\alpha_3 - n_1 - n_2}{K_s} + \dfrac{n_1}{K_f} \\ c_{12}^* = -\dfrac{\alpha_3 - n_1 - n_2}{K_{fr}} \\ c_{21}^* = \dfrac{\alpha_3 - n_1 - n_2}{K_{fr}} + \dfrac{n_2}{K_f} \\ c_{22}^* = -\dfrac{\alpha_3 - n_1 - n_2}{K_s}. \end{cases} \tag{3.137}$$

Again, assuming that the absolute velocity of the solid is relatively small when compared with the fluid velocity, assuming constant porosities, invoking Darcy's velocities for the dual-porosity medium as given in Eqs. (3.122) and (3.123), and adding the quasi-steady interporosity flow terms (treated as internal flow sources/sinks), Eqs. (3.135) and (3.136) can be further simplified as

$$\frac{k_1}{\mu}\nabla^2 p_1 = c_{11}^* \frac{\partial p_1}{\partial t} - c_{12}^* \frac{\partial p_2}{\partial t} + (\alpha_3 - n_2)\frac{\partial \varepsilon_{kk}^s}{\partial t} + \Gamma(p_1 - p_2) \tag{3.138}$$

$$\frac{k_2}{\mu}\nabla^2 p_2 = c_{21}^* \frac{\partial p_2}{\partial t} - c_{22}^* \frac{\partial p_1}{\partial t} + (\alpha_3 - n_1)\frac{\partial \varepsilon_{kk}^s}{\partial t} - \Gamma(p_1 - p_2). \tag{3.139}$$

For a more general formulation, the porosity should not be considered as a constant. As a result, the following analysis must be conducted. The total volume of the fractured porous medium, V, can be defined as

$$V = V_s + V_{p1} + V_{p2} \tag{3.140}$$

where V_s and V_p are the solid and fluid volumes, respectively. By definition, the porosity of each component is related to the volume fractions from Eq. (3.140), i.e.,

$$n_1 = \frac{V_{p1}}{V}, \quad n_2 = \frac{V_{p2}}{V}, \quad n_s = \frac{V_s}{V} = 1 - n_1 - n_2. \tag{3.141}$$

As a result,

$$\frac{Dn_1}{Dt} = \frac{1}{V}\left(\frac{DV_{p1}}{Dt} - n_1 \frac{DV}{Dt}\right) \tag{3.142}$$

$$\frac{Dn_2}{Dt} = \frac{1}{V}\left(\frac{DV_{p2}}{Dt} - n_2 \frac{DV}{Dt}\right). \tag{3.143}$$

From Eq. (3.140), one has

$$\frac{DV}{Dt} = \frac{DV_s}{Dt} + \frac{DV_{p1}}{Dt} + \frac{DV_{p2}}{Dt}. \tag{3.144}$$

Substituting Eq. (3.144) into Eqs. (3.142) and (3.143), with certain manipulations, yields

$$\frac{Dn_1}{Dt} = (1-n_1)n_1\frac{DV_{p1}}{V_1 Dt} - n_1 n_2\frac{DV_{p2}}{V_2 Dt} - n_1(1-n_1-n_2)\frac{DV_s}{V_s Dt} \tag{3.145}$$

$$\frac{Dn_2}{Dt} = (1-n_2)n_2\frac{DV_{p2}}{V_2 Dt} - n_1 n_2\frac{DV_{p1}}{V_1 Dt} - n_2(1-n_1-n_2)\frac{DV_s}{V_s Dt}. \tag{3.146}$$

Based on the definition:

$$\frac{1}{V_{p1}}\frac{DV_{p1}}{Dt} = -\frac{1}{\rho_{1f}}\frac{D\rho_{1f}}{Dt} = -\frac{1}{K_f}\frac{Dp_1}{Dt} \tag{3.147}$$

$$\frac{1}{V_{p2}}\frac{DV_{p2}}{Dt} = -\frac{1}{\rho_{2f}}\frac{D\rho_{2f}}{Dt} = -\frac{1}{K_f}\frac{Dp_2}{Dt} \tag{3.148}$$

and recalling the relationships in Eqs. (3.132) and (3.137), Eqs. (3.145) and (3.146) can be further modified as

$$\frac{Dn_1}{Dt} = -n_1\left(c_{22}^* + \frac{1-n_1}{K_f}\right)\frac{Dp_1}{Dt} + n_1 c_{21}^*\frac{Dp_2}{Dt} - n_1(1-\alpha_3)\dot{u}_{i,i}^s \tag{3.149}$$

$$\frac{Dn_2}{Dt} = -n_2\left(c_{12}^* + \frac{1-n_2}{K_f}\right)\frac{Dp_2}{Dt} + n_2 c_{11}^*\frac{Dp_1}{Dt} - n_2(1-\alpha_3)\dot{u}_{i,i}^s. \tag{3.150}$$

Substituting Eqs. (3.149) and (3.150) into Eqs. (3.135) and (3.136), the final flow equations for the dual-porosity poroelastic formulation can be written as

$$c_{11}^*\frac{Dp_1}{Dt} + c_{21}^*\frac{Dp_2}{Dt} + \frac{1}{1-n_2}\dot{v}_{1i,i}^f + \alpha_3\dot{u}_{i,i}^s = 0 \tag{3.151}$$

$$c_{21}^*\frac{Dp_2}{Dt} + c_{11}^*\frac{Dp_1}{Dt} + \frac{1}{1-n_1}\dot{v}_{2i,i}^f + \alpha_3\dot{u}_{i,i}^s = 0. \tag{3.152}$$

With similar assumptions to those made previously (i.e., Darcy's velocities, and negligible velocity of the moving solid), Eqs. (3.151) and (3.152) can be rewritten as

$$\frac{1}{(1-n_2)}\frac{k_1}{\mu}\nabla^2 p_1 = c_{11}^*\frac{\partial p_1}{\partial t} + c_{21}^*\frac{\partial p_2}{\partial t} + \alpha_3\frac{\partial \varepsilon_{ii}^s}{\partial t} \tag{3.153}$$

$$\frac{1}{(1-n_1)}\frac{k_2}{\mu}\nabla^2 p_2 = c_{21}^*\frac{\partial p_2}{\partial t} + c_{11}^*\frac{\partial p_1}{\partial t} + \alpha_3\frac{\partial \varepsilon_{ii}^s}{\partial t}. \tag{3.154}$$

The significant difference between Eqs. (3.153) and (3.154), where the porosities are not considered as constants, and Eqs. (3.138) and (3.139), where the porosities are assumed to be constants, can be readily identified, especially in the form of cross-influence terms.

Eqs. (3.153) and (3.154) have not been incorporated in any existing dual-porosity poroelastic models. In contrast to those models within the literature, which are mainly based on physical intuition, Eqs. (3.153) and (3.154) are derived according to rigorous mass conservation.

3.2.2.2 Two-Phase Fluid

In the following derivations, subscripts 1 and 2 are used to define matrix and fractures, while subscripts w and n denote wetting and nonwetting fluid phases, respectively.

Flow Through Fractured Porous Media

It is known that

$$S_{1w} + S_{1n} = 1, \qquad S_{2w} + S_{2n} = 1 \tag{3.155}$$

where S is the saturation. Assuming fluid pressures in the matrix and the fractures are the weighted values from the saturations, then

$$p_1 = S_{1w}p_{1w} + S_{1n}p_{1n} \tag{3.156}$$

$$p_2 = S_{2w}p_{2w} + S_{2n}p_{2n}. \tag{3.157}$$

Because both fluid pressures and saturations are a function of time, differentiating Eqs. (3.156) and (3.157) with respect to time leads to

$$\frac{\partial p_1}{\partial t} = S_{1w}\frac{\partial p_{1w}}{\partial t} + p_{1w}\frac{\partial S_{1w}}{\partial t} + S_{1n}\frac{\partial p_{1n}}{\partial t} + p_{1n}\frac{\partial S_{1n}}{\partial t} \tag{3.158}$$

$$\frac{\partial p_2}{\partial t} = S_{2w}\frac{\partial p_{2w}}{\partial t} + p_{2w}\frac{\partial S_{2w}}{\partial t} + S_{2n}\frac{\partial p_{2n}}{\partial t} + p_{2n}\frac{\partial S_{2n}}{\partial t}. \tag{3.159}$$

By definition, capillary pressures are given as

$$p_{1c} = p_{1n} - p_{1w}, \qquad p_{2c} = p_{2n} - p_{2w}. \tag{3.160}$$

In two-phase fluid flow, the solid mass conservation law defined in Eq. (3.126) remains unchanged except that the solid displacement is zero. However, the decoupled mass conservation of the fluid needs to be redefined as

$$\frac{\partial n_1 S_{1w}\rho_{1w}}{\partial t} + \dot{v}^f_{1wi}\cdot\vec{\nabla} n_1 S_{1w}\rho_{1w} + n_1 S_{1w}\rho_{1w}\nabla\cdot\dot{v}^f_{1wi} = 0 \tag{3.161}$$

$$\frac{\partial n_1 S_{1n}\rho_{1n}}{\partial t} + \dot{v}^f_{1ni}\cdot\vec{\nabla} n_1 S_{1n}\rho_{1n} + n_1 S_{1n}\rho_{1n}\nabla\cdot\dot{v}^f_{1ni} = 0 \tag{3.162}$$

$$\frac{\partial n_2 S_{2w}\rho_{2w}}{\partial t} + \dot{v}^f_{2wi}\cdot\vec{\nabla} n_2 S_{2w}\rho_{2w} + n_2 S_{2w}\rho_{2w}\nabla\cdot\dot{v}^f_{2wi} = 0 \tag{3.163}$$

$$\frac{\partial n_2 S_{2n}\rho_{2n}}{\partial t} + \dot{v}^f_{2ni}\cdot\vec{\nabla} n_2 S_{2n}\rho_{2n} + n_2 S_{2n}\rho_{2n}\nabla\cdot\dot{v}^f_{2ni} = 0. \tag{3.164}$$

Eqs. (3.161) through (3.164) can be written in a compact form as

$$\frac{\partial n_\pi S_{\pi\theta}\rho_{\pi\theta}}{\partial t} + \dot{v}^f_{\pi\theta i}\cdot\vec{\nabla} n_\pi S_{\pi\theta}\rho_{\pi\theta} + n_\pi S_{\pi\theta}\rho_{\pi\theta}\nabla\cdot\dot{v}^f_{\pi\theta i} = 0 \tag{3.165}$$

where subscripts $\pi = 1$ and 2, and $\theta = w$ and n, respectively.

Relative flow velocity, $\dot{v}^f_{\theta i}$, can be expressed as

$$\dot{v}^{f*}_{\pi\theta i} = n_\pi S_{\pi\theta}\dot{v}^f_{\pi\theta i}. \tag{3.166}$$

Coupled Processes

For two-phase fluid flow, solid mass conservation remains the same as for single-phase coupled processes in dual-porosity media, as shown in Eq. (3.126), while the fluid mass conservation relation is identical to that of Eq. (3.165), representing the abbreviated form of the flow equations.

Relative flow velocity, $\dot{v}^{f*}_{\theta i}$, can be expressed as

$$\dot{v}^{f*}_{\pi\theta i} = n_\pi S_{\pi\theta}(\dot{v}^f_{\pi\theta i} - \dot{u}^s_i) \tag{3.167}$$

or

$$\dot{u}^f_{\pi\theta i} = \frac{\dot{v}^f_{\pi\theta i}}{n_\pi S_{\pi\theta}} + n_\pi S_{\pi\theta}\dot{u}^s_i. \tag{3.168}$$

Applying the substantial derivatives and substituting Eq. (3.168) into Eq. (3.165), results in

$$n_\pi S_{\pi\theta}\frac{D\rho_{\pi\theta}}{Dt} + n_\pi\rho_{\pi\theta}\frac{DS_{\pi\theta}}{Dt} + S_{\pi\theta}\rho_{\pi\theta}\frac{Dn_\pi}{Dt} + \dot{v}^{f*}_{\pi\theta i}\frac{\rho_{\pi\theta}}{\partial x_i} + \rho_{\pi\theta}\dot{v}^{f*}_{\pi\theta i,i} + n_\pi S_{\pi\theta}\rho_{\pi\theta}\dot{u}^s_{i,i} = 0. \tag{3.169}$$

From Eq. (3.126), one has

$$\frac{Dn_1}{Dt} = \frac{(1-n_1-n_2)}{\rho_s}\frac{D\rho_s}{Dt} - \frac{Dn_2}{Dt} + (1-n_1-n_2)\dot{u}^s_{i,i} \tag{3.170}$$

$$\frac{Dn_2}{Dt} = \frac{(1-n_1-n_2)}{\rho_s}\frac{D\rho_s}{Dt} - \frac{Dn_1}{Dt} + (1-n_1-n_2)\dot{u}^s_{i,i}. \tag{3.171}$$

Flow equations with $\pi = 1$ and $\pi = 2$ can be obtained by substituting Eq. (3.170) into Eq. (3.169), and Eq. (3.171) into Eq. (3.169), respectively. For example, the flow equation with $\pi = 1$ is given as

$$\frac{n_1 S_{1\theta}}{\rho_{1\theta}}\frac{D\rho_{1\theta}}{Dt} + n_1\frac{DS_{1\theta}}{Dt} + \frac{(1-n_1-n_2)S_{1\theta}}{\rho_s}\frac{D\rho_s}{Dt} - S_{1\theta}\frac{Dn_2}{Dt} +$$

$$\dot{v}^{f*}_{1\theta i,i} + \frac{\dot{v}^{f*}_{1\theta i}}{\rho_{1\theta}}\frac{\partial \rho_{1\theta}}{\partial x_i} + (1-n_2)S_{1\theta}\dot{u}^s_{i,i} = 0. \tag{3.172}$$

Following a similar procedure to that for single-phase fluid flow, differentiations with respect to the porosity n_1 and n_2 are eliminated. The solid density is defined as a function of fluid pressures and solid strain using the effective stress law for dual-porosity media, except that the influence of the fluid compressibility on volume changes may be considered different between the matrix and the fractures. As a result, Eqs. (3.147) and (3.148) are redefined as

$$\frac{1}{V_{p1}}\frac{DV_{p1}}{Dt} = -\frac{1}{\rho_{1f}}\frac{D\rho_{1f}}{Dt} = -\frac{1}{K_{f_1}}\frac{Dp_1}{Dt} \tag{3.173}$$

$$\frac{1}{V_{p2}}\frac{DV_{p2}}{Dt} = -\frac{1}{\rho_{2f}}\frac{D\rho_{2f}}{Dt} = -\frac{1}{K_{f_2}}\frac{Dp_2}{Dt} \tag{3.174}$$

where the fluid bulk moduli K_{f_1} and K_{f_2} can be assessed by the following expressions using the concept of "saturation-averaging"

$$\frac{1}{K_{f_1}} = \frac{S_{1w}}{K_{1w}} + \frac{S_{1n}}{K_{1nw}}, \quad \frac{1}{K_{f_2}} = \frac{S_{2w}}{K_{2w}} + \frac{S_{2n}}{K_{2nw}}. \tag{3.175}$$

The flow equation (3.172) can then be rewritten as

$$\frac{n_1 S_{1\theta}}{\rho_{1\theta}}\frac{D\rho_{1\theta}}{Dt} + n_1\frac{DS_{1\theta}}{Dt} + S_{1\theta}\left[\frac{(1-n_2)(\alpha_3-n_1-n_2)}{K_s} - \frac{n_1 n_2}{K_{f_2}}\right]\frac{Dp_1}{Dt} +$$

$$S_{1\theta}(1-n_2)c^{**}_{21}\frac{Dp_2}{Dt} + \dot{v}^{f*}_{1\theta i,i} + \frac{\dot{v}^{f*}_{1\theta i}}{\rho_{1\theta}}\frac{\partial \rho_{1\theta}}{\partial x_i} + \alpha_3(1-n_2)S_{1\theta}\dot{u}^s_{i,i} = 0 \tag{3.176}$$

where c^{**}_{21} is slightly different from Eq. (3.137) in the definition of fluid compressibility, and can be written as

$$c^{**}_{21} = \frac{\alpha_3 - n_1 - n_2}{K_{fr}} + \frac{n_2}{K_{f_2}} \tag{3.177}$$

where K_{f_2} is the aggregate bulk modulus of fluids within the fractures, defined in Eq. (3.175). p_1 and p_2 are the fluid pressures in the matrix and fractures, which should be substituted with the saturation-weighted pressures, given in Eqs. (3.156) and (3.157).

For single-phase fluid flow, it is known that

$$S_{1\theta} = 1, \quad \frac{1}{\rho_{1\theta}}\frac{D\rho_{1\theta}}{Dt} = \frac{1}{K_f}\frac{Dp_1}{Dt}. \tag{3.178}$$

Substituting Eq. (3.178) into Eq. (3.176), and neglecting the convective terms in the governing equations as well as in the substantial time derivative, the equation for flow in the matrix for the single-phase fluid, Eq. (3.153), is recovered.

Eq. (3.176) represents two equations (for wetting and nonwetting fluid phases; i.e., $\theta = w$ and $\theta = n$) for the porous matrix (i.e., $\pi = 1$). Following a similar procedure, the governing flow equations for the fractures (i.e., $\pi = 2$) can be derived as

$$\frac{n_2 S_{2\theta}}{\rho_{2\theta}}\frac{D\rho_{2\theta}}{Dt} + n_2\frac{DS_{2\theta}}{Dt} + S_{2\theta}\left[\frac{(1-n_1)(\alpha_3 - n_1 - n_2)}{K_s} - \frac{n_1 n_2}{K_{f2}}\right]\frac{Dp_2}{Dt} +$$

$$S_{2\theta}(1-n_1)c_{11}^{**}\frac{Dp_1}{Dt} + \dot{v}_{2\theta i,i}^{f*} + \frac{\dot{v}_{2\theta i}^{f*}}{\rho_{2\theta}}\frac{\partial \rho_{2\theta}}{\partial x_i} + \alpha_3(1-n_1)S_{2\theta}\dot{u}_{i,i}^s = 0 \tag{3.179}$$

where $\theta = w$, $\theta = n$, and c_{11}^{**} should be expressed as

$$c_{11}^{**} = \frac{\alpha_3 - n_1 - n_2}{K_s} + \frac{n_1}{K_{f1}} \tag{3.180}$$

where K_{f1} is the fluid bulk modulus of the matrix blocks, defined in Eq. (3.175).

Again, for the single-phase fluid flow, one has

$$S_{2\theta} = 1, \qquad \frac{1}{\rho_{2\theta}}\frac{D\rho_{2\theta}}{Dt} = \frac{1}{K_f}\frac{Dp_2}{Dt}. \tag{3.181}$$

Substituting Eq. (3.181) into Eq. (3.179), with the omission of the convective terms, the equation for flow in the fractures for the single-phase fluid, Eq. (3.154) is recovered.

3.3 PARAMETRIC STUDY

This parametric investigation focuses on permeability, compressibility, and anisotropy.

3.3.1 Permeability

Permeability is arguably the most important parameter in evaluating fluid flow. Where the viscosity of the fluid remains sensibly constant, hydraulic conductivity may be used instead of the permeability. Permeability k is related to the hydraulic conductivity K_h by the following relation:

$$k = \frac{\mu K_h}{\rho_f g} \tag{3.182}$$

where μ is the fluid dynamic viscosity, ρ_f is the fluid density, and g is gravitational acceleration.

Due to this correlation, permeability and hydraulic conductivity may be exchanged, providing the appropriate substitutions are made. Permeability is the desired parameter where large changes in viscosity are expected, since permeability is a true material parameter.

3.3.1.1 Homogeneous Media

Two separate scenarios are analyzed: (a) the initial state with an invariant permeability, and (b) a subsequent stress-modified or stress-dependent state.

Invariant Permeability

For flow within a single capillary, or by superposition through a bundle of capillaries, permeability, k, may be defined proportionally to a characteristic length representative of hydraulic radius. This may be expressed as

$$k = \frac{c_g d_m^2}{F(n)} \tag{3.183}$$

where d_m is the "hydraulic radius" and is related to the grain diameter or pore dimension, $F(n)$ is the porosity factor, and c_g is a constant associated with grain packing and grain shape configuration.

Various attempts have been made to provide insight into the physical processes represented by Eq. (3.183) with substantial experimental work having been completed to further verify the expression (Hubbert 1940; Scheidegger 1957; Kozeny 1927; Krumbein and Monk 1943). Through experimental investigation, Hubbert (1940) substantiated Eq. (3.183) by presenting the relationship

$$k = N_g d_m^2 \left(\frac{\rho g}{\mu} \right) \tag{3.184}$$

where N_g is a dimensionless number associated with grain shape and packing and d_m is a size factor related either to the dimensions of the opening in the medium (pore space) or expressed as a mean size of the grain (Krumbein and Monk 1943). In addition, d_m can also be some function of the square root of the permeability k (Freeze and Cherry 1979). In this work, specifically, d_m is defined as the reciprocal of the specific surface which is the ratio of bulk volume to internal area of the porous medium (Kozeny 1927).

In the well-known theory of Kozeny (Kozeny 1927), the porous medium is represented by an assemblage of capillaries, of various cross-sections, where the Navier-Stokes equations are solved individually for each of the parallel arrangements. The permeability is given as

$$k = c_g^{\#} n^3 d_m^2 \tag{3.185}$$

where $c_g^{\#}$ is a constant representing the mean size and shape of the grains, and n is porosity.

Comparing Eq. (3.185) with Eq. (3.184), the importance of porosity is apparent. Real porous media comprise a mixture of large and small particles that may be uncemented or cemented. The porosity of consolidated materials depends mainly on the degree of cementation, while the porosity of unconsolidated materials depends on the packing of the grains, their shape, arrangement, and size distribution.

Porosity may have an important control over permeability, designated by the increase of permeability with an increase in porosity. Kozeny-Carman (Bear 1972) proposed an equation similar to Eq. (3.185), which includes a factor representing porosity as

$$k = \frac{n^3}{(1-n)^2} \frac{d_m^2}{180}. \tag{3.186}$$

This equation attempts to incorporate the influence of grain packing on hydraulic conductivity. Alternatively, packing may be incorporated directly for Hubbert's equation as shown in Eq. (3.184), enabling direct relationships to be established for permeability changes.

Stress-Dependent Permeability

Also related to Eq. (3.183), hydraulic radius is controlled by changes in effective or intergranular stresses (De Wiest 1969). Young et al. (1964) showed that an increase in stress acting on samples of argillaceous rocks produced a decrease in permeability of more than an order of magnitude in unconfined tests. In correlating confining stresses with permeability, the interaction between stresses and fluid pressures (i.e., effective stress law) has been intensively investigated (Brandt 1955; Gangi 1978; Walsh 1981; Gale 1982; Barton et al 1985; Jones 1975), in which one of the major indeterminates is the exact magnitude of effective stress (Terzaghi 1923; Walsh 1981; Skempton 1960; Geertsma 1957; Nur and Byerlee 1971; Robin 1973), as defined through total stresses and fluid pressures.

The influence of pore pressures on effective stresses is controlled through the Biot coefficient α, which has been experimentally defined by Geertsma (1957) and Skempton (1960) and rationalized by Nur and Byerlee (1971).

Scheidegger (1957) defined the permeability of a medium solely as a function of the total external stresses, expressed as a change of pore space under external loading. By carrying out laboratory tests on intact and jointed Barre granite, Kranz et al. (1979) proposed the following equation to predict the change of permeability, k, as a function of the difference between the confining stress σ_c and pore pressure p as

$$k \propto \left(\mid \sigma_c \mid - \frac{b_e^{\#}}{a_e^{\#}} \mid p \mid \right) \tag{3.187}$$

where $b_e^{\#}$ and $a_e^{\#}$ are constants representing the effective proportionality of the fluid pressure. Based on experimental results, Kranz et al. suggested that $b_e^{\#}/a_e^{\#} < 1$ for jointed rock and $b_e^{\#}/a_e^{\#} \approx 1$ for intact rock. They also reported a one to two order of magnitude difference between the permeability of jointed rock and that of intact rock at low effective stresses, and almost no difference at high effective stresses.

An empirical formula relating permeability to effective stress is given by Walsh (1981) as

$$k = k_0 \left\{ 1 - (\sqrt{2}\xi_f) \ln \left(\frac{\sigma_e}{\sigma_e^0} \right) \right\}^3 \tag{3.188}$$

where k_0 is the initial permeability, σ_e^0 is the initial effective stress, ξ_f is a factor related to the fracture geometry, and acting effective stress, σ_e, may be estimated from the following equation:

$$\sigma_e = \sigma_c + \alpha p \tag{3.189}$$

where σ_c is the mean stress evaluated by $\sigma_c = \frac{\sigma_{kk}}{3}$ and σ_{kk} is the total stress.

Louis (1974) studied the influence of overburden stress on the permeability measured at various depths in boreholes drilled in fractured formations. After examination of numerous test results, he concluded that the relationship between permeability and normal stress follows a negative exponential function, i.e.,

$$k = k_0 e^{-\beta_s^* \Delta\sigma_n} \tag{3.190}$$

where, k_0 is the initial permeability, β_s^* is a site parameter based on the particulars of the test site, and $\Delta\sigma_n$ is the normal stress. This relationship has been widely used to analyze the coupling of permeability and stresses, due to its simplicity. However, numerous permeability tests showed that this relationship may not always match field data (Witherspoon 1981).

To express hydraulic conductivity as a function of stress conditions, Gangi (1978) derived an equation based on Hertzian contact theory (Hertz 1895), which may be expressed as

$$k = k_0 \left\{ 1 - \frac{1}{2} \left(\frac{\Delta\sigma_c + \Delta\sigma_i}{E_0} \right)^{2/3} \right\}^4 \tag{3.191}$$

where σ_i is the equivalent cementing pressure, E_0 is the effective modulus of the grains, and σ_c is the confining stress given in Eq. (3.189). This relationship identifies an inverse proportionality between permeability and effective stresses. Experimental investigation of the coupling between the permeability and stresses was also completed by Gale (1982) and by Barton et al. (1985).

Alternatively, Bai and Meng (1997) assumed that permeability variations may be attributed to the changes in void spaces or as a result of grain compression, which can be calculated by

$$\Delta k_j = k_0 \left[\sum_{i=1,i\neq j}^{3} \left\{ 1 - \frac{1}{2} \left[\frac{9(1-\nu^2)^2}{2} (\pi \Delta\varepsilon_i)^2 \right]^{1/3} \right\}^2 - 1 \right] \tag{3.192}$$

where k_0 is the original permeability, ν is Poisson ratio, and the normal strain vector $\Delta\varepsilon_i$ can be converted to the normal stress vector $\Delta\sigma_i$ through the following expression:

$$\Delta\varepsilon_i = \frac{1}{E}[\Delta\sigma_i - \nu(\Delta\sigma_j + \Delta\sigma_k)] \tag{3.193}$$

where E is the elastic modulus.

3.3.1.2 Heterogeneous Media

Since scale effects are significant, study of the permeability of heterogeneous media, including fractured media, is especially relevant.

Invariant Permeability

One of the most popular analogs in evaluating fluid flow through fractures is the "parallel plate analog". Understanding the assumptions invoked in the development of this analog is helpful in understanding its applicability to "fracture flow" problems.

Parallel Plate Analog: Assuming that viscous forces are much greater than inertial forces, terms on the left-hand side of the Navier-Stokes equation (3.8) can be neglected. Further assuming a no-flow boundary condition at the plate walls, and noting zero velocity gradients orthogonal to the principal flow direction, Eq. (3.8) reduces to

$$\frac{\partial p}{\partial x} = \mu \frac{\partial^2 v_x}{\partial z^2} \tag{3.194}$$

$$\frac{\partial p}{\partial y} = \mu \frac{\partial^2 v_y}{\partial z^2} \tag{3.195}$$

$$\frac{\partial p}{\partial z} = 0. \tag{3.196}$$

The following boundary conditions can be satisfied if the flow is laminar:

$$\frac{\partial v_x}{\partial z} = 0, \qquad \frac{\partial v_y}{\partial z} = 0 \qquad (z = 0). \tag{3.197}$$

Integrating Eqs. (3.194), (3.195), and (3.196) with respect to the z direction and implementing the boundary condition in Eq. (3.197), yields

$$\begin{cases} z\dfrac{\partial p}{\partial x} = \mu\dfrac{\partial v_x}{\partial z} \\ \\ z\dfrac{\partial p}{\partial y} = \mu\dfrac{\partial v_y}{\partial z} \end{cases} \tag{3.198}$$

$$v_x = v_y = 0 \qquad\qquad (z = \pm\frac{b}{2}) \tag{3.199}$$

where b is the gap between parallel plates (aperture). A no-slip boundary condition is implied by Eq. (3.199), where the geometry of Figure 3-4 is applied.

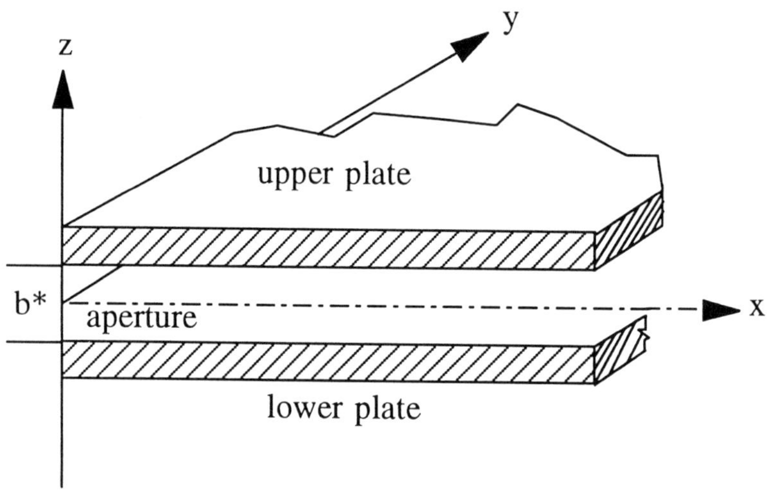

Figure 3-4. Schematic flow between parallel plates.

Integrating Eq. (3.198) with respect to z, again while incorporating Eq. (3.199), gives

$$\begin{cases} v_x = \dfrac{1}{2\mu}\left(z^2 - \dfrac{b^2}{4}\right)\dfrac{\partial p}{\partial x} \\ \\ v_y = \dfrac{1}{2\mu}\left(z^2 - \dfrac{b^2}{4}\right)\dfrac{\partial p}{\partial y}. \end{cases} \tag{3.200}$$

The specific discharge, which is defined as the flow rate through a unit flow area, can be determined from the integration with respect to the z direction in Eq. (3.200). For example, the specific discharge in the x direction, q_x, with the fracture spacing s^* for multiple parallel fractures can be expressed as

$$q_x = \frac{1}{s^*} \int_{-\frac{b}{2}}^{\frac{b}{2}} v_x(z) dz. \tag{3.201}$$

Substituting $v_x(z)$ in Eq. (3.200) into Eq. (3.201) and performing the integration, yields

$$q_x = \frac{1}{s^*} \frac{\partial p}{\partial x} \left[\frac{1}{2\mu} \left(\frac{z^3}{3} - \frac{b^2 z}{4} \right)_{-\frac{b}{2}}^{\frac{b}{2}} \right] = -\frac{b^3}{12\mu s^*} \frac{\partial p}{\partial x}. \tag{3.202}$$

Similarly,

$$q_y = -\frac{b^3}{12\mu s^*} \frac{\partial p}{\partial y}. \tag{3.203}$$

Comparing Eqs. (3.201) and (3.202) with the Darcy velocity in Eq. (3.15), the permeability can be deduced as

$$k = \frac{b^3}{12 s^*}. \tag{3.204}$$

Based on the morphological similarity, flow within planar fractures is analogous to flow between parallel plates. As a result, Eq. (3.204) is applicable for describing fracture flow in which the permeability is related to the cube of fracture gap (or aperture) over a unit fracture spacing. Eq. (3.204) is well-known as the cubic law of fracture flow, the application of which has been verified through experimental study (Witherspoon et al. 1980).

From the definition of specific discharge, q_s, it is known that flow rate, q, can be expressed as

$$q = A_f q_s \tag{3.205}$$

where A_f is the flow area. With reference to Figure 3-4, and Eqs. (3.204) and (3.205), flow rate in laminar flow along a single fracture, q_1, can be written as

$$q_1 = A_f q_s = -(A_f) \left(\frac{k}{\mu} \right) \vec{\nabla} p = -(s^* l) \left(\frac{b^3}{12 s^* \mu} \right) \left(\frac{\Delta p}{L} \right) = -\frac{b^3 l}{12\mu} \frac{\Delta p}{L} \tag{3.206}$$

For n^* fractures, the flow rate is

$$q_{n^*} = -n^* \frac{b^3 l}{12\mu} \frac{\Delta p}{L}. \tag{3.207}$$

Because the flow rate can be generally expressed from Darcy's law as

$$q_{n^*} = -A_f \frac{k}{\mu} \frac{\Delta p}{L}, \tag{3.208}$$

equating Eqs. (3.207) and (3.208), enables the permeability in Eq. (3.208) to be represented by

$$k = f_s \frac{b^3}{12} \tag{3.209}$$

where f_s is

$$f_s = \frac{n^* l}{A_f} \tag{3.210}$$

which represents the total fracture length per unit cross-sectional flow area.

For multiple pairs of parallel fractures,

$$f_s = \frac{n^* l}{n^* l s^*} = \frac{1}{s^*} \tag{3.211}$$

and Eq. (3.204) is recovered by substituting Eq. (3.211) into Eq. (3.209).

Permeability and Mechanical Properties of Fractured Media: In the groundwater literature, hydraulic head is frequently adopted as the independent variable, as an alternative to pore fluid pressure. Under this situation, Darcy's velocity can be expressed as

$$q_i = -K_h \frac{\partial h}{\partial x_i} \tag{3.212}$$

where q_i is the velocity component in a Cartesian coordinate system (x_i, $i = 1, 2, 3$), K_h is the hydraulic conductivity introduced previously [refer to Eq. (3.182) for its relation with the permeability], and h is the hydraulic head, which can be given as

$$h_j^* = \frac{p}{\rho g} + z \tag{3.213}$$

where ρ is the fluid density, g is gravitational acceleration, and z is the elevation with respect to an arbitrary datum.

The permeability, k, is defined as an effective area of flow. In other words, the permeability is only associated with the active flow conduits by excluding the "dead ends," filled pore spaces, vugs, or cavities completely separated from flow pathways, and any cross-sections in the flow regions which do not contribute to the fluid flow. In view of the "effective flow area," this definition is defined differently for a homogeneous porous medium and for a fractured medium. Specifically, the area is defined as the effective void space for the former and the effective fracture apertures for the latter.

To introduce the concept of hydraulic aperture, Eq. (3.204) is rewritten as

$$k = \frac{b_h^3}{12 s^*} \tag{3.214}$$

where b_h is the fracture hydraulic aperture that represents the effective flow area, and is also equivalent to the original hydraulic aperture b_0. s^* is the fracture spacing. However, the hydraulic aperture is likely affected by the mechanical properties of the fracture. In most cases, a reduction of effective aperture is the result of a decrease in the effective flow area within the fracture due to infilling and roughness. Therefore, a modification to the hydraulic aperture must be made through the following relation:

$$b_h = F_l b_m \tag{3.215}$$

where b_m is the mechanical aperture, which is in general larger than b_h (Piggott and Elsworth 1993) since, in general, $F_l \leq 1$. It is apparent that F_l incorporates the influence of a number of rock mechanical properties.

Significant efforts have been made to identify the form of F_l in Eq. (3.215). Louis (1969) related F_l to the fracture roughness by the following equation:

$$F_l = \left\{[1 + 8.8(R_l/D_h)^{1.5}]^{-1}\right\}^{1/3} \tag{3.216}$$

where R_l is the absolute roughness and D_h is the hydraulic diameter of the fracture.

From correlations with experimental data, Witherspoon et al. (1980) proposed the concept of a *correction factor* by considering the effective flow within a fracture which deviates from the ideal parallel plate conditions, i.e.,

$$F_l = \left(\frac{1}{f^{\#}}\right)^{1/3} \tag{3.217}$$

where $f^{\#}$ is the "corrective factor" and $f^{\#} \geq 1$ in general.

An equation similar to Eq. (3.217) was used by Elliot et al. (1985) but with a different intent; they attributed the "correction factor" to a degree of surface matching, interlocking, roughness, deposits, loading history, and sample size effects.

Based on numerous tests, Barton et al. (1985) provided an empirical relationship between the hydraulic and mechanical apertures, such as

$$F_l = \frac{1}{JRC^{2.5}} \tag{3.218}$$

where JRC is the joint roughness coefficient.

Incorporating the effects of tortuosity and contact areas, Cook (1988) developed an alternative relationship:

$$F_l = \left\{[1 + \log(b_m^*)]\frac{(b_m^*)^2}{2 - b_m^*}\right\}^{1/3} \tag{3.219}$$

where b_m^* is the dimensionless mechanical aperture (with reference to original aperture).

In comparison, many of these relations show nonlinear characteristics between hydraulic and mechanical apertures.

Stress-Dependent Permeability

Darcy's velocity, described in Eq. (3.212), may not be easily determined if the hydraulic conductivity K_h or the permeability k is not a constant. As a result of stress changes, the effective flow area, i.e., the permeability, is no-longer a constant but a function of the variation in stress. In other words, the fracture aperture may contract due to compressive loading, and expand as a result of extensional stresses.

Indirect relationships between stress and fracture hydraulic aperture, predicated on the direct link between normal stress and mechanical aperture, have been proposed by several researchers. Goodman (1976) proposed the following expression:

$$\frac{\Delta\sigma_n - \Delta{\sigma_n}^0}{\Delta{\sigma_n}^0} = \zeta_1 \left(\frac{\Delta u_n}{b_0 - \Delta u_n}\right)^{\zeta_2} \tag{3.220}$$

where $\Delta\sigma_n$ and $\Delta{\sigma_n}^0$ are the ambient normal and initial normal stresses, respectively; Δu_n is the normal displacement; ζ_1 and ζ_2 are empirical parameters; and b_0 is the original hydraulic aperture.

Lamas (1995) presented a similar relationship:

$$\Delta\sigma_n = \left[-\log\left(\frac{b_0 - \Delta u_n}{10 b_0}\right)\right]^{\xi_l} - 1 \tag{3.221}$$

where ξ_l is a parameter determined from experimental data.

For fracture-dominated flow, the stress-permeability relationship becomes more important. Based on statistical analyses, Oda (1986) proposed a relationship between permeability and the stress tensor, according to fracture structure and crack geometry. This statistical study showed that these tensors can be determined in terms of measurable in situ quantities such as crack orientations and crack traces on rock exposures.

With reference to the analysis in the previous section, the hydraulic aperture, b_h, may be modified as

$$b_h = b_0 + \Delta b_s \tag{3.222}$$

where b_0 is the original hydraulic aperture and Δb_s is the change in aperture due to a change in stress.

An explicit linearized relationship between the hydraulic aperture and stress status was provided by Bai and Elsworth (1994); i.e.,

$$b_h = b_0 + \Delta b_s = b_0 + \frac{E(s^* + b_0)\Delta\varepsilon}{s^* K_n + E} \tag{3.223}$$

where s^* is the fracture spacing, $\Delta\varepsilon$ is the normal strain, K_n is the joint normal stiffness, and E is the elastic modulus. The hydraulic aperture discussed in the previous section does not include the portion of the perturbed aperture Δb_s. Specifically, Darcy's velocity in the previous section is a fixed value, while for stress-dependent permeability it is a variable due to the variation in permeability with stress changes.

Incorporating the influences of both normal deformation and shear dilatation on the effect of fluid flow in orthogonally fractured media, the dimensionless permeability changes due to solid deformation may be expressed as (Elsworth and Xiang 1989; Bai and Elsworth 1994)

$$\frac{k}{k_0} = \left[1 + \frac{\Delta\varepsilon}{b}\left(\frac{K_n}{E} + \frac{1}{s^*}\right)^{-1} + \frac{\Delta\gamma}{b}\left(\frac{K_{sh}}{G} + \frac{1}{s^*}\right)^{-1}(\tan\phi_d)\right]^3 \tag{3.224}$$

where k_0 is the initial permeability, $\Delta\varepsilon$ and $\Delta\gamma$ are the body and shear strains, G is the shear modulus, b is the fracture aperture, and ϕ_d is the fracture dilatational angle. This follows directly from consideration of the partitioning of strains within a fractured medium, where shear and normal stiffnesses, K_{sh} and K_n, represent the stress-deformation response of the individual fractures. This is equivalent to behavior defined in terms of a modulus reduction ratio, $R_m = E_{mass}/E_{intact}$ (Ouyang and Elsworth 1993; subscripts *mass* and *intact* represent the quantities in terms of rock mass or intact rock, respectively), such that

$$\frac{k}{k_0} = \left\{1 + \Delta\varepsilon\left[\frac{s^*}{b}(1 - R_m^n) + 1\right] + \Delta\gamma\left[\frac{s^*}{b}(1 - R_m^s) + 1\right](\tan\phi_d)\right\}^3 \tag{3.225}$$

where

$$R_m^n = \frac{1}{1 + \dfrac{E}{K_n s^*}} \tag{3.226}$$

$$R_m^s = \frac{1}{1 + \dfrac{G}{K_{sh} s^*}}. \tag{3.227}$$

This enables permeability change behavior to be related directly to popular rock mass rating schemes (Liu et al. 1999), where the standard repetitive length of individual matrix blocks, including one fracture, is taken as s^* instead of $s^* + b$ (since $s^* \gg b$), then Eq. (3.225) collapses to Eq. (3.224). The integer bracketed term of Eq. (3.225) drops out under this requirement. Figure 3-5 depicts a schematic description of a three-dimensional fracture set in a local coordinate system $(Ox'y'z')$ and subjected to external load.

The orientations of the fracture sets for a flow direction that is not orthogonal to the fractures can be determined from the relation of the global coordinate (XYZ) to the directional cosines of the fracture vector normal, i.e.,

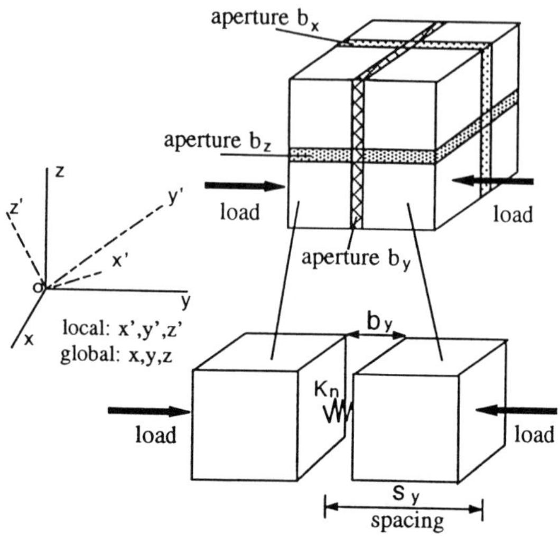

Figure 3-5. Schematic 3-D fractured block.

(l_1^*, m_1^*, n_1^*), (l_2^*, m_2^*, n_2^*), and (l_3^*, m_3^*, n_3^*). As a result, the relationship in Eq. (3.224) can be written more generally as

$$\frac{k_i}{k_{0i}} = \frac{1}{(b_j)^3 + (b_k)^3} \left\{ b_j^3 \left[1 + \frac{\Delta\sigma_j}{(K_n)_j b_j} + \frac{s_j^* - b_j}{E_j b_j} \left(\Delta\sigma_j - \nu_j \left(\Delta\sigma_i + \Delta\sigma_k \right) \right) + \right. \right.$$

$$\left. (\Delta\tau_{jk} + \Delta\tau_{ji}) \left(\frac{G_j}{s_j^*} + (K_{sh})_j \right)^{-1} \frac{\tan(\phi_d)_j}{b_j} \right]^3 +$$

$$b_k^3 \left[1 + \frac{\Delta\sigma_k}{(K_n)_k b_k} + \frac{s_k^* - b_k}{E_k b_k} \left(\Delta\sigma_k - \nu_k \left(\Delta\sigma_i + \Delta\sigma_j \right) \right) + \right.$$

$$\left. \left. (\Delta\tau_{kj} + \Delta\tau_{ki}) \left(\frac{G_j}{s_k^*} + (K_{sh})_k \right)^{-1} \frac{\tan(\phi_d)_k}{b_k} \right]^3 \right\} \tag{3.228}$$

where the subscripts $i, j, k = x', y', z'$; $i \neq j \neq k$; ν is Poisson ratio; and $\Delta\sigma$ and $\Delta\tau$ are the normal and shear stresses.

As a specific example, the dimensionless permeability change in the z' direction may be defined as

$$\frac{k_{z'}}{k_{0z'}} = \frac{1}{(b_{x'})^3 + (b_{y'})^3} \left\{ b_{x'}^3 \left[1 + \frac{\Delta\sigma_{x'}}{(K_n)_{x'} b_{x'}} + \frac{s_{x'}^* - b_{x'}}{E_{x'} b_{x'}} \left(\Delta\sigma_{x'} - \nu_{x'} \left(\Delta\sigma_{y'} + \Delta\sigma_{z'} \right) \right) \right. \right.$$

$$+ (\Delta\tau_{x'y'} + \Delta\tau_{x'z'}) \left(\frac{G_{x'}}{s_{x'}} + (K_{sh})_{x'}\right)^{-1} \frac{\tan(\phi_d)_{x'}}{b_{x'}}\Bigg]^3 +$$

$$b_{y'}^3 \left[1 + \frac{\Delta\sigma_{y'}}{(K_n)_{y'}b_{y'}} + \frac{s_{y'}^* - b_{y'}}{E_{y'}b_{y'}}(\Delta\sigma_{y'} - \nu_{y'}(\Delta\sigma_{x'} + \Delta\sigma_{z'})) + \right.$$

$$\left.(\Delta\tau_{y'x'} + \Delta\tau_{y'z'}) \left(\frac{G_{y'}}{s_{y'}^*} + (K_{sh})_{y'}\right)^{-1} \frac{\tan(\phi_d)_{y'}}{b_{y'}}\right]^3 \Bigg\}. \tag{3.229}$$

The opening or closure of the fracture when it is subject to extension or compression is attributed to the normal stresses only. The dilatation due to shear displacements always tends to increase the fracture aperture, and correspondingly increase the permeability magnitude (Meng and Bai 1998).

Transformation between local stresses and permeabilities and their global counterparts may be achieved by using the tensorial transformation properties (e.g., Jaeger and Cook 1979). For example, the following relations can be derived for the case in which the global coordinate system (xyz) is coincident with the principal stresses:

$$\Delta\sigma_{x'} = (l_1^*)^2\Delta\sigma_1 + (m_1^*)^2\Delta\sigma_2 + (n_1^*)^2\Delta\sigma_3 \tag{3.230}$$

$$\Delta\sigma_{y'} = (l_2^*)^2\Delta\sigma_1 + (m_2^*)^2\Delta\sigma_2 + (n_2^*)^2\Delta\sigma_3 \tag{3.231}$$

$$\Delta\sigma_{z'} = (l_3^*)^2\Delta\sigma_1 + (m_3^*)^2\Delta\sigma_2 + (n_3^*)^2\Delta\sigma_3 \tag{3.232}$$

$$\Delta\tau_{x'y'} = l_1^*l_2^*\Delta\sigma_1 + m_1^*m_2^*\Delta\sigma_2 + n_1^*n_2^*\Delta\sigma_3 \tag{3.233}$$

$$\Delta\tau_{x'z'} = l_3^*l_1^*\Delta\sigma_1 + m_3^*m_1^*\Delta\sigma_2 + n_3^*n_1^*\Delta\sigma_3 \tag{3.234}$$

$$\Delta\tau_{y'z'} = l_2^*l_3^*\Delta\sigma_1 + m_2^*m_3^*\Delta\sigma_2 + n_2^*n_3^*\Delta\sigma_3. \tag{3.235}$$

Conversion of the permeabilities between the local and global systems follows a similar process.

Alternative Permeability Forms

For simplicity of illustration, the stresses and strains in the following are denoted as scalars. Applying a uniaxial load to the fracture network, Elsworth (1989) derived the permeability change in relation to Eq. (3.214) by assuming the individual fractures are distinctly soft with respect to the porous matrix, as

$$\Delta k_2 = \frac{1}{12s^*}(b + s^*\Delta\varepsilon)^3 \tag{3.236}$$

where $\Delta\varepsilon$ is the elastic strain orthogonal to the fracture network due to application of the load. For relatively small fracture spacing, Eq. (3.236) can be more precisely written as

$$\Delta k_2 = \frac{1}{12s^*}[b + (s^* + b)\Delta\varepsilon]^3. \tag{3.237}$$

The porous matrix is assumed to be stiff with negligible deformation in the above two equations. If this restriction is released and the contribution of deformations from both fracture and matrix are distinguished and incorporated, the permeability change can be expressed as (Bai and Elsworth 1994)

$$\Delta k_2 = \frac{1}{12s^*}\left[b + \frac{E(s^*+b)\Delta\varepsilon}{sK_n + E}\right]^3 \tag{3.238}$$

where E is the elastic modulus and K_n is the fracture normal stiffness.

In all cases mentioned above, the stress dependent permeability is defined either within the matrix blocks or within the fractures, but not in both. In reality, rock deformation and fluid flow contribute mutually to behavior. In combining these effects, three conditions must be simultaneously accommodated:

- The effective areas of flow are summed for both fracture and matrix, i.e., permeability results from the cumulative effect of fracture and matrix permeabilities
- Elastic strains are individually calculated and superposed through the summation of the respective permeabilities
- Load or stress acting on either fracture or matrix is uniform, as required by equilibrium considerations.

To consider the influence of dual-porosity behavior on effective permeability, fracture and porous medium permeabilities are rationed according to their respective initial in situ volumes. This rationale is justified since the flow conduits in both porous matrix and fractures are strongly related to the ratio of respective void volume to the total volume (porosity). Using a similar concept as proposed by Kozeny-Carman (Bear 1972), the effective permeability for the dual-porosity medium can be expressed as

$$\Delta k_{12} = \frac{n_1}{n}\Delta k_1 + \frac{n_2}{n}\Delta k_2 \tag{3.239}$$

where subscripts 1 and 2 represent matrix and fracture, respectively; and the total porosity n is assessed as $n = n_1 + n_2$. Further, in Eq. (3.239),

$$\Delta k_1 = k_0\left\{1 - \frac{1}{2}\left[\frac{9(1-\nu^2)^2}{2}(\pi\Delta\varepsilon_1)^2\right]^{1/3}\right\}^2 \tag{3.240}$$

where

$$\Delta\varepsilon_1 = \frac{\Delta\sigma}{E} \tag{3.241}$$

and $\Delta\sigma$ is the total stress change, with E and ν requiring to be determined from intact rock containing no fractures. Also,

$$\Delta k_2 = \frac{1}{12s^*}[b + (s^* + b)\Delta\varepsilon_2]^3 \tag{3.242}$$

where

$$\Delta\varepsilon_2 = \frac{\Delta\sigma}{K_n(s^* + b)} \tag{3.243}$$

and similarly, K_n must be determined from the fractured rock where deformation of the solid matrix is negligible.

3.3.2 Compressibility

For transient fluid flow, the storativity of the porous medium is a vital parameter that quantifies the duration of the transient portion of flow. The compressibility of the porous medium, together with that of the fluid, contributes to the storativity.

3.3.2.1 Constitutive Compressibility

Under isothermal conditions, the general (or total) compressibility, c^*, can be evaluated through experimental tests where the density changes with respect to the applied pressure per unit density is measured; i.e.,

$$c^* = \frac{1}{\rho}\frac{\partial\rho}{\partial p}. \tag{3.244}$$

Eq. (3.244) can be rearranged as

$$c^*\partial p = \frac{\partial\rho}{\rho}. \tag{3.245}$$

Differentiating with respect to time on both sides of Eq. (3.245), yields the familiar form of the storativity relation as

$$c^*\frac{\partial p}{\partial t} = \frac{1}{\rho}\frac{\partial\rho}{\partial t}. \tag{3.246}$$

Because of Eq. (3.75), pore pressure-induced changes in compressibility also indicate volume changes, popularly defined as changes in storage. Assuming that a general bulk modulus, K^*, is the inverse of the compressibility, Eq. (3.246) can also be written as

$$\frac{1}{K^*}\frac{\partial p}{\partial t} = \frac{1}{\rho}\frac{\partial\rho}{\partial t}. \tag{3.247}$$

For nonisothermal cases, the following equation of state must be satisfied:

$$d\rho = \frac{\partial \rho}{\partial p} dp + \frac{\partial \rho}{\partial T} dT \tag{3.248}$$

where T is the temperature. Eq. (3.248) can also be written as

$$\frac{d\rho}{\rho} = c^* dp - \beta dT \tag{3.249}$$

where c^* is defined in Eq. (3.244) and β is described by

$$\beta = -\frac{1}{\rho} \frac{\partial \rho}{\partial T}. \tag{3.250}$$

3.3.2.2 Compressibility in the Field Equations

The compressibility term appearing in the governing flow equations can be considered as the macroscopic compressibility. Mass conservation in a porous medium, in its simplest form, may be expressed as

$$\frac{n\rho v_{xi} - n\rho v_{(x+\Delta x)i}}{\Delta x_i} = \frac{\partial}{\partial t}(n\rho) \tag{3.251}$$

where v_{xi} is fluid velocity term in the x_i direction.

The right-hand side of Eq. (3.251) represents the rate change of fluid accumulation in the control volume. It is known that

$$n\rho v_{x_i+\Delta x_i} \geq n\rho v_{xi} \tag{3.252}$$

therefore, Eq. (3.251) should be rewritten as

$$-\nabla \cdot n\rho \mathbf{v} = \frac{\partial}{\partial t}(n\rho) \tag{3.253}$$

where the negative sign is to ensure that the rate change is always positive.

Substituting Darcy's velocity in Eq. (3.15) into Eq. (3.253) leads to

$$\nabla \cdot (n\rho \frac{k}{\mu} \vec{\nabla} p) = \frac{\partial}{\partial t}(n\rho). \tag{3.254}$$

Rearranging the compressibility, defined in Eq. (3.244), while assuming the density is only a function of pressure, one has

$$\frac{d\rho}{\rho} = c^* dp. \tag{3.255}$$

Integrating Eq. (3.255) on both sides, and solving with the given initial conditions, yields the following relation:

$$\rho = \rho_0 e^{[c^*(p-p_0)]}. \tag{3.256}$$

Because c^* in Eq. (3.256) is usually small, the exponential term can be expanded using Taylor's series. After the truncation of the higher order terms in the function series, one has

$$\rho \approx \rho_0[1 + c^*(p - p_0)]. \tag{3.257}$$

Assuming p_0 and n are constants, and substituting Eq. (3.257) into (3.254), the following equation results:

$$\frac{k}{\mu}\nabla^2 p + c^*\frac{k}{\mu}(\nabla p)^2 = c^*\frac{\partial p}{\partial t}. \tag{3.258}$$

The second term on the left-hand side of Eq. (3.258) is apparently nonlinear. This term may be omitted if its magnitude is much smaller than that of the first term. If so, Eq. (3.16) is recovered from Eq. (3.258).

There are other alternative forms of compressibility. Recalling Biot's (1941) poroelastic approach, the total compressibility was expressed as

$$c^* = \frac{1}{R} - \frac{\alpha}{H} \tag{3.259}$$

where R and H are Biot's constants. If Terzaghi's effective stress law is followed, c^* is most widely suggested to be (Verruijt 1969; Bear 1972; Huyakorn and Pinder 1983):

$$c^* = \frac{n}{K_f} \tag{3.260}$$

where K_f is the fluid bulk modulus.

If the compressibility of the solid grains is not negligible, however, then total compressibility is the arithmetically averaged value of the respective porosity (Liggett and Liu 1983); i.e.,

$$c^* = \frac{n}{K_f} + \frac{1-n}{K_s} \tag{3.261}$$

where K_s is the bulk modulus of the solid grains. The form of c^* in this equation is the dominant one in the current literature. If equating Eqs. (3.259) and (3.261), then R may be derived for Biot's theory as follows:

$$\frac{1}{R} = \frac{n}{K_f} + \frac{1-n}{K_s} + \frac{1}{K}\left(1 - \frac{K}{K_s}\right)^2. \tag{3.262}$$

c^* was suggested by Rice and Cleary (1976) as

$$c^* = \frac{n}{K_f} - \frac{1+n}{K_s} + \frac{1}{K}. \tag{3.263}$$

Geertsma (1957) suggested adding the influence of the compressibility of the solid skeleton, then Eq. (3.261) can be modified as

$$c^* = \frac{n}{K_f} + \frac{\alpha - n}{K_s}. \tag{3.264}$$

where c^* in this equation appears to be the most sensible form of the total compressibility. Since Eq. (3.264) can be rigorously derived rather than merely proposed, as shown in Chapter 2 on deformation, it is used in the present analysis.

3.3.3 Anisotropic Effect

The foregoing analysis is restricted to the definition that the properties of the medium are independent of direction. This is typically a simplification of real behavior; the effects of anisotropy may be adequately incorporated.

For anisotropic media, the flow equation (3.16) becomes

$$\nabla \cdot (\frac{\tilde{k}}{\mu} \cdot \vec{\nabla} p) = \tilde{\delta} : \tilde{c}^* \frac{\partial p}{\partial t}. \tag{3.265}$$

The difference between this equation and Eq. (3.16) is that permeability k and compressibility c^* are now tensorial, rather than scalar, quantities.

The coupled flow equation for homogeneous media (Biot's formulation), Eq. (3.32), can be rewritten as

$$\nabla \cdot (\frac{\tilde{k}}{\mu} \cdot \vec{\nabla} p) = \tilde{\delta} : \tilde{c}^* \frac{\partial p}{\partial t} + \tilde{\alpha} : \frac{\partial \tilde{\varepsilon}}{\partial t}. \tag{3.266}$$

The Biot coefficient α and volumetric strain component ε_{kk} defined in the previous Eq. (3.32) are now no longer scalars, but tensors.

In a homogeneous, isotropic porous medium where the flow is assumed to be laminar, Darcy's law can be defined as

$$\vec{v} = -\frac{k}{\mu} \vec{J} = -\frac{k}{\mu} \begin{pmatrix} J_1 \\ J_2 \\ J_3 \end{pmatrix} \tag{3.267}$$

where $\vec{J}$ is the hydraulic gradient. It must be noted that the permeability is a single valued constant in an isotropic medium. However, in anisotropic rock formations the permeability is no longer a scalar. In general, the permeability is a tensor, defined as

$$\tilde{k} = \begin{bmatrix} k_{11} & k_{12} & k_{13} \\ k_{21} & k_{22} & k_{23} \\ k_{31} & k_{32} & k_{33} \end{bmatrix}. \tag{3.268}$$

Assuming symmetric characteristics of the permeability tensor, which gives

$$k_{ij} = k_{ji} \qquad (i, j = 1, 2, 3) \tag{3.269}$$

Eq. (3.269) can be interpreted that the flow in one direction is equal to flow in the opposite direction. This symmetric property of the permeability tensor was challenged by Sagar and Runchal (1982). However, as disputed by Chen et al. (1999), Sagar and Runchal's claim may not be sensible due to a possible error in their derivation.

Reformulating Darcy's velocity as

$$\vec{v} = -\frac{1}{\mu}\tilde{k}\cdot\vec{J} \tag{3.270}$$

and assuming the flow to be caused by a pressure gradient $\vec{J}$ in the rectangular Cartesian coordinates, the expanded form of Darcy's velocity can be written as

$$v_i = -\frac{k_{ij}}{\mu}J_i \quad \Leftrightarrow \quad \begin{cases} v_1 = \frac{1}{\mu}(k_{11}J_1 + k_{12}J_2 + k_{13}J_3) \\ \\ v_2 = \frac{1}{\mu}(k_{12}J_1 + k_{22}J_2 + k_{23}J_3) \\ \\ v_3 = \frac{1}{\mu}(k_{13}J_1 + k_{23}J_2 + k_{33}J_3). \end{cases} \tag{3.271}$$

Manipulation of Eq. (3.271) may result in the following five special cases.

- Case 1: k_{ij} when $i \neq j$

  The non-zero diagonal terms imply cross-directional influences. For instance, the first equation of Eq. (3.271) indicates that the flow velocity v_1 is affected by the pressure gradients in the other two directions, i.e., J_2 and J_3.

- Case 2: $k_{ii} \neq 0$, $k_{ij} = 0$ when $i \neq j$

  In this case, the permeability tensor has diagonal terms only; i.e, the fluid flows in three principal directions only:

$$v_i = -\frac{k_{ii}}{\mu}J_i \tag{3.272}$$

- Case 3: $k_{11} = k_{22} \neq k_{33}$, $k_{ij} = 0$ when $i \neq j$

  This is a typical case of cross-anisotropy where layered materials show dominant flow directions in their bedding planes whereas the flow in the orthogonal (e.g., $\vec{x}_3$) direction is frequently negligible.

- Case 4: $k_{11} = k_{22} = k_{33}$, $k_{ij} = 0$ when $i \neq j$

  Fluid flow occurs in three principal directions with equal access (permeability). This is the orthogonally isotropic case.

- Case 5: $\tilde{k} = k$

  When the permeability tensor reduces to a constant, fluid flow is the same in all directions, and complete isotropic flow prevails.

Chapter 4

TRANSPORT

4.1 INTRODUCTION

The well-known conservation laws are related to three balanced physical systems where momentum, mass, or energy are conserved. For the applications addressed in this text, the physical interactions represented in these systems are force equilibrium, flow continuity, and mass and thermal transfer. Transport processes are considered as part of a mass balance system where solute transport is tracked, or as an energy balance where heat, or enthalpy, are tracked. For solute transport, the dependent variable is mass concentration, defined as solute mass per unit volume. Apart from the difference in dependent variables, the major difference between flow and transport is that the former is an active potential (diffusive) process driven by the spatial variations of fluid pressure or hydraulic head, while the latter is a combination of an active potential (dispersion) process determined by spatial variations of solute concentration and a passive carry-away (convection) process controlled by the bulk velocity of regional flow. Where diffusion dominates, the dispersive transport equation reduces to the same form as the pressure diffusion equation, albeit with different dependent variables. Where the convective term becomes significant, the represented processes are quite different, and solution of the governing equations by Eulerian methods becomes more difficult.

4.2 MATHEMATICAL FORMULATION

The discussion of the equations governing transport is divided between homogeneous and heterogeneous media, together with consideration of stochastic processes.

4.2.1 Homogeneous Media

The correlation between the flow processes described in the previous chapter and those of transport is investigated here, followed by a discussion of certain fundamental concepts affecting transport in porous media.

4.2.1.1 Relation Between Flow and Transport

Although transport phenomena are mainly associated with chemical (or solute) migration, a separate physical process from fluid flow, the equations governing transport and flow are strongly correlated.

Nonlinear Flow and Transport-Type Equations

For single-phase fluid flow, mass conservation may be expressed as (Bear 1972)

$$\frac{\partial n\rho}{\partial t} + \mathbf{v}^{*}\cdot\vec{\nabla} n\rho + n\rho\nabla\cdot\mathbf{v}^{*} = 0 \tag{4.1}$$

where n is the porosity, ρ is the fluid density, and $\mathbf{v}^*$ is the intrinsic flow velocity. A homogeneous fluid is a single-species fluid or a homogeneous mixture. If the fluid is compressible, the fluid density is assumed to depend only on the pressure.

By definition, the apparent velocity of the fluid can be expressed as

$$\mathbf{v} = \frac{\mathbf{v}^{*}}{n} \tag{4.2}$$

where $\mathbf{v}^*$ is commonly referred to as the Darcy velocity.

Using the apparent velocity in Eq. (4.2) to express Eq. (4.1), and assuming the porosity n as a constant after expansion of the gradient, yields

$$\frac{\partial \rho}{\partial t} + \mathbf{v}\cdot\vec{\nabla}\rho + \rho\nabla\cdot\mathbf{v} = 0. \tag{4.3}$$

Recalling Darcy's law and neglecting gravitational effects, the fluid apparent velocity may be given as

$$\mathbf{v} = -\frac{k}{n\mu}\vec{\nabla} p \tag{4.4}$$

where k is the permeability and μ is the fluid dynamic viscosity.

For isothermal conditions, the total compressibility is defined in Chapter 3 on flow, which can be further expressed in more general terms as

$$c^{*}\rho\vec{\nabla} p = \vec{\nabla}\rho \tag{4.5}$$

or more specifically

$$c^{*}\rho\partial p = \partial\rho. \tag{4.6}$$

Taking the derivative with respect to time t in Eq. (4.6),

$$\frac{\partial \rho}{\partial t} = c^* \rho \frac{\partial p}{\partial t} \tag{4.7}$$

and substituting Eqs. (4.4) and (4.5) into (4.3), while assuming k, μ and c are constants, yields

$$\nabla^2 \rho = \frac{n c^* \mu}{k} \frac{\partial \rho}{\partial t}. \tag{4.8}$$

From Eq. (4.5), one has

$$\nabla^2 \rho = c^* \vec{\nabla} \rho \vec{\nabla} p + c^* \rho \nabla^2 p = c^{*2} \rho (\nabla p)^2 + c^* \rho \nabla^2 p. \tag{4.9}$$

Substituting Eqs. (4.7) and (4.9) into Eq. (4.8) yields

$$\frac{k}{\mu} \nabla^2 p + c^* \frac{k}{\mu} (\vec{\nabla} p)^2 = n c^* \frac{\partial p}{\partial t} \tag{4.10}$$

and incorporating Eq. (4.4), Eq. (4.10) can be rewritten as

$$D_p \nabla^2 p - \mathbf{v} \vec{\nabla} p = \frac{\partial p}{\partial t} \tag{4.11}$$

where

$$D_p = \frac{k}{n \mu c^*} \tag{4.12}$$

which is equivalent to the dispersion coefficient adopted in transport models. Using Eq. (4.12), Eq. (4.10) can be further simplified as

$$D_p \nabla^2 p + c^* D_p (\vec{\nabla} p)^2 = \frac{\partial p}{\partial t}. \tag{4.13}$$

Compared to the traditional flow equation, Eq. (4.13) is frequently referred to as the "nonlinear" flow equation. The linearization of the nonlinear flow equation, shown in Eq. (4.13), is usually based on the assumption that inertial forces (both convective and local) are smaller than the viscous ones. Prandtl (1952) referred to such a motion as "creep," which is characterized by a low Reynolds number, indexing the ratio between the inertial and the viscous (frictional) forces.

A Conversion Between Flow and Transport

Substituting the pressure, p, by concentration, c, in Eq. (4.11), the equation resembles the dispersion-convection transport equation, sometimes referred to as the advection-dispersion equation. However, since no direct expression exists between p and c, a conversion must be sought between the two mechanisms.

In general, this conversion is sought through the density-concentration ($\rho - c$) relation because both carry identical units.

Before deriving the ρ-c relationship, it may be helpful to review some basic concepts of solute transport:

- The *species* represents a component of the solute, such as CH_4, SiO_2, NaCl, etc.

- In contrast, the *phase* denotes the type of the fluids, such as oil, gas, air, and water.

- There are two kinds of concentrations, defined in the following:

  - *mass concentration* is a measure of density in terms of particulate matters, such as precipitates, sand, clay, or acid, for example. Its mathematical definition is

$$\rho_\omega = \frac{m_\omega}{V_p} \tag{4.14}$$

    where m_ω is the mass of species ω, and V_p is the fluid volume.

  - *molar concentration* is a measure of density with reference to the dissolved species, such as any types of chemical solution, which can be expressed as

$$c_\omega = \frac{m_\omega}{M_\omega^* V_p} \tag{4.15}$$

    where M_ω is the number of moles of species ω, which is related to the molecular weight.

The density-concentration relation can be readily derived from Eqs. (4.14) and (4.15) as

$$\rho_\omega = c_\omega M_\omega^* \tag{4.16}$$

or, alternatively,

$$c_\omega = \frac{\rho_\omega}{M_\omega^*}. \tag{4.17}$$

These relations are suitable for single-phase fluid flow.

From Eqs. (4.2), (4.4), and (4.5), the governing equation of mass conservation, Eq. (4.5), can be rewritten as

$$\frac{\partial n\rho}{\partial t} + \mathbf{v}^* \cdot \vec{\nabla} n\rho - n\rho \nabla \cdot \left(\frac{k}{\mu c^*} \frac{\vec{\nabla}\rho}{\rho} \right) = 0 \tag{4.18}$$

where n is the porosity, ρ is the fluid density, and $\mathbf{v}^*$ is the intrinsic flow velocity.

Taking the divergence of the last term of the left-hand side of Eq. (4.18), and incorporating Eq. (4.13), one has

$$\frac{\partial n\rho}{\partial t} + \mathbf{v}^{*}\cdot\vec{\nabla} n\rho - nD_p^{*}\nabla^2\rho + \frac{nD_p^{*}(\vec{\nabla}\rho)^2}{\rho} = 0 \tag{4.19}$$

where

$$D_p^{*} = \frac{k}{\mu c^{*}}. \tag{4.20}$$

Assuming that the ratio of the quadratic term in Eq. (4.19), $(\vec{\nabla}\rho)^2$, to the density, ρ, is relatively small and may therefore be neglected, Eq. (4.19) reduces to

$$\frac{\partial n\rho}{\partial t} + \mathbf{v}^{*}\cdot\vec{\nabla} n\rho - nD_p^{*}\nabla^2\rho = 0. \tag{4.21}$$

For the mass conservation of species ω, the above equation can also be written as

$$\frac{\partial n\rho_\omega}{\partial t} + \mathbf{v}^{*}\cdot\vec{\nabla} n\rho_\omega - nD_{p_\omega}^{*}\nabla^2\rho_\omega = 0. \tag{4.22}$$

Substituting the relation of Eq. (4.16) into Eq. (4.22), and recalling that M_ω^{*} is a constant, Eq. (4.22) can be transformed to a transport equation as

$$\frac{\partial nc_\omega}{\partial t} + \mathbf{v}^{*}\cdot\vec{\nabla} nc_\omega - nD_p^{*}\nabla^2 c_\omega = 0. \tag{4.23}$$

For the case of multi-phase multiple species, Eq. (4.23) is modified as

$$\frac{\partial nS_\psi c_{\psi\omega}}{\partial t} + \mathbf{v}^{*}\cdot\vec{\nabla} nS_\psi c_{\psi\omega} - nS_\psi D_{p\omega}^{*}\nabla^2 c_{\psi\omega} = 0 \tag{4.24}$$

where subscripts ψ and ω denote the *phase* and *species*, respectively.

For single-phase single species transport, and assuming that n is a constant, Eq. (4.23) becomes a standard dispersion-convection transport equation:

$$D_p^{*}\nabla^2 c - \mathbf{v}\cdot\vec{\nabla} c = \frac{\partial c}{\partial t}. \tag{4.25}$$

For cases involving multiple species in a single-phase, the transport process is the mass migration of a component of the fluid phase. In contrast, the flow process is the mass migration of a fluid phase (Bear 1993). In the absence of chemical reactions, the mass conservation of a species is defined as (Bear 1972)

$$\frac{\partial n\rho_\omega}{\partial t} + \nabla\cdot(n\mathbf{v}^{*}) + \nabla\cdot J_\omega^{*} = 0 \tag{4.26}$$

where J^*_ω is the diffusive flux for species ω. Bear (1972) further defined J^*_ω as

$$J^*_\omega = -n\rho D^*_{p\omega}\vec{\nabla}\left(\frac{\rho_\omega}{\rho}\right) = -nD^*_{p\omega}\left(\vec{\nabla}\rho_\omega - \frac{\rho_\omega\vec{\nabla}\rho}{\rho}\right) \tag{4.27}$$

where ρ is the density of the fluid system and ρ_ω is the density of the species.

Based on the divergence operation, one has

$$\nabla\cdot J^*_\omega = -nD^*_{p\omega}\left(\nabla^2\rho_\omega - \frac{\vec{\nabla}\rho_\omega\vec{\nabla}\rho}{\rho}\right). \tag{4.28}$$

Expanding the second term on the left-hand side of Eq. (4.26), and substituting Eq. (4.28), Eq. (4.26) can be written as

$$\frac{\partial n\rho_\omega}{\partial t} + \mathbf{v}^*\cdot\vec{\nabla}(n\rho_\omega) + n\rho_\omega\nabla\cdot\mathbf{v}^* - nD^*_{p\omega}\nabla^2\rho_\omega + nD^*_p\frac{\vec{\nabla}\rho_\omega\vec{\nabla}\rho}{\rho} = 0. \tag{4.29}$$

For transport of a single species in single-phase flow, and also assuming that the fluid average flow velocity $\mathbf{v}^*$ in Eq. (4.29) is a constant, Eq. (4.29) reduces to

$$\frac{\partial n\rho}{\partial t} + \mathbf{v}^*\cdot\vec{\nabla}(n\rho) - nD^*_p\nabla^2\rho + nD^*_p\frac{(\vec{\nabla}\rho)^2}{\rho} = 0 \tag{4.30}$$

which has identical form to Eq. (4.19).

4.2.1.2 A Generalized Dispersion-Convection Model

To illustrate, the generalized dispersion-convection model may be solved using the method of Laplace transforms. Eq. (4.25) is a general dispersion-convection equation. In one-dimensional cases, and dropping both subscript and superscript for the dispersion coefficient, Eq. (4.25) reduces to

$$D\frac{\partial^2 c}{\partial x^2} - \mathbf{v}\frac{\partial^2 c}{\partial x} = \frac{\partial c}{\partial t} \tag{4.31}$$

where $\mathbf{v}$ is the vector of convective flow velocity.

If the flow velocity, $\mathbf{v}$, is assessed as an aggregate magnitude, it can be assumed to be a constant value; as a result, Eq. (4.31) is simplified as

$$D\frac{\partial^2 c}{\partial x^2} - v\frac{\partial^2 c}{\partial x} = \frac{\partial c}{\partial t}. \tag{4.32}$$

Eq. (4.32) represents the majority of transport models in the one-dimensional domain.

The initial and inlet boundary conditions are given as

$$\begin{cases} c = 0 & t = 0 \\ c = c_0 & x = 0 \end{cases} \tag{4.33}$$

where c_0 is the constant concentration at the inlet boundary.

Three situations may exist for the outlet boundary conditions:

$$\begin{cases} c = 0 \quad x = L^* & \text{(concentration)} \\ c = 0 \quad x = \infty & \text{(concentration)} \\ \dfrac{dc}{dx} = 0, \quad x = L^* & \text{(flux)} \end{cases} \tag{4.34}$$

where L^* can be taken as the length of a laboratory sample column or the longest traveling distance of the solute.

The advantage of using dimensionless terms is represented by the flexibility of choosing parametric ranges arbitrarily. The following dimensionless terms are defined:

$$\begin{cases} x_d = \dfrac{x}{L^*} \\ \\ t_d = \dfrac{vt}{L^*} \\ \\ \gamma^* = \dfrac{vL^*}{D} \end{cases} \tag{4.35}$$

where t_d is the dimensionless time or pore volume injected and γ^* is the equivalent Peclet number expressing the ratio of the rate of transport by convection to that by dispersion.

Eq. (4.32) can be modified as

$$\frac{1}{\gamma^*}\frac{\partial^2 c}{\partial x_d^2} - \frac{\partial c}{\partial x_d} = \frac{\partial c}{\partial t_d}. \tag{4.36}$$

The initial and inlet boundary conditions are:

$$\begin{cases} c = c^0 & t = 0 \\ c = c_0 & x_d = 0 \end{cases} \tag{4.37}$$

where c^0 is the initial concentration and c_0 is the inlet concentration.

Three different outlet boundary conditions at $x_d = 1$ are:

$$\begin{cases} c = 0 & x_d = 1 \\ c = 0 & x_d = \infty \\ \dfrac{dc}{dx_d} = 0 & x_d = 1. \end{cases} \tag{4.38}$$

Applying a Laplace transform to Eq. (4.32) with the above initial, inlet, and the third-type outlet boundary conditions, Eq. (4.36) becomes an ordinary differential equation:

$$\frac{1}{\gamma^*}\frac{d^2\bar{c}}{dx_d^2} - \frac{d\bar{c}}{dx_d} = s\bar{c} - c^0. \tag{4.39}$$

The inlet and third-type outlet boundary condition in the Laplace domain become

$$\begin{cases} \bar{c} = \frac{c_0}{s}, & x_d = 0 \\ \frac{d\bar{c}}{dx_d} = 0 & x_d = 1. \end{cases} \tag{4.40}$$

Solution of Eq. (4.39), subject to the boundary conditions in the Laplace domain, Eq. (4.40), can be given as

$$\bar{c} = e_1^{\#} e^{\phi_1^* x_d} + e_2^{\#} e^{\phi_2^* x_d} - \frac{\gamma^* c^0}{\phi_1^* \phi_2^*} \tag{4.41}$$

where

$$\begin{cases} \phi_1^* = \frac{\gamma^*}{2}\left[1 + \sqrt{1 + \frac{4s}{\gamma^{*2}}}\right] \\ \phi_2^* = \frac{\gamma^*}{2}\left[1 - \sqrt{1 + \frac{4s}{\gamma^{*2}}}\right] \\ \varrho^* = \frac{c_0}{s} - \frac{\gamma^* c^0}{\phi_1^* \phi_2^*} \\ R_a = \frac{\phi_1^*}{\phi_2^*} \\ e_1^{\#} = \frac{\varrho^*}{1 - R_a e^{(\phi_1^* - \phi_2^*)}} \\ e_2^{\#} = \frac{\varrho^*}{1 - R_a^{-1} e^{(\phi_2^* - \phi_1^*)}}. \end{cases} \tag{4.42}$$

The concentration in real space can be recovered from the inversion of the solution in Eq. (4.41). A popular inversion technique is to employ a numerical inversion method, such as that proposed by Stehfest (1970).

4.2.2 Heterogeneous Media

The difference between fluid flow and solute transport is that, in fluid flow, the transported medium is a fluid, and in solute transport, it is a component of the fluid (Bear 1993). Aside from this difference, the Coats and Smith (1964) model for transport with stagnant regions is similar to the Warren and Root (1963) model, which is primarily used to define fluid flow through fractured porous media. In many cases, Coats and Smith's (1964) model has been broadly referenced in the modeling of fluid flow and contaminant transport through fractured porous media (Tang et al. 1981; Bibby 1981; Huyakorn et al. 1983; Nilson and Lie 1990; Rowe and Booker 1990; Sudicky and McLaren

1992; Harrison et al. 1992; Leo and Booker 1993). In contrast, however, Warren and Root's (1963) model has been rarely mentioned in the literature of solute transport through micropore-macropore regions (Passioura 1971; Passioura and Rose 1971; Joy and Kouwen 1991; Koenders and Williams 1992; Joy et al. 1993; Piquemal 1992, 1993) with an exception that Sahimi (1993) provided an implicit link between the two models. Correspondingly, models based on the same conceptualization have been developed independently (e.g., spherical block model for transport through fractured porous media by Huyakorn et al. 1983, and for transport through micropore-macropore regions by Correa et al. 1987).

In the following description, the subscripts 1 and 2 represent macropores and micropores, respectively.

4.2.2.1 Coats and Smith's Model

Based on similar concepts to the dual-porosity model proposed by Warren and Root (1963) for fluid flow, the transport model proposed by Coats and Smith (1964) has attracted significant interest due to its simplicity in representing one-dimensional transport. In their model, solute transport mainly occurs in connected macropore regions, while the solute in "dead end" micropores may be interchanged within the two porous spaces depending upon the concentration difference.

The Coats and Smith model can be expressed by a pair of differential equations:

$$D_1 \frac{\partial^2 c_1}{\partial x^2} - v \frac{\partial c_1}{\partial x} = f^* \frac{\partial c_1}{\partial t} + (1 - f^*) \frac{c_2}{\partial t} \tag{4.43}$$

$$0 = (1 - f^*) \frac{c_2}{\partial t} - \Omega (c_1 - c_2) \tag{4.44}$$

where c is the solute concentration, D is the dispersion coefficient, v is the average flow velocity in the macropore region, f^* is the fraction of the pore space contained within the macropore region, Ω is the rate of mass transfer between stagnant and flowing fluids, x is the distance from the source and t is the time. Eq. (4.44) implies that in the micropores, the fluid is stagnant and no diffusion occurs.

Defining the rate coefficient of transport between micro- and macropore regions, Ω^*, as

$$\Omega^* = \frac{\Omega L^*}{v}, \tag{4.45}$$

together with selecting the dimensionless parameters defined in Eq. (4.35), Eqs. (4.43) and (4.44) can be rewritten in terms of the dimensionless quantities as

$$\frac{1}{\gamma^*} \frac{\partial^2 c_1}{\partial x_d^2} - \frac{\partial c_1}{\partial x_d} = f^* \frac{\partial c_1}{\partial t_d} + (1 - f^*) \frac{\partial c_2}{\partial t_d} \tag{4.46}$$

$$0 = (1 - f^*)\frac{\partial c_2}{\partial t_d} - \Omega^*(c_1 - c_2). \tag{4.47}$$

The initial and inlet boundary conditions are

$$\begin{cases} c_1 = c_1^0 & t = 0 \\ c_2 = c_2^0 & t = 0 \\ c_1 = c_0 & x_d = 0 \end{cases} \tag{4.48}$$

where c_1^0 and c_2^0 are the initial concentrations of macropores and micropores, respectively.

Two different outlet boundary conditions at $x_d = 1$ are

$$\begin{cases} c_1 = 0 & x_d = 1 \\ \\ \dfrac{dc_1}{dx_d} = 0 & x_d = 1. \end{cases} \tag{4.49}$$

The first boundary condition at $x_d = 1$ in Eq. (4.49) is frequently used for modeling field situations, while the second condition in Eq. (4.49) is usually applied to laboratory core plug experiments.

Applying a Laplace transform to Eqs. (4.46) and (4.47) with the above initial, inlet, and second outlet boundary conditions, Eqs. (4.46) and (4.47) become ordinary differential equations:

$$\frac{1}{\gamma^*}\frac{d^2\bar{c}_1}{dx_d^2} - \frac{d\bar{c}_1}{dx_d} = f^*(s\bar{c}_1 - c^0) + (1 - f^*)(s\bar{c}_2 - c_2^0) \tag{4.50}$$

$$(1 - f^*)(\bar{c}_2 s - c_2^0) = \Omega^*(\bar{c}_1 - \bar{c}_2) \tag{4.51}$$

where s is a Laplace parameter.

From Eq. (4.51), one has

$$\bar{c}_2 = \frac{c_2^0(1 - f^*) + \Omega^*\bar{c}_1}{\Omega^* + s(1 - f^*)}. \tag{4.52}$$

The boundary conditions in the Laplace domain can be written as

$$\begin{cases} \bar{c} = \frac{c_0}{s}, & x_d = 0 \\ \dfrac{d\bar{c}}{dx_d} = 0 & x_d = 1. \end{cases} \tag{4.53}$$

Substituting Eq. (4.52) into Eq. (4.50), yields

$$\frac{d^2\bar{c}_1}{\partial x_d^2} - \gamma^*\frac{d\bar{c}_1}{dx_d} - f(s)\bar{c}_1 = \Pi \tag{4.54}$$

where

$$f(s) = \gamma^* s \left[f^* + \frac{\Omega^*(1-f^*)}{\Omega^* + s(1-f^*)} \right] \tag{4.55}$$

$$\Pi = -\gamma^* \left[f^* c_1^0 + \frac{\Omega^*(1-f^*)c_2^0}{\Omega^* + s(1-f^*)} \right] \tag{4.56}$$

Solution of Eq. (4.54) subject to the boundary condition expressed in Eq. (4.53) can be given as

$$\bar{c}_1 = e_1 e^{\phi_1 x_d} + e_2 e^{\phi_2 x_d} + \frac{\Pi}{\phi_1 \phi_2} \tag{4.57}$$

where

$$e_1 = \frac{-\varrho \phi_2 e^{\phi_2}}{\phi_1 e^{\phi_1} - \phi_2 e^{\phi_2}} \tag{4.58}$$

$$e_2 = \frac{\varrho \phi_1 e^{\phi_1}}{\phi_1 e^{\phi_1} - \phi_2 e^{\phi_2}} \tag{4.59}$$

$$\varrho = \frac{c_0}{s} - \frac{\Pi}{\phi_1 \phi_2} \tag{4.60}$$

$$\phi_1 = \frac{\gamma^*}{2} \left[1 + \sqrt{1 + \frac{4f(s)^*}{\gamma^{*2}}} \right] \tag{4.61}$$

$$\phi_2 = \frac{\gamma^*}{2} \left[1 - \sqrt{1 + \frac{4f(s)^*}{\gamma^{*2}}} \right]. \tag{4.62}$$

Although based on a phenomenological conceptualization, Coats and Smith model (1964) is considered one of the most popular models of solute transport due to its inherent simplicity.

4.2.2.2 Piquemal Model

Based on a more rigorous volume averaging method, Piquemal (1992) presented a slightly different model to that of Coats and Smith (1964). The first equation of Piquemal (1992) can be expressed as

$$f^* D_1 \frac{\partial^2 c_1}{\partial x^2} - f^* v \frac{\partial c_1}{\partial x} = f^* \frac{\partial c_1}{\partial t} + (1 - f^*) \frac{c_2}{\partial t} \tag{4.63}$$

which can be compared with Eq. (4.43) of the Coats and Smith model (1964).

The second equation, representing transport in micropores, is identical to Eq. (4.44). Substituting identical dimensionless parameters, as given in Eq. (4.35), Eq. (4.63) becomes

$$\frac{1}{\gamma^*} \frac{\partial^2 c_1}{\partial x_d^2} - \frac{\partial c_1}{\partial x_d} = \frac{\partial c_1}{\partial t_d} + \frac{1 - f^*}{f^*} \frac{\partial c_2}{\partial t_d}. \tag{4.64}$$

Again, the second equation in dimensionless form, defining transport in micropores, is identical to Eq. (4.47). Following a similar procedure to that applied in the previous section, using the identical initial and boundary conditions [second outlet condition in Eq. (4.49)] as in the Coats and Smith (1964) model, the transport equation for macropores can be expressed in Laplace space as

$$\frac{d^2\bar{c}_1}{\partial x_d^2} - \gamma^* \frac{d\bar{c}_1}{dx_d} - f(s)^* \bar{c}_1 = \Pi^* \tag{4.65}$$

where

$$f(s)^* = \gamma^* s \left\{ 1 + \frac{\Omega^*(1-f^*)}{f^*[\Omega^* + s(1-f^*)]} \right\} \tag{4.66}$$

$$\Pi^* = -\gamma^* \left\{ c_1^0 + \frac{\Omega^*(1-f^*)c_2^0}{f^*[\Omega^* + s(1-f^*)]} \right\}. \tag{4.67}$$

Assuming the following boundary conditions in the Laplace domain,

$$\begin{cases} \bar{c}_1 = \frac{c_0}{s}, & x_d = 0 \\ \frac{d\bar{c}_1}{dx_d} = 0 & x_d = 1 \end{cases} \tag{4.68}$$

solution of Eq. (4.65) subjected to the boundary condition in the Laplace domain can be given as

$$\bar{c}_1 = e_1^* e^{\phi_1 x_d} + e_2^* e^{\phi_2 x_d} + \frac{\Pi^*}{\phi_1 \phi_2} \tag{4.69}$$

where

$$e_1^* = \frac{-\varrho \phi_2 e^{\phi_2}}{\phi_1 e^{\phi_1} - \phi_2 e^{\phi_2}} \tag{4.70}$$

$$e_2^* = \frac{\varrho \phi_1 e^{\phi_1}}{\phi_1 e^{\phi_1} - \phi_2 e^{\phi_2}} \tag{4.71}$$

$$\varrho = \frac{c_0}{s} - \frac{\Pi^*}{\phi_1 \phi_2}. \tag{4.72}$$

4.2.2.3 Correa et al. Model

The Correa et al. (1987) model is similar to the model for fluid flow introduced by Moench (1984) where the domain of the porous matrix was envisioned as spherical balls, and the mass interchange between the matrix and fluid conducting fractures occurs at their interfaces, known as the "fracture skins." However, spherical transport in the matrix in the Correa et al. model was simplified to one-dimensional linear transport, which may only strictly represent cross-layer transport, even through the term representing radius "r" is retained in their model. The governing expression of the Correa et al. (1987) model is expressed as

$$D_1 \frac{\partial^2 c_1}{\partial x^2} - v \frac{\partial c_1}{\partial x} = f^* \frac{\partial c_1}{\partial t} + 3(1 - f^*) \frac{D_s}{R^\#} \left[\frac{\partial c_2}{\partial r_\#} \right]_{r_c = R^\#} \tag{4.73}$$

$$D_2 \frac{\partial c_2}{\partial r_c^2} = \frac{\partial c_2}{\partial t} \tag{4.74}$$

where D_2 is the microscopic dispersion, r_c is the arbitrary distance from the center of the matrix block to the fracture interface, and $R^\#$ is the radius of the spherical matrix block.

Using x_d and t_d defined in Eq. (4.35) and the following dimensionless parameters:

$$\gamma_1^* = \frac{vL^*}{D_1}, \quad \gamma_2^* = \frac{vL^*}{D_2}, \quad x_z = \frac{r_c}{L^*}, \quad \kappa^\# = \frac{3(1 - f^*)}{R^\# v} \tag{4.75}$$

and employing the following initial and boundary conditions:

$$\begin{cases} c_1 = c_1^0 & t = 0 \\ c_2 = c_2^0 & t = 0 \\ c_1 = c_0 & x_d = 0 \\ \dfrac{dc_1}{dx_d} = 0, & x_d = 1 \end{cases} \tag{4.76}$$

further, substituting the dimensionless terms in Eq. (4.75) into Eqs. (4.73) and (4.74), one obtains

$$\frac{1}{\gamma_1^*} \frac{\partial^2 c_1}{\partial x_d^2} - \frac{\partial c_1}{\partial x_d} = f^* \frac{\partial c_1}{\partial t_d} + \kappa^\# \left[\frac{\partial c_2}{\partial x_z} \right]_{x_z = R^\# / L^*} \tag{4.77}$$

$$\frac{1}{\gamma_2^*} \frac{\partial^2 c_2}{\partial x_z^2} = \frac{\partial c_2}{\partial t_d}. \tag{4.78}$$

Applying a Laplace transform to Eqs. (4.77) and (4.78), along with the initial and boundary conditions in Eq. (4.76), yields the following ordinary differential equations:

$$\frac{1}{\gamma_1^*} \frac{d^2 \bar{c}_1}{dx_d^2} - \frac{d\bar{c}_1}{dx_d} = f^*(s\bar{c}_1 - c_1^0) + \kappa^\# \left[\frac{d\bar{c}_2}{dx_z} \right]_{x_z = R^\# / L^*} \tag{4.79}$$

$$\frac{1}{\gamma_2^*} \frac{d^2 \bar{c}_2}{dx_z^2} = s\bar{c}_2 - c_2^0. \tag{4.80}$$

The boundary conditions in the Laplace domain become

$$\begin{cases} \bar{c}_1 = \frac{c_0}{s}, & x_d = 0 \\ \frac{d\bar{c}_1}{dx_d} = 0, & x_d = 1 \\ \bar{c}_2 = \bar{c}_1, & x_z = R^{\#}/L^{*} \\ \frac{d\bar{c}_2}{dx_z} = 0, & x_d = 0 \end{cases} \tag{4.81}$$

The solution of Eq. (4.80), incorporating the boundary conditions expressed in Eq. (4.81), results in

$$\kappa^{\#}\left[\frac{d\bar{c}_2}{dx_z}\right]_{x_z=R^{\#}/L^{*}} = \psi_1^{\#}(\bar{c}_1 - \psi_2^{\#}) \tag{4.82}$$

where

$$\begin{cases} \phi^{\#} = \sqrt{\gamma_2^{*} s} \\ \psi_1^{\#} = \dfrac{\kappa^{\#}\phi^{\#}\left[1 - \exp\left(\dfrac{2\phi^{\#}R^{\#}}{L^{*}}\right)\right]}{1 + \exp\left(\dfrac{-2\phi^{\#}R^{\#}}{L^{*}}\right)} \\ \psi_2^{\#} = -\dfrac{2\gamma_2^{*}c_2^{0}}{\phi^{\#}(1+\phi^{\#})}. \end{cases} \tag{4.83}$$

Substituting Eq. (4.82) into Eq. (4.79), one has

$$\frac{1}{\gamma_1^{*}}\frac{d^2\bar{c}_1}{dx_d^2} - \frac{d\bar{c}_1}{dx_d} = f^{*}(s\bar{c}_1 - c_1^{0}) + \psi_1^{\#}(\bar{c}_1 - \psi_2^{\#}). \tag{4.84}$$

Solution of Eq. (4.84) with the boundary conditions in Eq. (4.81) gives:

$$\bar{c}_1 = e_1^{\#}\exp(\rho_1^{\#}x_d) + e_2^{\#}\exp(\rho_2^{\#}x_d) + \bar{c}^{\#} \tag{4.85}$$

where

$$\begin{cases} \rho_1^{\#} = \frac{\gamma_1^{*}}{2}\left[1 + \sqrt{1 + \frac{4\delta_1^{*}}{\gamma_1^{*2}}}\right] \\ \rho_2^{\#} = \frac{\gamma_1^{*}}{2}\left[1 - \sqrt{1 + \frac{4\delta_1^{*}}{\gamma_1^{*2}}}\right] \\ \delta_1^{*} = \gamma_1^{*}(f^{*}s + \psi_1^{\#}) \\ \delta_2^{*} = \gamma_1^{*}(f^{*}c_1^{0} + \psi_1^{\#}\psi_2^{\#}) \\ \bar{c}^{\#} = \dfrac{-\delta_2^{*}}{\rho_1^{\#}\rho_2^{\#}} \end{cases} \tag{4.86}$$

Coefficients $e_1^{\#}$ and $e_2^{\#}$ in Eq. (4.85) can be obtained by solving the nonhomogeneous differential equation (4.84) as

$$e_1^{\#} = \frac{\varrho_1^* e^{\rho_2^{\#}} - \varrho_2^*}{e^{\rho_2^{\#}} - e^{\rho_1^{\#}}}, \quad e_2^{\#} = \frac{\varrho_2^* - \varrho_1^* e^{\rho_1^{\#}}}{e^{\rho_2^{\#}} - e^{\rho_1^{\#}}} \tag{4.87}$$

$$\varrho_1^* = \frac{c_0}{s} + \varrho_2^*, \quad \varrho_2^* = \bar{c}^{\#}. \tag{4.88}$$

4.2.2.4 Modified Correa et al. Model

The format of the first equation in the modified Correa et al. (1987) model is basically unchanged compared with the original formulation. As a result, this discussion focuses on the derivation of the second equation. Applying the spherical coordinate system for the transport within the spherical matrix blocks, the second equation in the Correa et al. (1987) model is modified as

$$\frac{D_2}{r_c^2}\frac{\partial}{\partial r_c}\left(r_c^2 \frac{\partial c_2}{\partial r_c^2}\right) = \frac{\partial c_2}{\partial t}. \tag{4.89}$$

Using the previous dimensionless parameters and the initial and boundary conditions, Eq. (4.89) becomes

$$\frac{1}{\gamma_2^*}\left(\frac{\partial^2 c_2}{\partial x_z^2} + \frac{2}{x_z}\frac{\partial c_2}{\partial x_z} =\right)\frac{\partial c_2}{\partial t_d}. \tag{4.90}$$

Applying a Laplace transform to Eq. (4.90), along with the initial and boundary conditions in Eq. (4.76), yields the following ordinary differential equations:

$$\frac{1}{\gamma_2^*}\left(\frac{d^2\bar{c}_2}{dx_z^2} + \frac{2}{x_z}\frac{d\bar{c}_2}{dx_z}\right) = s\bar{c}_2 - c_2^0. \tag{4.91}$$

The boundary conditions in the Laplace domain become

$$\begin{cases} \bar{c}_1 = \frac{c_0}{s} & x_d = 0 \\ \frac{d\bar{c}_1}{dx_d} = 0 & x_d = 1 \\ \bar{c}_2 = \bar{c}_1 & x_z = R^{\#}/L^* \\ \bar{c}_2 = \text{finite} \end{cases} \tag{4.92}$$

The solution of Eq. (4.91), incorporating the boundary condition in Eq. (4.92), results in

$$\kappa^{\#}\left[\frac{d\bar{c}_2}{dx_z}\right]_{x_z=R^{\#}L^*} = \kappa^{\#}\left(\bar{c}_1 - \frac{c_2^0}{s}\right)\left[\phi^{\#}\coth\frac{\phi^{\#}R^{\#}}{L^*} - \frac{L^*}{R^{\#}}\right] \tag{4.93}$$

where $\phi^{\#}$ is defined as before.

Substituting Eq. (4.93) into Eq. (4.79), yields

$$\frac{1}{\gamma_1^*}\frac{d^2\bar{c}_1}{dx_d^2} - \frac{d\bar{c}_1}{dx_d} = f^*(s\bar{c}_1 - c_1^0) + \psi_1^{\#}\left(\bar{c}_1 + \frac{c_2^0}{s}\right) \tag{4.94}$$

where $\psi_1^{\#}$ is defined in Eq. (4.83).

Solution of Eq. (4.94) with the boundary conditions of Eq. (4.92) gives

$$\bar{c}_1 = e_1^{\#}\exp(\rho_1^{\#}x_d) + e_2^{\#}\exp(\rho_2^{\#}x_d) + \bar{c}^{\#} \tag{4.95}$$

where $\rho_1^{\#}$ and $\rho_2^{\#}$ are as defined before [i.e., Eq. (4.86]. However, the related definition for $\psi_2^{\#}$ is now substituted by the following expression:

$$\psi_2^{\#} = -\frac{c_2^0}{s}. \tag{4.96}$$

Coefficients $e_1^{\#}$ and $e_2^{\#}$ in Eq. (4.95) can be obtained by solving the nonhomogeneous differential equation (4.94) as

$$e_1^{\#} = -\frac{\varrho_1^*\rho_2^{\#}e^{\rho_2^{\#}}}{\rho_1^{\#}e^{\rho_1^{\#}} - \rho_2^{\#}e^{\rho_2^{\#}}} \tag{4.97}$$

$$e_2^{\#} = \frac{\varrho_1^*\rho_1^{\#}e^{\rho_1^{\#}}}{\rho_1^{\#}e^{\rho_1^{\#}} - \rho_2^{\#}e^{\rho_2^{\#}}}. \tag{4.98}$$

4.2.2.5 Alternative Capacitance Models

Based on the capacitance concept proposed by Deans (1963), Coats and Smith (1964) presented a phenomenological model that has been widely applied to address abnormal tracer breakthrough in geologic media. Abnormal breakthrough refers to the presence of extended breakthrough tails in the time domain, commonly attributed to the effect of media heterogeneities. Physically, dead-end pores, or stagnant regions within pores, act as capacitances to the flowing areas, resulting in an increase in the length of the mixing zone, thus creating an extended tail. Mathematically, an additional equation is added to specify the mode of mass transfer between the flowing (macropores) and stagnant regions (micropores). The format of the second equation can vary from quasi-steady rate changes (Coats and Smith, 1964) to unsteady diffusion (Correa et al. 1987). Among numerous parameters, dispersion coefficient D, flowing fraction of pore volume f^*, and mass transfer coefficient Ω are the most influential factors in the classical capacitance model (Coats and Smith 1964).

Excluding consideration of the convective mechanism, Coats and Smith's (1964) capacitance model is a direct counterpart model to the flow model of Warren and Root (1963). However, through comparison between Warren and Root's (1963) model [Eqs. (3.91) and (3.92) in Chapter 2 and Coats and Smith's (1964) model [Eqs. (4.43) and (4.44)], it is noted that, unlike in the former model, the mass interchange terms are not symmetric; the interchange term only appears in the micropore domain. Since both models are based on different phenomenological conceptualizations, these differences are attributed to the different physical processes represented.

A further careful comparison between the Coats and Smith's (1964) model and the Warren and Root's (1963) model indicates that it is easier to extend the latter model to multi-porosity cases than the former, due to the symmetry of the interchange terms. To minimize this deficiency, an alternative capacitance model is proposed as follows:

$$D_1 \frac{\partial^2 c_1}{\partial x^2} - v \frac{\partial c_1}{\partial x} = f^* \frac{\partial c_1}{\partial t} + \Omega(c_1 - c_2) \tag{4.99}$$

$$0 = (1 - f^*) \frac{c_2}{\partial t} - \Omega(c_1 - c_2) \tag{4.100}$$

This alternative capacitance model contains the symmetric mass interchange terms, similar to the Warren and Root model for fluid flow. Using identical dimensionless parameters to those given in Eq. (4.35), Eqs. (4.99) and (4.100) can be rewritten as

$$\frac{1}{\gamma^*} \frac{\partial^2 c_1}{\partial x_d^2} - \frac{\partial c_1}{\partial x_d} = f^* \frac{\partial c_1}{\partial t_d} + \Omega^*(c_1 - c_2). \tag{4.101}$$

$$0 = (1 - f^*) \frac{\partial c_2}{\partial t_d} - \Omega^*(c_1 - c_2) \tag{4.102}$$

The initial and boundary condition are given as

$$\begin{cases} c_1 = c_1^0 & t = 0 \\ c_2 = c_2^0 & t = 0 \\ c_1 = c_0 & x_d = 0 \\ \dfrac{dc_1}{dx_d} = 0, & x_d = 1. \end{cases} \tag{4.103}$$

In the Laplace domain, Eqs. (4.101) and (4.102) become the following ordinary differential equations:

$$\frac{1}{\gamma^*} \frac{d^2 \bar{c}_1}{dx_d^2} - \frac{d\bar{c}_1}{dx_d} = f^*(s\bar{c}_1 - c^0) + \Omega^*(\bar{c}_1 - \bar{c}_2) \tag{4.104}$$

$$(1 - f^*)[\bar{c}_2 s - c_2^0] = \Omega^*(\bar{c}_1 - \bar{c}_2) \tag{4.105}$$

where s is a Laplace parameter. The boundary conditions in the Laplace domain are described by Eq. (4.48).

Since Eq. (4.105) is identical to (4.51), $\bar{c}_2$ in Eq. (4.52) can be used again. Substitution of $\bar{c}_2$ into Eq. (4.104) yields:

$$\frac{d^2\bar{c}_1}{dx_d^2} - \gamma^* \frac{d\bar{c}_1}{dx_d} - f(s)^* \bar{c}_1 = \Pi^* \tag{4.106}$$

where

$$f(s)^* = \gamma^* \left[f^* s + \Omega^* - \frac{\Omega^{*2}}{(1-f^*)s + \Omega^*} \right] \tag{4.107}$$

$$\Pi^* = -\gamma^* \left[f^* c_1^0 + \frac{\Omega^*(1-f^*)c_2^0}{(1-f^*)s + \Omega^*} \right]. \tag{4.108}$$

Solution of Eq. (4.106), which satisfies the boundary conditions in Eq. (4.48), can be expressed as

$$\bar{c}_1 = e_1^* e^{\phi_1^* x_d} + e_2^* e^{\phi_2^* x_d} + \frac{\Pi^*}{\phi_1^* \phi_2^*} \tag{4.109}$$

where

$$e_1^* = \frac{-\varrho^* \phi_2^* e^{\phi_2^*}}{\phi_1^* e^{\phi_1^*} - \phi_2^* e^{\phi_2^*}} \tag{4.110}$$

$$e_2^* = \frac{\varrho^* \phi_1^* e^{\phi_1^*}}{\phi_1^* e^{\phi_1^*} - \phi_2^* e^{\phi_2^*}} \tag{4.111}$$

$$\varrho^* = \frac{c_0}{s} - \frac{\Pi^*}{\phi_1^* \phi_2^*} \tag{4.112}$$

$$\phi_1^* = \frac{\gamma^*}{2} \left[1 + \sqrt{1 + \frac{4f(s)^*}{\gamma^{*2}}} \right] \tag{4.113}$$

$$\phi_2^* = \frac{\gamma^*}{2} \left[1 - \sqrt{1 + \frac{4f(s)^*}{\gamma^{*2}}} \right]. \tag{4.114}$$

4.2.2.6 Multi-Porosity Models

The use of a dual-porosity approach to represent transport behavior (envisioning two regions, e.g., mobile and immobile regions) may be insufficient to characterize highly heterogeneous porous media containing pore structures at a variety of length scales. In this, mesopores of intermediate scale may exist as a buffer zone to modify mass transfer between macropores and micropores.

Characteristic response of a triple-porosity medium yields a spectrum of breakthrough curves, typically with enhanced tailing (Abdassah and Ershaghi 1986; Bai et al. 1993). As pointed out by Gwo et al. (1996), it is appropriate to view soils as consisting of a continuous distribution of pore sizes that may be segregated into macro-, meso-, and micropore regions, analogous to particle-size distribution being segregated into sand, silt, and clays. Luxmoore et al. (1990) defined macropores and micropores as pores with "equivalent pore diameters" (EPD) greater than 1 mm and less than 0.01 mm, respectively. Gwo et al. (1995) alternatively divided the pore structures into three regions by adding a mesopore region with a corresponding EPD between 0.01 mm and 1 mm.

General Triple-Porosity Model

Assuming constant porosity and using the average convective velocity v ($v = v^*/n$, where v^* is the intrinsic interstitial or Darcy velocity), and assuming that the quasi-steady rate of solute exchange is proportional to the concentration gradients between the three continua (Bear 1972; Bai et al. 1993; Gwo et al. 1995), the general governing equations of solute transport for a triple-porosity system can be described as

$$\frac{\partial}{\partial x_i}\left(D_{1ij}\frac{\partial c_1}{\partial x_j}\right) - v_{1i}\frac{\partial c_1}{\partial x_i} = \frac{\partial c_1}{\partial t} + \frac{\xi_{12}}{n_1}(c_1 - c_2) + \frac{\xi_{13}}{n_1}(c_1 - c_3) \tag{4.115}$$

$$\frac{\partial}{\partial x_i}\left(D_{2ij}\frac{\partial c_2}{\partial x_j}\right) - v_{2i}\frac{\partial c_2}{\partial x_i} = \frac{\partial c_2}{\partial t} + \frac{\xi_{21}}{n_2}(c_2 - c_1) + \frac{\xi_{23}}{n_2}(c_2 - c_3) \tag{4.116}$$

$$\frac{\partial}{\partial x_i}\left(D_{3ij}\frac{\partial c_3}{\partial x_j}\right) - v_{3i}\frac{\partial c_3}{\partial x_i} = \frac{\partial c_3}{\partial t} + \frac{\xi_{31}}{n_3}(c_3 - c_1) + \frac{\xi_{32}}{n_3}(c_3 - c_2) \tag{4.117}$$

where subscripts 1, 2, and 3 represent macro-, meso-, and micropores for unconsolidated porous media, and fractures, micro-fractures, and porous matrix for rock masses, respectively; subscripts i and j are coordinate indices; c is the solute concentration; t is the time after inception of the transport, x_i is the coordinate system; D_{ij} is the hydrodynamic dispersion tensor; v_i is the average flow velocity; n is the porosity; and ξ is a concentration exchange coefficient characterizing mass transfer between the pores of various scales.

Incorporating the effects of linear sorption, Eqs. (4.115) through (4.117) may be further modified and written in short form as

$$\frac{\partial}{\partial x_i}\left(D_{mij}\frac{\partial c_m}{\partial x_j}\right) - v_{mi}\frac{\partial c_m}{\partial x_i} = \frac{\partial c_m}{\partial t} + \frac{\xi_{mm_1}}{n_m}(c_m - c_{m_1}) + \frac{\xi_{mm_2}}{n_m}(c_m - c_{m_2}) + \frac{\partial \varsigma_m^*}{\partial t} \tag{4.118}$$

$$\frac{\partial \varsigma_m^*}{\partial t} = K_{pm}^{\#}(c_m - \iota_m^* \varsigma_m^*) \tag{4.119}$$

where subscripts m, m_1, and m_2 follow rotational order, i.e., $m = 1, 2, 3$; $m_1 = 2, 3, 1$; and $m_2 = 3, 1, 2$, respectively. ς^* is the sorption isotherm that is represented by the product of solid-phase density and sorbed mass per unit mass of solids over the porosity, $K_p^{\#}$ is the sorption intensity factor, and ι^* is the process factor denoting whether the process is irreversible ($\iota^* = 0$), or reversible (e.g., $\iota^* \varsigma^* > c$).

Depending on physical interpretation, two different "triple-porosity" models may be recovered from the degenerate forms of this general "triple-porosity" model.

Triple-Porosity Model with Negligible Mesopore Dispersion

Eqs. (4.115) through (4.117) incorporate the comprehensive coupling but can be difficult to solve analytically. A first approximation may be made by considering that: (a) the solute exchange between macropores and micropores is both indirect and insignificant; (b) mechanical dispersion is the predominant transport mechanism within the macropores; (c) convection is the dominant transport mechanism within mesopores (McKibbin 1985; Houseworth 1988; Bouhroum 1993); and (d) diffusion or replenishment is pervasive in the micropores. As a result, Eqs. (4.115) through (4.117) can be simplified as (Bai and Roegiers 1997)

$$\frac{\partial}{\partial x_i}\left(D^m_{1\,ij}\frac{\partial c_1}{\partial x_j}\right) - v_{1i}\frac{\partial c_1}{\partial x_i} = \frac{\partial c_1}{\partial t} + \frac{\xi_{12}}{n_1}(c_1 - c_2) \tag{4.120}$$

$$-v_{2i}\frac{\partial c_2}{\partial x_i} = \frac{\partial c_2}{\partial t} + \frac{\xi_{21}}{n_2}(c_2 - c_1) + \frac{\xi_{23}}{n_2}(c_2 - c_3) \tag{4.121}$$

$$\frac{\partial}{\partial x_i}\left(D^d_{3ij}\frac{\partial c_3}{\partial x_j}\right) = \frac{\partial c_3}{\partial t} + \frac{\xi_{32}}{n_3}(c_3 - c_2) \tag{4.122}$$

where D_1^m is the mechanical dispersion coefficient, and D_3^d is the effective diffusion coefficient incorporating the tortuosity factor.

Further approximations can be made by assuming: (a) one-dimensional transport, (b) constant dispersivity and diffusivity, and, (c) spherical block structure of the micropore aggregate (Huyakorn et al. 1983; Correa et al. 1987). As shown in Figure 4-1, this spherical block represents an equivalent micropore domain with the outer bounding surface being defined as the mesopore skin. After dropping the superscripts for D_1 and D_3, Eqs. (4.120), (4.121), and (4.122) can be modified as

$$D_1\frac{\partial^2 c_1}{\partial x^2} - v_1\frac{\partial c_1}{\partial x} = \frac{\partial c_1}{\partial t} + \frac{\xi_{12}}{n_1}(c_1 - c_2) \tag{4.123}$$

$$-v_2\frac{\partial c_2}{\partial x} = \frac{\partial c_2}{\partial t} + \frac{\xi_{21}}{n_2}(c_2 - c_1) + \frac{3n_3D_3}{n_2R_b}\left[\frac{\partial c_3}{\partial r}\right]_{r_b=R_b} \tag{4.124}$$

$$\frac{D_3}{r_b^2}\frac{\partial}{\partial r_b}\left(r_b^2\frac{\partial c_3}{\partial r_b}\right) = \frac{\partial c_3}{\partial t} \tag{4.125}$$

where r_b is the radial distance from the center and R_b is the radius of the equivalent spherical micropore block. It may be noted from Eq. (4.125) that transport in the micropores is decoupled from that in the mesopores.

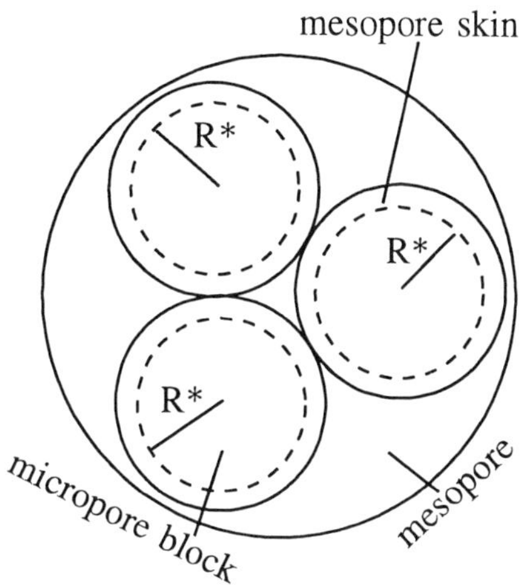

Figure 4-1. Spherical block description for micropores.

For a more general solution, the following dimensionless terms are introduced:

$$\begin{cases} x_d = \frac{x}{L^*}, & \tau^* = \frac{v_1 t}{L^*}, & \gamma_1^* = \frac{v_1 L^*}{D_1}, & \gamma_3^* = \frac{v_1 L^*}{D_3}, & x_z = \frac{r_b}{L^*} \\ a_{ij} = \frac{\xi_{ij} L^*}{n_i v_i}, & a^* = \frac{3 n_3 D_3}{n_2 R_b v_2}, & b_i^* = \frac{v_1}{v_i}, & R^* = \frac{r_b}{L^*} & \end{cases} \tag{4.126}$$

where subscripts i and j = 1, 2; but $j \neq i$; and L^* is an arbitrary length which may represent the length of the core sample or the distance of pollutant migration from the source.

Incorporating all dimensionless terms, Eqs. (4.123), (4.124), and (4.125) are rewritten as

$$\frac{1}{\gamma_1^*}\frac{\partial^2 c_1}{\partial x_d^2} - \frac{\partial c_1}{\partial x_d} = b_1^*\frac{\partial c_1}{\partial \tau^*} + a_{12}(c_1 - c_2) \tag{4.127}$$

$$-\frac{\partial c_2}{\partial x_d} = b_2^* \frac{\partial c_2}{\partial \tau^*} + a_{21}(c_2 - c_1) + a^* \left[\frac{\partial c_3}{\partial x_z} \right]_{x_z = R^*/L^*} \tag{4.128}$$

$$\frac{1}{\gamma_3^*} \left(\frac{\partial^2 c_3}{\partial x_z^2} + \frac{2}{x_z} \frac{\partial c_3}{\partial x_z} \right) = \frac{\partial c_3}{\partial \tau^*}. \tag{4.129}$$

For step injection at the inlet and constant flux (zero flux for this case) at the outlet, boundary and initial conditions may be described by

$$\begin{cases} c_1 = c_2 = c_0 & (x_d = 0) \\ \dfrac{\partial c_1}{\partial x_d} = 0 & (x_d = 1) \\ c_3 = \text{finite} & (x_z = 0) \\ c_3 = c_2 & (x_z = R^*/L^*) \\ c_1 = c_1^0, \quad c_2 = c_2^0, \quad c_3 = c_3^0 & (\tau^* = 0) \end{cases} \tag{4.130}$$

where c_0 is the concentration at the source and c_1^0, c_2^0 and c_3^0 are the initial concentrations in macropores, mesopores, and micropores, respectively. The boundary conditions expounded in the above equations are applicable to the injection of solute through a finite core comprising a triple-porosity medium.

Applying a Laplace transform to Eqs. (4.127), (4.128), and (4.129), one has

$$\frac{1}{\gamma_1^*} \frac{d^2 \bar{c}_1}{dx_d^2} - \frac{d\bar{c}_1}{dx_d} = b_1^*(s\bar{c}_1 - c_1^0) + a_{12}(\bar{c}_1 - \bar{c}_2) \tag{4.131}$$

$$-\frac{d\bar{c}_2}{dx_d} = b_2^*(s\bar{c}_2 - c_2^0) + a_{21}(\bar{c}_2 - \bar{c}_1) + a^* \left[\frac{d\bar{c}_3}{dx_z} \right]_{x_z = R^*/L^*} \tag{4.132}$$

$$\frac{1}{\gamma_3^*} \left(\frac{d^2 \bar{c}_3}{dx_z^2} + \frac{2}{x_z} \frac{d\bar{c}_3}{dx_z} \right) = s\bar{c}_3 - c_3^0 \tag{4.133}$$

where s is the Laplace transform parameter. Boundary conditions in the Laplace domain are changed to

$$\begin{cases} \bar{c}_1 = \bar{c}_2 = \frac{c_0}{s} & (x_d = 0) \\ \dfrac{d\bar{c}_1}{dx_d} = 0 & (x_d = 1) \\ \bar{c}_3 = \text{finite} & (x_z = 0) \\ \bar{c}_3 = \bar{c}_2 & (x_z = R^*/L^*) \end{cases} \tag{4.134}$$

and $\bar{c}_3$ can be derived independently from Eq. (4.133). Following a similar procedure proposed by Moench (1984), one has

$$\left[\frac{d\bar{c}_3}{dx_z}\right]_{x_z=R^*/L^*} = \left(\bar{c}_2 - \frac{c_3^0}{s}\right)\left[\phi^{\#}\coth(\phi^{\#}R^*/L^*) - \frac{L^*}{R^*}\right] \tag{4.135}$$

where $\phi^{\#} = \sqrt{\gamma_3^* s}$.

Substituting Eq. (4.135) into (4.132), then

$$-\frac{d\bar{c}_2}{dx_d} = (b_2^* s + a_{21} + \psi_1^{\#})\bar{c}_2 - a_{21}\bar{c}_1 + \psi_1^{\#}\psi_2^{\#} - b_2 c_2^0 \tag{4.136}$$

where

$$\begin{cases} \psi_1^{\#} = a^*\left[\phi^{\#}\coth(\phi^{\#}R^*/L^*) - \frac{L^*}{R^*}\right] \\ \psi_2^{\#} = -\frac{c_3^0}{s} \end{cases} \tag{4.137}$$

Eq. (4.131) can be rewritten as

$$\frac{1}{\gamma_1^*}\frac{d^2\bar{c}_1}{dx_d^2} - \frac{d\bar{c}_1}{dx_d} = (b_1^* s + a_{12})\bar{c}_1 - a_{12}\bar{c}_2 - b_1^* c_1^0. \tag{4.138}$$

The solutions of the coupled Eqs. (4.136) and (4.138) are reported in Bai and Roegiers (1997).

Triple-Porosity Model with Sorption

The triple-porosity responses of the porous medium may not only provide a rationale to replicate the complex physical phenomena related to transport in heterogeneously structured soils, but also suggest an alternative tool in matching unusual response in physical experiments. As illustrated in Figure 4-2, a structured soil may be divided into three distinct porous domains (macro-, meso-, and micropores) that may be alternately envisioned as consisting of fractures, micro-fractures, and matrix pores with three different velocity and dispersion profiles. While macropores act as preferential channels for mass transport, mesopore regions are considered as branch paths to divert solute either out-from or into the least permeable micropore aggregates.

Eqs. (4.118) and (4.119) may be a relatively complete formulation of the general triple-porosity system. However, it is difficult to circumvent the complexity of this system, especially due to its comprehensive mass interchanges among various pore structures. To enable the application of analytical approaches, Eqs. (4.118) and (4.119) can be simplified using the following assumptions: (a) direct solute exchange between macropores and micropores is insignificant; (b) solute within the micropores is immobile (Coats and Smith

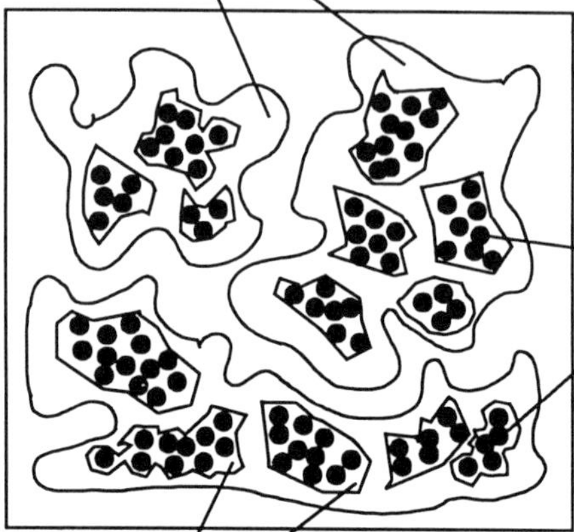

Figure 4-2. A triple-porosity system.

1964); (c) the dispersion coefficient D is constant; and (d) solute transport is one-dimensional. As a result, Eqs. (4.118) and (4.119), in an expanded form, become (Bai et al. 1997a)

$$D_1 \frac{\partial^2 c_1}{\partial x^2} - v_1 \frac{\partial c_1}{\partial x} = \frac{\partial c_1}{\partial t} + \frac{\xi_{12}}{n_1}(c_1 - c_2) + \frac{\partial \varsigma_1^*}{\partial t} \tag{4.139}$$

$$D_2 \frac{\partial^2 c_2}{\partial x^2} - v_2 \frac{\partial c_2}{\partial x} = \frac{\partial c_2}{\partial t} + \frac{\xi_{21}}{n_2}(c_2 - c_1) + \frac{\xi_{23}}{n_2}(c_2 - c_3) + \frac{\partial \varsigma_2^*}{\partial t} \tag{4.140}$$

$$0 = \frac{\partial c_3}{\partial t} + \frac{\xi_{32}}{n_3}(c_3 - c_2) + \frac{\partial \varsigma_3^*}{\partial t} \tag{4.141}$$

$$\frac{\partial \varsigma_1^*}{\partial t} = K_{p1}^{\#}(c_1 - \iota_1^* \varsigma_1^*) \tag{4.142}$$

$$\frac{\partial \varsigma_2^*}{\partial t} = K_{p2}^{\#}(c_2 - \iota_2^* \varsigma_2^*) \tag{4.143}$$

$$\frac{\partial \varsigma_3^*}{\partial t} = K_{p3}^{\#}(c_3 - \iota_3^* \varsigma_3^*). \tag{4.144}$$

The dimensionless terms can be used for more general solution. Expressions for x_d, b_i^*, and τ^* are identical to those in Eq. (4.126). Other dimensionless parameters are

$$\gamma_i^* = \frac{v_i L^*}{D_i}, \qquad a_{ij}^* = \frac{\xi_{ij} L^*}{n_i v_1}, \qquad \eta_i^* = \frac{L^*}{v_1} K_{p_i}^{\#} \tag{4.145}$$

where subscripts $i = 1, 2, 3$; $j = 1, 2, 3$ but $i \neq j$; $i = 1$ and 2 for γ_i^* and b_i^* only; L^* is an arbitrary length that may represent sample or domain length. Since b_1^* is always equal to 1, it is omitted in the following formulation.

Incorporating all dimensionless terms, Eqs. (4.139) through (4.144) are rewritten as

$$\frac{1}{\gamma_1^*}\frac{\partial^2 c_1}{\partial x_d^2} - \frac{\partial c_1}{\partial x_d} = \frac{\partial c_1}{\partial \tau^*} + a_{12}^*(c_1 - c_2) + \frac{\partial \varsigma_1^*}{\partial \tau^*} \tag{4.146}$$

$$\frac{1}{\gamma_2^*}\frac{\partial^2 c_2}{\partial x_d^2} - \frac{\partial c_2}{\partial x_d} = b_2^*\left[\frac{\partial c_2}{\partial \tau^*} + a_{21}^*(c_2 - c_1) + a_{23}^*(c_2 - c_3) + \frac{\partial \varsigma_2^*}{\partial \tau^*}\right] \tag{4.147}$$

$$0 = \frac{\partial c_3}{\partial \tau^*} + a_{32}^*(c_3 - c_2) + \frac{\partial \varsigma_3^*}{\partial \tau^*} \tag{4.148}$$

$$\frac{\partial \varsigma_1^*}{\partial \tau^*} = \eta_1^*(c_1 - \iota_1^* \varsigma_1^*) \tag{4.149}$$

$$\frac{\partial \varsigma_2^*}{\partial t} = \eta_2^*(c_2 - \iota_2^* \varsigma_2^*) \tag{4.150}$$

$$\frac{\partial \varsigma_3^*}{\partial \tau^*} = \eta_3^*(c_3 - \iota_3^* \varsigma_3^*). \tag{4.151}$$

For step injection at the inlet and vanishing flux at the outlet, boundary and initial conditions may be described by

$$\begin{cases} c_1 = c_2 = c_0 & (x_d = 0) \\ \dfrac{\partial c_1}{\partial x_d} = \dfrac{\partial c_2}{\partial x_d} = 0 & (x_d = 1) \\ c_1 = c_1^0 \quad c_2 = c_2^0 \quad c_3 = c_3^0 & (\tau^* = 0) \\ \varsigma_1^* = \varsigma_1^{*0},\ \varsigma_2^* = \varsigma_2^{*0},\ \varsigma_3^* = \varsigma_3^{*0} & (\tau^* = 0) \end{cases} \tag{4.152}$$

where c_0 is the concentration at the source; c_1^0, c_2^0, and c_3^0 are the initial concentrations; and ς_1^{*0}, ς_2^{*0}, and ς_3^{*0} are the initial sorbent concentrations in macropores, mesopores, and micropores, respectively. The field equations and boundary conditions, represented in the previous, are applicable to solute injection within a finite column of a triple-porosity medium.

Applying a Laplace transform to Eqs. (4.146) through (4.151), yields

$$\frac{1}{\gamma_1^*}\frac{d^2 \bar{c}_1}{dx_d^2} - \frac{d\bar{c}_1}{dx_d} = s\bar{c}_1 - c_1^0 + a_{12}^*(\bar{c}_1 - \bar{c}_2) + s\bar{\varsigma}_1^* - \varsigma_1^{*0} \tag{4.153}$$

$$\frac{1}{\gamma_2^*}\frac{d^2\bar{c}_2}{dx_d^2} - \frac{d\bar{c}_2}{dx_d} = b_2^*\left[(s\bar{c}_2 - c_2^0) + a_{21}^*(\bar{c}_2 - \bar{c}_1) + a_{23}^*(\bar{c}_2 - \bar{c}_3) + s\bar{\varsigma^*}_2 - \varsigma_2^{*0}\right] \tag{4.154}$$

$$0 = s\bar{c}_3 - c_3^0 + a_{32}^*(\bar{c}_3 - \bar{c}_2) + s\bar{\varsigma}_3^* - \varsigma_3^{*0} \tag{4.155}$$

$$s\bar{\varsigma}_1^* - \varsigma_1^{*0} = \eta_1^*(\bar{c}_1 - \iota_1^*\bar{\varsigma}_1^*) \tag{4.156}$$

$$s\bar{\varsigma}_2^* - \varsigma_2^{*0} = \eta_2^*(\bar{c}_2 - \iota_2^*\bar{\varsigma}_2^*) \tag{4.157}$$

$$s\bar{\varsigma}_3^* - \varsigma_3^{*0} = \eta_3^*(\bar{c}_3 - \iota_3^*\bar{\varsigma}_3^*) \tag{4.158}$$

where s is the Laplace transform parameter.

Boundary conditions in the Laplace domain are transformed as

$$\begin{cases} \bar{c}_1 = \bar{c}_2 = \frac{c_0}{s} & (x_d = 0) \\ \frac{d\bar{c}_1}{dx_d} = \frac{d\bar{c}_2}{dx_d} = 0 & (x_d = 1). \end{cases} \tag{4.159}$$

The relationship between $\bar{\varsigma}_i^*$ and $\bar{c}_i$ (i=1,2,3) can be derived from Eqs. (4.156), (4.157), and (4.158). The subsequent results can be substituted into Eqs. (4.153), (4.154), and (4.155) to eliminate unknowns $\bar{\varsigma}_i^*$ (i=1,2,3). As a result, Eqs. (4.153) through (4.158) reduce to

$$\frac{1}{\gamma_1^*}\frac{d^2\bar{c}_1}{dx_d^2} - \frac{d\bar{c}_1}{dx_d} = s\bar{c}_1 - c_1^0 + a_{12}(\bar{c}_1 - \bar{c}_2) + \eta_1^{\#}\bar{c}_1 + \eta_1^{*0} \tag{4.160}$$

$$\frac{1}{\gamma_2^*}\frac{d^2\bar{c}_2}{dx_d^2} - \frac{d\bar{c}_2}{dx_d} = b_2^*[(s\bar{c}_2 - c_2^0) + a_{21}^*(\bar{c}_2 - \bar{c}_1) + a_{23}^*(\bar{c}_2 - \bar{c}_3)] + \eta_2^{\#}\bar{c}_2 + \eta_2^{*0} \tag{4.161}$$

$$0 = s\bar{c}_3 - c_3^0 + a_{32}^*(\bar{c}_3 - \bar{c}_2) + \eta_3^{\#}\bar{c}_3 + \eta_3^{*0} \tag{4.162}$$

where

$$\eta_1^{\#} = \frac{s\eta_1^*}{s + \eta_1^*\iota_1^*}, \quad \eta_2^{\#} = \frac{sb_2^*\eta_2^*}{s + \eta_2^*\iota_2^*}, \quad \eta_3^{\#} = \frac{s\eta_3^*}{s + \eta_3^*\iota_3^*} \tag{4.163}$$

$$\eta_1^{*0} = \varsigma_1^{*0}\left(\frac{s}{s + \eta_1^*\iota_1^*} - 1\right) \tag{4.164}$$

$$\eta_2^{*0} = b_2^*\varsigma_2^{*0}\left(\frac{s}{s + \eta_2^*\iota_2^*} - 1\right) \tag{4.165}$$

$$\eta_3^{*0} = \varsigma_3^{*0}\left(\frac{s}{s + \eta_3^*\iota_3^*} - 1\right). \tag{4.166}$$

The relationship between $\bar{c}_2$ and $\bar{c}_3$ can be derived from Eq. (4.162) as

$$\bar{c}_3 = \frac{c_3^0 + a_{32}^*\bar{c}_2 - \eta_3^{*0}}{s + a_{32}^* + \eta_3^{\#}}. \tag{4.167}$$

Substituting Eq. (4.167) into (4.161), then

$$\frac{1}{\gamma_2^*}\frac{d^2\bar{c}_2}{dx_d^2} - \frac{d\bar{c}_2}{dx_d} - \phi_{21}^*\bar{c}_2 = -b_2^* a_{21}^*\bar{c}_1 + \phi_{22}^* \tag{4.168}$$

where

$$\phi_{21}^* = b_2^*\left(s + a_{21}^* + a_{23}^* - \frac{a_{23}^* a_{32}^*}{s + a_{32}^* + \eta_3^{\#}}\right) + \eta_2^{\#} \tag{4.169}$$

$$\phi_{22}^* = -b_2^*\left[c_2^0 + \frac{a_{23}^*(c_3^0 - \eta_3^{*0})}{s + a_{32}^* + \eta_3^{\#}}\right] + \eta_2^{*0}. \tag{4.170}$$

Eq. (4.160) can be rewritten as

$$\frac{1}{\gamma_1^*}\frac{d^2\bar{c}_1}{dx_d^2} - \frac{d\bar{c}_1}{dx_d} - \phi_{11}^*\bar{c}_1 = -a_{12}^*\bar{c}_2 + \phi_{12}^* \tag{4.171}$$

where

$$\phi_{11}^* = s + a_{12}^* + \eta_1^{\#}, \qquad \phi_{12}^* = -c_1^0 + \eta_1^{*0}. \tag{4.172}$$

For brevity and completeness, the detailed analytical procedure and entire solutions of the coupled equations (4.168) and (4.171) are given in Chapter 5 on analytical solution.

4.2.3 Comparative Analysis

Six models discussed in this section are selected for comparison. They are: (a) single-porosity model [Eq. (4.32)], (b) dual-porosity model [Eqs. (4.99) and (4.100)], (c) triple-porosity model [Eqs. (4.123) through (4.125)], (d) Coats and Smith's (1964) model [Eqs. (4.43) and (4.44)], (e) Piquemal's (1992) model [Eqs. (4.63) and (4.44)], and (f) Correa et al.'s (1987) model [Eqs. (4.73) and (4.74)]. In the comparison, it is difficult to assume identical parameters due to the differences in conceptualizations. However, the examinations are made with identical dimensionless time and distance for all the models.

For simplicity, the units for all selected parameters in the triple-porosity model are omitted, since only the dimensionless parameters are used in the equations. However, a uniform mass-length-time system is used in the analysis. The unchanged parameters for analyzing spatial and temporal concentrations are assumed as $D_1 = 5$, $v_1 = 1$, $c_1^0 = 0.007$, $c_2^0 = 0.008$, $c_3^0 = 0.09$, $n_1 = 0.3$, $n_2 = 0.1$, and $n_3 = 0.05$. The rest of the parameters are listed in Table 4.1.

Table 4.1. Parameters in comparative study.

Figure	L^*	D_3	v_2	v_3	R^*	ξ_{12}	ξ_{21}	τ^*	x_d
Spatial	15	0.1	0.5	0.2	10^{-8}	0.02	0.02	0.5	0-1
Temporal	5	0.2	0.4	0.1	10^{-6}	0.01	0.01	0-1	0.5

Figure 4-3 shows comparisons between all six models for the spatial distribution of concentration. The models yielding similar results are the single- and dual-porosity, Coats and Smith (1964), and Piquemal (1992) models. The triple-porosity and Correa et al. (1987) models deviate from the other models with more dramatic changes in the upstream regions. Correa et al.'s model generates a dominant diffusive flow while the triple-porosity model shows the dominance of convection near the source and pervasive dispersion in the downstream region. Because Correa et al.'s model assumes a similar micropore transport process as the triple-porosity model, this figure reveals the important influence of coupling flow in the micropores.

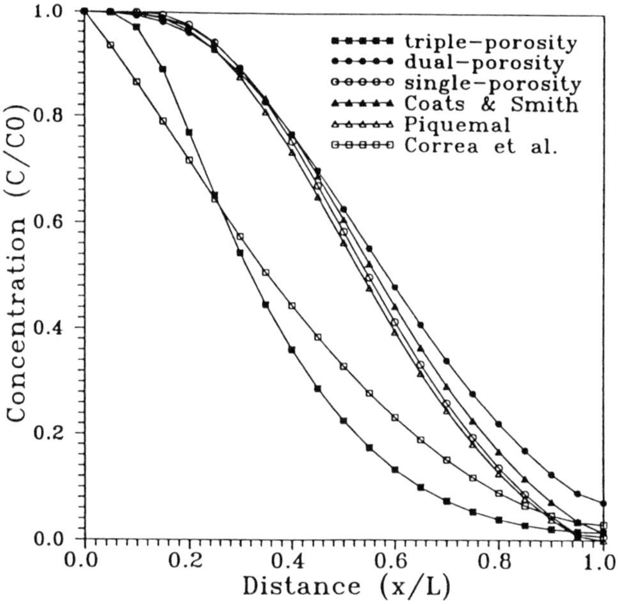

Figure 4-3. Comparison of spatial concentrations.

The comparison of breakthrough curves for the six models is given in Figure 4-4. Again, the Correa et al. (1987) and the triple-porosity models deviate from the other four models which show close resemblance in the breakthrough profiles. However, Correa et al.'s (1987) model maintains diffusive characteristics with extensive tailing, while the triple-porosity model represents a prominent convective response with abrupt plug-like solute arrival and drastic concentration change. Because of its use of additional parameters, the triple-porosity model is better in representing observational data.

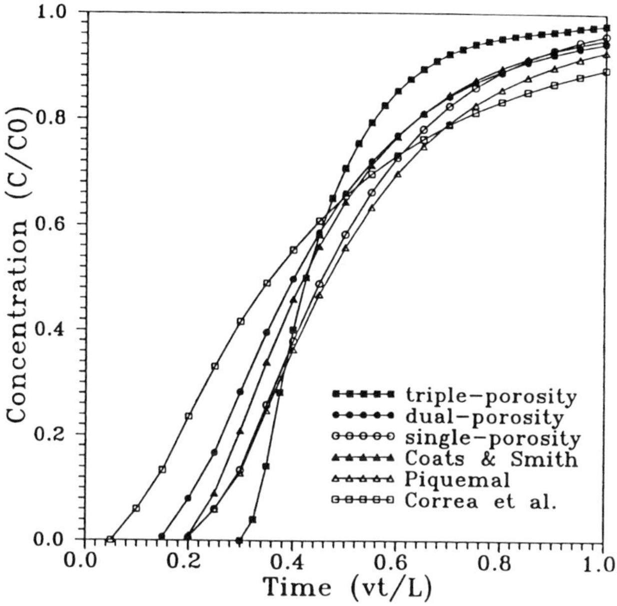

Figure 4-4. Comparison of temporal concentrations.

4.2.4 Stochastic Processes

Based on a stochastic representation of the void structure and void sizes, Aberg (1992) presented a method to calculate the grain size distribution in relation to the effect of gravel packs. Joy et al. (1993) studied particulate transport in porous media using a stochastic model, constrained by experimental data, to evaluate the influence of the random movements of the particles and the contribution of the fluid in the development of these trajectories. As defined by Imdakm and Sahimi (1991), the continuum approaches are purely phenomenological and may be considered as trajectory analysis models where the paths of the particles within a single pore are computed using force balances. Morita (1993) developed such an approach, applied to the numerical study of gravel pack damage. Statistical models, in contrast, assume particle transport to be a random process where the probability of having the particles move in any direction is essentially uniform for the homogeneous pore distribution and biased for heterogeneous pore distribution. Todd et al. (1984) studied formation damage in oil reservoirs using a random-walk technique to move the particles through an assumed square network. Houi and Lenormand (1986) developed a similar model for studying filtration processes.

Bouhroum and Civan (1994) observed in an experimental study that the presence of a sharp particle front cannot be explained by the variety of current models that predict a smooth increase of particle concentration within the gravel pack. Physically, this phenomenon may be the result of clogging due to the process of particle exclusion in passing the pore throat.

The random walk method (RWM) may be an ideal choice if particle transport is envisioned as a random process instead of a deterministic process, such as described by all other methods.

Brownian motion defines the ceaseless irregular motion of a particle in a liquid or gas. Assuming that u is the displacement of a particle in Brownian motion from its starting point after time t, it can be shown that u has the probability density function

$$\Theta_p = \frac{1}{\sqrt{4\pi t D}} \exp\left[-\frac{(x)^2}{4tD}\right]$$

where x is the distance, D is the diffusion coefficient, and u is normally distributed with mean 0, and variance

$$\varsigma^2 = 2tD$$

where ς is the standard deviation. This equation states that the mean square displacement of a particle in Brownian motion is proportional to the time t. A model for Brownian motion is provided by a particle undergoing a random walk. Assuming $x_i, i = 1, 2, ...n^*$ as identically distributed random variables with mean 0 and finite variance, the summation $u_{n^*} = x_1 + x_2 + ... + x_{n^*}$ represents the displacement from its starting position of a particle performing the n^*th step random walk with mean 0 and variance proportional to the number of steps, or

$$\Theta_{p_{n^*}} = \frac{1}{\sqrt{4\pi t D n^*}} \exp\left[-\frac{(x)^2}{4tDn^*}\right]. \tag{4.173}$$

The traditional dispersion-convection transport equation is given in Eq. (4.25). A more general dispersion-convection equation incorporating heterogeneities in permeability and storage capacities over a representative elementary volume (REV) may be written as

$$D^{\#}\frac{\partial^2 c}{\partial x^2} - v\frac{\partial c}{\partial x} = \frac{\partial c}{\partial t} \tag{4.174}$$

where c is the solute concentration, v is the average flow velocity, and $D^{\#}$ is the equivalent dispersion coefficient, considering the permeability contrast between highly conductive macropores and less permeable micropores. Similar to Schestakow (1979), $D^{\#}$ can be expressed as

$$D^{\#} = \alpha^{\#} v = \frac{L^{\#}(1-\theta^{\#})k_{mi}}{k_{ma}} \tag{4.175}$$

where $\alpha^{\#}$ is the heterogeneous dispersivity, $L^{\#}$ is the characteristic length of the micropore aggregate, $\theta^{\#}$ is the relative saturation of the macropores, k_{mi} and k_{ma} are the permeabilities of micropores and macropores, respectively. Inclusion of the parameter $D^{\#}$ makes the analysis suitable for both homogeneous and heterogeneous media.

For a constant concentration at the source and free spreading of solute, boundary and initial conditions may be given as

$$\begin{cases} c(x,0) = 0 \\ c(0,t) = c_0 \\ c(\infty,t) = 0 \end{cases} \tag{4.176}$$

Using a pulse step function, Eq. (4.176) may be written as

$$\begin{cases} c(x,0) = c_0\delta(x) \\ c(0,t\to 0) = \dfrac{c_0\delta(t)}{v} \\ c(x\to\infty,t) = 0 \end{cases} \tag{4.177}$$

where c_0 is the initial concentration, which can be expressed as (Luckner and Schestakow 1991)

$$c_0 = \frac{m_\omega}{a^{\#}\varphi} \tag{4.178}$$

where m_ω is the mass in moles, φ is the storage coefficient, $a^{\#}$ is the cross-sectional area of transport, $\delta(t)$ is the delta function, and v is the velocity of the front.

Assuming that the solute per unit mass is injected at x=0 and at time t, then

$$c(x,t) = \tilde{c}(x^*,t) \tag{4.179}$$

where

$$x^* = x - vt \tag{4.180}$$

and

$$\begin{cases} \dfrac{\partial c}{\partial x} = \dfrac{\partial \tilde{c}}{\partial x^*} \\ \\ \dfrac{\partial^2 c}{\partial x^2} = \dfrac{\partial^2 \tilde{c}}{\partial x^{*2}}. \end{cases} \tag{4.181}$$

Also from Eqs. (4.179) and (4.180), one obtains

$$\frac{\partial c}{\partial t} = \frac{\partial \tilde{c}}{\partial x^*}\frac{\partial x^*}{\partial t} + \frac{\partial \tilde{c}}{\partial t} = -v\frac{\partial \tilde{c}}{\partial x^*} + \frac{\partial \tilde{c}}{\partial t}. \tag{4.182}$$

Substituting Eq. (4.179) into Eq. (4.174), yields

$$D^{\#}\frac{\partial^2 \tilde{c}}{\partial x^{*2}} = \frac{\partial \tilde{c}}{\partial t}. \tag{4.183}$$

Boundary and initial conditions are consequently modified to

$$\begin{cases} \tilde{c}(x^*,0) = 0 \\ \tilde{c}(x^* \to 0, t \to 0) = \dfrac{c_0\delta(t)}{v^\#} \\ \tilde{c}(x^* \to \infty, t) = 0 \end{cases} \tag{4.184}$$

where $v^\#$ is the velocity of the front and $v^\# = x^*/t$.

Applying a Laplace transform to Eq. (4.183) and omitting the tilde over the c, the following ordinary differential equation results:

$$\frac{d^2\bar{c}}{dx^{*2}} - \frac{1}{D^\#}s\bar{c} = 0. \tag{4.185}$$

The following boundary conditions are applied in the Laplace domain:

$$\begin{cases} \bar{c}(x^* \to \infty, s) = 0 \\ \bar{c}(x^* \to 0, s) = \frac{c_0}{v^\#} \end{cases} \tag{4.186}$$

where it should be noted that the front velocity is considered a constant in the present case. Solution of Eq. (4.185) can be given as

$$\bar{c}(x^*, s) = g_1^\# \exp(\gamma x^*) + g_2^\# \exp(-\gamma^* x^*) \tag{4.187}$$

where $\gamma^* = \sqrt{\frac{s}{D^\#}}$.

From the boundary condition in Eq. (4.186), it is known that $g_1^\# = 0$, and

$$\bar{c}(x^* \to 0, s) = \frac{c_0}{v^\#}. \tag{4.188}$$

As a result,

$$\bar{c}(x^*, s) = \frac{c_0}{v^\#} \exp(-\gamma x^*). \tag{4.189}$$

Again, assuming that the front velocity $v^\#$ is approximately a constant value, the following Laplace inversion can be performed on Eq. (4.189)(Luckner and Schestakow 1991):

$$L[\Upsilon \exp(-\varpi\sqrt{s})] = \frac{\Upsilon\varpi}{2t\sqrt{\pi t}} \exp\left(-\frac{\varpi^2}{4t}\right) \tag{4.190}$$

where Υ and ϖ are parameters.

Recalling the definition of the front velocity as $v^\# = \frac{x^*}{t}$, the concentration in real space can be obtained as

$$c(x^*, t) = \frac{c_0}{2\sqrt{\pi t D^\#}} \exp\left(-\frac{x^{*2}}{4tD^\#}\right). \tag{4.191}$$

Substituting the relation in Eq. (4.180), Eq. (4.191) can be rewritten as

$$\frac{c}{c_0} = \frac{1}{\sqrt{4\pi t D^{\#}}} \exp\left[-\frac{(x - vt)^2}{4tD^{\#}}\right]. \tag{4.192}$$

Considering a process in which a particle is traveling in discrete steps in the x-direction, it is known from the theory of stochastic processes that the distribution function for the probability of traveling a certain distance after a great number of independent steps is *Gaussian*, as described in Eq. (4.173). This *Gaussian* or normal distribution function can be written as (Bear 1993):

$$\Theta(x) = \frac{1}{\sqrt{2\pi\chi_2}} \exp\left[-\frac{(x - \chi_1)^2}{2\chi}\right] \tag{4.193}$$

where

$$\begin{cases} \chi_1 = N^{\#} m_d \\ \chi_2 = N^{\#} \varsigma^2 \end{cases} \tag{4.194}$$

where m_d is the average distance covered in each step, which is commonly known as the *mean*; ς is the standard deviation, a measure of the spreading; $N^{\#}$ is the total number of steps in which each step may be considered to consist of a deterministic part of magnitude χ_1, and random part of maximum magnitude χ_2.

Comparing Eq. (4.192) with Eq. (4.193), it is noted that they are in the same form and become identical if

$$\chi_1 = vt, \quad \chi_2 = 2tD^{\#}. \tag{4.195}$$

The consequent parametric relationships relate χ_1 to the effect of convection and χ_2 to that of dispersion. As a result, the random walk process can be considered to be a representation of transport by both dispersion and convection. The simulation attempts to describe the random walk process for a single particle. To simulate dispersion, the process has to be repeated for a large number of particles. In fact, the use of a random process in the description of dispersive transport can be envisioned to be representative of the physical background of dispersion due to the random characteristics of the structure of the porous medium.

For pulse-type injection at a point x_0, an accumulated number of elements are allowed to "walk" simultaneously over the time interval Δt, and over the fixed distance $x_1 = vt$ via convection (similar to the "method of characteristics"), along with an additional random distance via dispersion $x_2 = \varsigma = \sqrt{2D^{\#}t}$. This scenario becomes apparent from the relationships given in Eqs. (4.194) and (4.195), and can be expressed as

$$x(\Delta t) = x_0 + x_1 + x_2 = x_0 + vt + \sqrt{2D^{\#}t}. \tag{4.196}$$

4.3 PARAMETRIC STUDY

Discussion on certain transport parameters, sensitivity analysis, and convection-dominated transport are the focus of this section.

4.3.1 Parameters for Homogeneous Media

The purpose of a transport model is to enable the simulation of solute transport and to compute the concentration of either a dissolved or a reactive chemical species at any specified place and time in the subsurface. Discussion of reactive processes is beyond the scope of this treatise.

Considering the balance of the net rate changes in mass and mass flux changes in the REV, as well as the variations of mass due to chemical reactions within the REV, the conservation of mass in solute transport can be generally expressed as (Bear 1972; Bear and Verruijt 1987)

$$\frac{\partial m}{\partial t} = \nabla\cdot\mathbf{J}_t \pm R^* \pm \lambda^* m \tag{4.197}$$

where m is the total mass of solute (absorbed and liquid phase) per unit volume, λ^* is a general rate constant of such reactions as radioactive and/or biological decay, and R_g is a general term of all chemical and biological reactions. A positive sign indicates sources while a negative signs denote sinks. $\mathbf{J}_t$ represents the total mass flux, which contains all the major physical, chemical, and mechanical concepts of the transport mechanisms.

In general, $\mathbf{J}_t$ consists of three basic components:

$$\mathbf{J}_t = \mathbf{J}_t^m + \mathbf{J}_t^d + \mathbf{J}_t^c \tag{4.198}$$

where the superscripts m, d, and c denote mechanical dispersion, diffusion, and convection, respectively.

Specifically, Eq. (4.198) can be expressed as

$$\mathbf{J}_t = \tilde{D}^m n\vec{\nabla}c + \tilde{D}^d n\vec{\nabla}c - \mathbf{v}^* c \tag{4.199}$$

where $\mathbf{v}^*$ is a vector defining the intrinsic flow velocity, and $\tilde{D}^m$ and $\tilde{D}^d$ are the tensors of the mechanical dispersion coefficient and of diffusion coefficient, respectively. Since the formats of the first and second terms on the right-hand side of Eq. (4.199) are very similar, it is customary to express these terms together as

$$\mathbf{J}_t = \mathbf{J}_t^h + \mathbf{J}_t^c = \tilde{D}^h n\vec{\nabla}c - \mathbf{v}^* c \tag{4.200}$$

where $\tilde{D}^h$ is the tensor of the hydrodynamic dispersion coefficient, written as

$$\tilde{D}^h = \tilde{D}^m + \tilde{D}^d. \tag{4.201}$$

Taking the divergence with respect to $\mathbf{J}_t$ in Eq. (4.199), and substituting the result into Eq. (4.197), the traditional dispersion-convection transport equation [i.e., Eq. (4.25)] is obtained.

Mechanical dispersion, also known as "mixing", represents the nonsteady, irreversible solute spreading process. The term "mechanical" refers to the external influence, i.e., seepage flow, which is often considered as an external force. In contrast, convection describes spreading as a result of velocity variations.

The mechanical dispersion tensor for isotropic media may be expressed as

$$D_{ij}^m = \alpha_t \mathbf{v} \delta_{ij} + (\alpha_l - \alpha_t) \frac{\mathbf{v}_i \mathbf{v}_j}{\mathbf{v}}$$

where α_l and α_t are the longitudinal and transverse dispersivities, respectively; $\mathbf{v}$ is the vector of flow velocity; and $\mathbf{v}_i$ and $\mathbf{v}_j$ are the components of flow velocities. Where the coordinate directions are aligned parallel and perpendicular to the principal flow orientations, the matrix form of $\tilde{D}_{ij}^m$ can be given as

$$D_{ij}^m = \begin{bmatrix} \alpha_l \mathbf{v} & 0 & 0 \\ 0 & \alpha_t \mathbf{v} & 0 \\ 0 & 0 & \alpha_t \mathbf{v} \end{bmatrix}. \tag{4.202}$$

Conversely, molecular diffusion occurs in the absence of flow, which is also unsteady, but more significant at low flow velocities. Diffusive phenomena are usually embedded in the dispersion process, but are separate from the convection process. As a result of their significantly different behaviors, the dominance of one respective process frequently leads to very different transport consequences. Molecular diffusion ameliorates variations in tracer concentration within the liquid phase. Molecular diffusion produces an additional flux of tracer. The tensor representing the diffusion coefficient can be written as

$$D_{ij}^d = D^d \tau_{ij}^{\#} \tag{4.203}$$

where D^d is the coefficient of molecular diffusion and $\tau_{ij}^{\#}$ is the tortuosity tensor. Although the actual driving force for diffusive transport is the gradient in chemical potential of the solute, a hydraulic gradient is not required for transport of the contaminant.

Mechanical dispersion and molecular diffusion occur simultaneously, and their combined effects are called "hydrodynamic dispersion." Convection is the process whereby solutes are transported along with the flowing fluid or solvent, such as water, in response to a gradient in total hydraulic head, which is different from the concentration controlled dispersive process. In Eq. (4.32), it is known that when the impact of flow velocity is much larger than that of hydrodynamic dispersion, the equation degenerates into the following form:

$$-v \frac{\partial^2 c}{\partial x} = \frac{\partial c}{\partial t}. \tag{4.204}$$

This convection-driven transport is controlled by the regional flow velocity, rather than by Fick's dispersive principle. Mathematically, Eq. (4.204) is a first-order hyperbolic equation.

On the other hand, if the transport is a diffusion-controlled process in which the effect of flow velocity is negligible, Eq. (4.32) can be approximated as

$$D\frac{\partial^2 c}{\partial x^2} = \frac{\partial c}{\partial t}. \tag{4.205}$$

Fickian diffusion is the dominant process in Eq. (4.205). Mathematically, it is a second-order parabolic equation, which carries the same form as the classical diffusion equation.

Applying the same dimensionless terms as defined in Eq. (4.35), Eq. (4.32) is rewritten as Eq. (4.36) where the equivalent Peclet number, γ^*, is defined as the ratio of flow convection (velocity) to hydrodynamic dispersion, multiplied by the length of the transport domain, L^*. The range of this equivalent Peclet number represents the spectrum from dispersion-dominated to convection-dominated transport processes. In view of the formulation of Eq. (4.36), Eqs. (4.204) and (4.205) are two limiting cases.

The original Peclet number, as defined by Bear (1972), is written as

$$Pe = \frac{lv}{D^d} \tag{4.206}$$

where l is a characteristic length that may be defined as (Bear 1972): (a) an individual channel length for fractured media, (b) a pore length for porous media, (c) mean grain size, or (d) other characteristic medium length; D^d is the molecular diffusion. Bear (1972) considered the Peclet number to be a microscopic parameter, which characterizes transport at the pore scale. However, this definition has very little use in engineering applications, where the characteristic length must be specifically defined at the field scale, and the diffusion coefficient should be replaced by the dispersion coefficient to cover a broader spectrum of processes.

4.3.2 Sensitivity Analysis for Heterogeneous Media

The aim of developing dual-porosity or multi-porosity models is to replicate the behavior of structured porous media that have multiple characteristic scales, while providing a more flexible tool in predicting and matching actual measurements. As defined in the dimensionless groups of Eqs. (4.35), (4.126), and (4.145), principal parameters of a triple-porosity model [Eqs. (4.146) - (4.151)] for a designated location x_d and time t_d are: (a) equivalent Peclet number (EPN) γ_i^* ($i = 1, 2$) for macro- and mesopores; (b) solute exchange intensity factor a_{ij}^* ($ij = 12, 21, 23, 32$); (c) flow velocity ratio b_i^* ($i = 1, 2$); and (d) equivalent sorption intensity factor η_i^* (i=1,2,3). The dimensionless initial solute concentrations $c_i^* = c_i^0/c^0$ and the process factors ι_i^* (i=1,2,3) are predetermined for all porous phases as $c_1^* = 0.1$, $c_2^* = 0.3$, $c_3^* = 0.5$, $\iota_1^* = 0.1$, $\iota_2^* = 0.2$ and $\iota_3^* = 0.3$.

All parameters that are changed among the various figures are listed in Table 4.2, while the invariant parameters for all the figures are: $\gamma_2^* = 10$, $a_{21}^* = 0.4$, $\varsigma_2^{*0} = 0.01$, $\varsigma_3^{*0} = 0.001$, in which the dimensionless solid-phase concentrations are defined as $\varsigma_i^{\#} = \varsigma_i^{*0}/c^0$. In the figures illustrated, all concentrations are normalized by the maximum value at the injection point and are referred to those of macropores. The analysis is limited to the study of temporal concentration changes (breakthrough) at a specified location ($x_d = 0.5$).

Table 4.2. Selected dimensionless parameters.

Figure	γ_1^*	a_{12}^*	a_{23}^*	a_{32}^*	b_2^*	η_1^*	η_2^*	η_3^*	ς_1^{*0}
4.5	.1-100	0.5	0.3	0.2	2	0.001	0.01	0.1	0.1
4.6	1-20	0	0.3	0.2	2	0	0.01	0.1	0
4.7	20	0.5	0.3	0.2	2-50	0.001	0.01	0.1	0.1
4.8	20	0.5	0.3	0.2	2	0.001-0.5	0.01-0.6	0.1-0.7	0.1
4.9	20	0.5	5	4	2	0.001	0.01	0.1	0.1

Determination of sensible parametric ranges for Table 4.2 are obtained from: (a) previous experimental data, (b) related literature search, and (c) physical intuition and judgment. Even though the grouped dimensionless parameters are used, reasonable magnitudes of original parameters still need to be considered and accommodated because the variation of one parameter may affect other parameters due to the chain relationships, shown in the dimensionless expressions. Certain general rules may be derived from the literature, in view of the parametric ranges.

Among all parameters, γ_i^* may be the most important to be defined since it relates to the reverse relationship between the flow velocity and hydrodynamic dispersion. For the case of the "capacitance" concept, γ_i^* for the macropores is typically in the range 50 to 780 for the experimental setup by Coats and Smith (1964). However, the span of this range may drop to as low as 10 to 18 as evident in the test data of Bouhroum and Bai (1996). The breakthrough profiles for the limiting cases of γ^* in a single-porosity scenario can be obtained from Passioura (1971). In view of the relative magnitudes of velocity v_i and dispersion D_i (i=1,2,3) for each porous phase, the general rules of $v_1 > v_2 > v_3$ and $D_1 < D_2 < D_3$ for the triple-porosity model can be found in Gwo et al. (1995). Applying the definition that $\gamma_i^* = \frac{v_i L^*}{D_i}$, it appears that $\gamma_1^* > \gamma_2^*$ when L^* is a constant for the present case. However, these rules are not universal, and exceptions should be permitted (e.g., $D_1 > D_2 > D_3$; also see Gwo et al. 1995). Due to the greater flow cross-sectional area for the larger pore phase, $v_1 > v_2$ (i.e., $b_2^* > b_1^*$) for dual-porosity situations may be commonly defined (Gerke and van Genuchten 1993).

With regard to the solute exchange factor a_{ij}^*, a broad range is defined as 0.017 to 1.54 by Coats and Smith (1964), while this range is narrowed to 0.34 to 0.59 by Bouhroum and Bai (1996). For the dimensional parameter ξ_{ij}, which is a concentration exchange coefficient, it is known that $\xi_{12} > \xi_{23}$ (Gwo et al. 1995). Using physical intuition and considering the principle of mass balance, the following equality results: $\xi_{12} = \xi_{21}$ and $\xi_{23} = \xi_{32}$. Based on the literature (Warren and Root 1963; Bai et al. 1993), higher conductivity media, in general, occupy a smaller volume (i.e., have smaller overall porosity), one has $n_3 > n_2 > n_1$. As a result, and recalling that $a_{ij}^* = \frac{\xi_{ij} L^*}{n_i v_1}$, the following relationship becomes a natural consequence when L^* and v_1 are held constant: $a_{12}^* > a_{21}^* > a_{23}^* > a_{32}^*$. Again, exceptions do exist.

Fewer references may be made to the parameters representing sorption in multi-porosity media. However, Mannhardt and Nasr-El-Din (1994) and Ogata (1964) indicated that sorption mechanisms act as a supplemental form of the capacitance effect in mobile and immobile zones proposed by Coats and Smith (1964). The parametric ranges for sorption are chosen from the literature and based on physical intuition.

The following sensitivity study involves the change of one parameter at a time, while all other parameters are held constant.

Figure 4-5 depicts the effects of increasing the equivalent Peclet number (EPN) of macropores (γ_1^*) on the temporal variations of concentration. As γ_1^* increases, the only possible accompanying phenomenon is the decreasing of D_1, since other parameters are held constant. According to the traditional concept in a single-porosity medium, decreasing dispersion would lead to truncated tailing and late breakthrough. This seems to be the only case in Figure 4-5, when γ_1^* changes from 1 to 10. The remaining cases contradict the normal observation. In other words, a decrease of γ_1^* results in an increase in tailing. This unusual phenomenon is the result of the "capacitance" effect, as expounded by Mannhardt and Nasr-El-Din (1994). Indeed, an enhanced solute exchange between macropores and mesopores would restrict the local solute concentration (at a point) in reaching its maximum value, and as such lead to extended tailing. This response is aggravated when the ratio of γ_1^* to γ_2^* becomes quite large (e.g., the ratio$=\frac{v_1 D_2}{v_2 D_1} = 10$) where the contribution of solute exchange becomes substantial. Interestingly, a change in the slope of the concentration response is noted for this case at the initial breakthrough. This behavior may reflect the local replenishment of mass from the mesopores to the macropores when the storage of the macropores becomes exhausted. This type of slope change, demonstrated as a fluctuation on the profiles of the breakthrough curves, was observed experimentally by Neretnieks (1993). This unusual dispersive response is difficult to decipher in transport modeling because the process may be overshadowed by simultaneous convective processes.

To further illustrate that the behavior of the single-porosity model is different from that of the triple-porosity model subjected to different EPN γ^*,

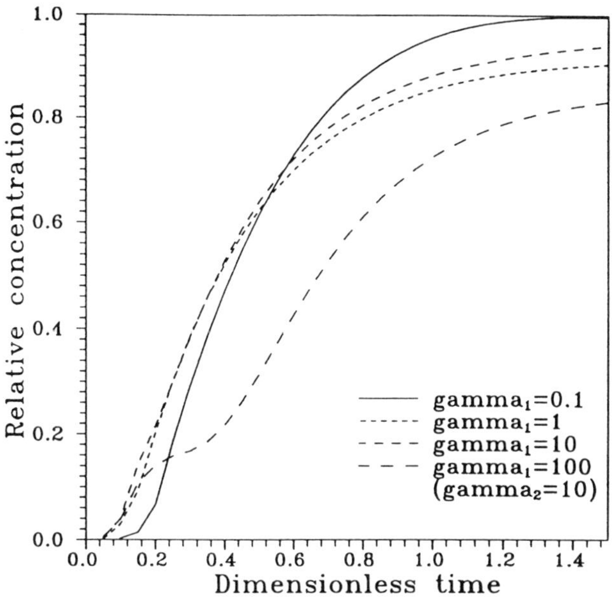

Figure 4-5. Breakthrough curves for various EPNs of macropores (γ_1^*). (Reprinted from Bai et al. 1997a, with permission from ASCE.)

Figure 4-6 depicts the breakthrough curves for the single-porosity model (refer to Table 4.2) with different γ^*. As expected in traditional transport modeling, a smaller γ^* corresponds to earlier breakthrough and tailing, while the opposite is true for the larger γ^*.

The velocity ratio b_i^* between the macropores and mesopores provides a reasonable benchmark to evaluate the impact of flow velocity alone. Figure 4-7 depicts significant differences in temporal concentration when the velocity contrast between macro- and mesopores increases. The smaller velocity ratio (b_2^*=2) signifies less dominant transport within the macropores, but a more significant influence from the mass exchange between macropores and mesopores, leading to an early breakthrough and extended tailing. With the increase of the velocity ratios (equivalent to the decrease of v_2 since v_1 is fixed), the dominance of macropore transport becomes increasingly apparent. This results in the progressively delayed but convection-controlled breakthrough curves as an indication of the increasing velocity contrast between the main flow region and the less permeable regions. Similar phenomena have been observed by McKibbin (1985), Houseworth (1988), and Bouhroum and Bai (1996). The velocity contrast is attributed to the permeability contrast between the different pore phases. The convective component becomes dominant as the permeability contrast increases.

The previous calculation does not consider the strong effects from the sorp-

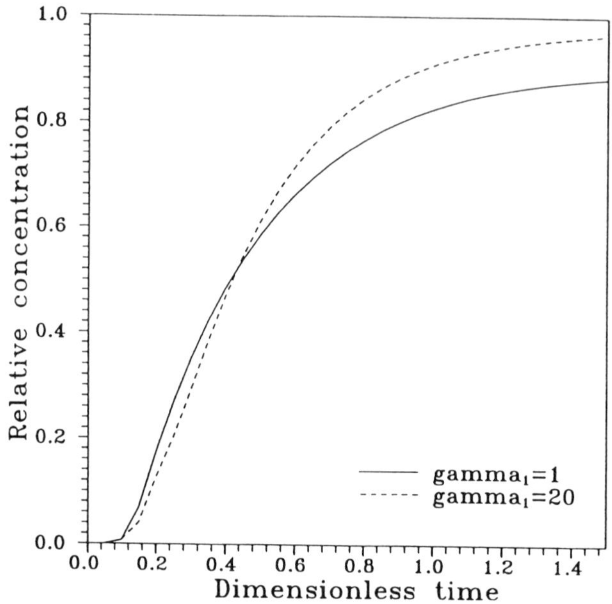

Figure 4-6. Breakthrough curves for two EPNs of macropores (γ_1). (Reprinted from Bai et al. 1997a, with permission from ASCE.)

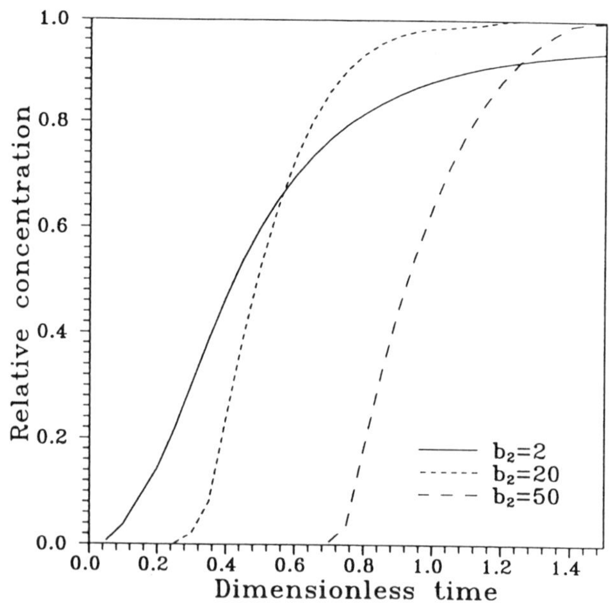

Figure 4-7. Breakthrough curves for various velocity ratios (b_i^*). (Reprinted from Bai et al. 1997a, with permission from ASCE.)

tion process because relatively small equivalent sorption intensity factors (ESIFs) are used. Figure 4-8 reveals the significant differences in temporal concentrations with the increased magnitudes of ESIF in the macropores, especially during the late evolution of the breakthrough curves. Different from the impact of interporosity mass exchange due to the increase in macropore adsorption, this retardation process results in progressive tailing without the exhibition of early abrupt breakthrough. Even though the solute exchange between the various porous spaces also creates retardation in the concentration changes, the differences between these two processes are that the retardation by fluid-solid sorption occurs primarily in the individual porous space (e.g., macropores), while the retardation by interporosity mass exchange occurs in the multiple porous spaces simultaneously. Consequently, the former process results in direct concentration modification, such as a reduction in concentration. Ogata (1964) verified the retardation effect by sorption.

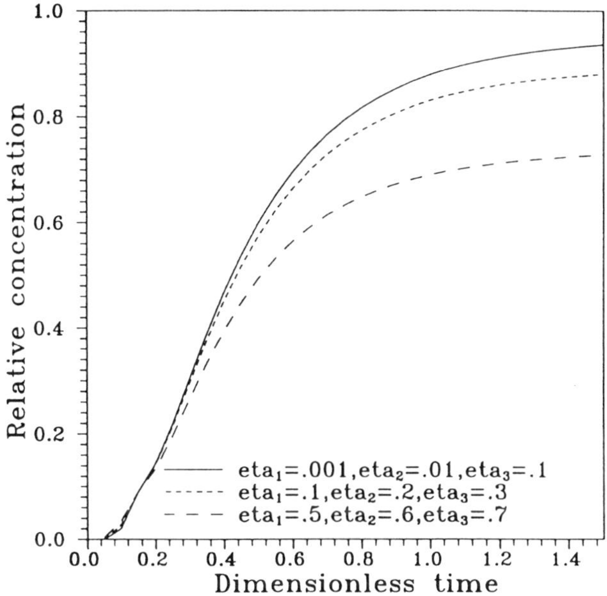

Figure 4-8. Effect of adsorption on temporal concentration. (Reprinted from Bai et al. 1997a, with permission from ASCE.)

4.3.3 Convection-Dominated Transport

Convection-dominated solute transport in the subsurface is an important phenomenon where, for example, fracture dominated flows become important in addressing wellhead protection concerns. Significant numerical difficulties are encountered where convection dominates the flow response. The problem arises in representing a convective phenomenon with a static mesh.

4.3.3.1 Problem Definition

Mathematically, the problem is a consequence of instability due to convective domination of the first order hyperbolic equation. The problem can be circumvented by applying any of the following methods: (a) method of characteristics (Huyakorn and Pinder 1983), (b) Eulerian-Lagrangian moving coordinates (Zhang et al. 1993), (c) function transformation (Bai et al. 1994d), and (d) random walk method (Bear and Verruijt 1987). The common nature of these methods is to eliminate the convective term and to transform the transport equation into a more stable parabolic type of equation. Solution of the dispersion-convection equation [e.g., Eq. (4.32)], is particularly challenging as Peclet number, indexing the ratio of convective to diffusive fluxes, increases. At high Peclet numbers, one is usually forced to choose between accepting the presence of nonphysical oscillations within the solution or suffering unwanted numerical dispersion. Of key importance is awareness of the changing nature of the governing equation. Where dispersion dominates [e.g., Eq. (4.205)], the equation is parabolic and causes no particular problem in numerical solution. Where convection dominates [e.g., Eq. (4.204)], the behavior is analogous to a first-order hyperbolic partial differential equation that exhibits a frontal character and creates difficulties in its numerical solution.

Applying the dimensionless concentration and changing only the definition for the dimensionless time t_d in Eq. (4.35) as

$$\left\{\begin{array}{l} c_d = \dfrac{c(x,t)}{c(0,t)} \\ \\ t_d^* = \dfrac{Dt}{L^{*2}} \end{array}\right. \tag{4.207}$$

Eq. (4.36) can be reformulated as

$$\frac{\partial^2 c}{\partial x_d^2} - \gamma^* \frac{\partial c}{\partial x_d} = \frac{\partial c}{\partial t_d^*} \tag{4.208}$$

with the following initial and boundary conditions:

$$\left\{\begin{array}{ll} c(x_d, 0) = 0 & \\ c(0, t_d^*) = 1 & \\ c(L^*, t_d^*) = 0 & \text{(for prescribed concentration)} \\ \dfrac{\partial c(L^*, t_d^*)}{\partial x_d} = 0 & \text{(for prescribed flux)} \end{array}\right. \tag{4.209}$$

Any attempts to solve Eq. (4.208) for a large equivalent Peclet number (EPN),

γ^*, by any numerical technique lead to oscillatory results (Gladwell and Wait 1979).

4.3.3.2 Alternative Methodologies

The most successful technique in eliminating numerical oscillations may, however, be attributed to the application of the upwinding method (Christie et al. 1976) modified from the finite difference iteration. The method is also referred to as upstream weighting (Huyakorn and Pinder 1983). However, despite the utility of this method in reducing spurious oscillations, excessive smearing or numerical dispersion is arbitrarily added to the solutions. Noorishad et al. (1992) provided a review of the effect of the upwind method.

Results for higher dimensional elements in two and three dimensions, where low element continuity is maintained, results in no net improvement (Heinrich et al. 1977). Alternatively, higher order elements using cubic or bicubic Hermitian interpolating functions, together with collocation finite element methods, may concurrently minimize oscillation and smearing (Mohsen 1984; Pinder and Shapiro 1979; van Genuchten and Pinder 1977). However, the computational costs incurred in using higher order elements are high and formulation of the problem often turns out to be cumbersome.

By examining Eq. (4.208), it is known that the first term on the left-hand side of the equation is symmetric in nature, which generally does not induce stability problems in numerical schemes. However, the asymmetry of the numerical formulation, which is reported to cause stability problems, is due to the existence of the other two terms. In consequence, alternative methodologies for minimizing numerical dispersion and oscillation are frequently divided into two groups according to the negative impact due to the existence of: (a) the first term on the right-hand side of Eq. (4.208), and (b) the second term on the left-hand side of Eq. (4.208), respectively. The former is related to the temporal treatment, while the latter is associated with the spatial treatment. The latter method is more popular because it is more effective in reducing the numerical dispersion and oscillation. Occasionally, methods are developed for reducing the negative impacts of both terms in Eq. (4.208) by combined temporal and spatial treatments.

Temporal Treatment

Leismann and Frind (1989) attempted to achieve matrix symmetry by placing the convective term in the previous time level in time marching. The resulting numerical errors are minimized by introducing an artificial dispersion term and by optimal time weighting of all terms on the basis of a Taylor expansion of the governing equation.

Sudicky (1989) indicated that improper selection of a time step size could lead to artificial smearing (i.e., numerical dispersion) or oscillations in the solution. He used Laplace transformation to eliminate the temporal derivative term from the dispersion-convection equation and solved the ordinary differential equation in the Laplace domain using the conventional Galerkin finite element technique.

Spatial Treatment

In general, the spatial treatment is divided into two groups: moving coordinate methods and function transformation methods.

Moving Coordinate Methods: This method converts a dispersion-convection equation into a hyperbolic-type equation in order to achieve numerical stability (Zhang et al. 1993). Assuming that the medium is fully saturated and that retardation is negligible, the governing equation proposed by Zhang et al. (1993) collapses to the same format as in Eq. (4.208). The classical dispersion-convection transport equation is evaluated in the fixed Eulerian coordinate system.

Using the concept of the total derivative (or Lagrangian derivative), one has

$$\frac{dc}{dt_d^*} = \frac{\partial c}{\partial t_d^*} + \gamma^* \frac{\partial c}{\partial x_d}. \tag{4.210}$$

Substituting Eq. (4.210) into Eq. (4.208), a parabolic-like equation is formed:

$$D\frac{\partial^2 c}{\partial x_d} = \frac{dc}{dt_d^*}. \tag{4.211}$$

The concentration c in Eq. (4.211) no longer represents the concentration at a point in space and in time, but rather the concentration of a fluid particle moving along the characteristic path described by the following equation:

$$\gamma^* = \frac{dx}{dt}. \tag{4.212}$$

This approach is similar in concept and procedure to the "random walk method" in Section 4.2.4 describing stochastic processes, however, with a different explanation. In Eq. (4.211), a Lagrangian moving coordinate system, defined in Eq. (4.210), is used to replace the traditional fixed Eulerian coordinate system. As noted by Zhang et al. (1993), the Lagrangian formulation eliminates the convective term so the governing equation takes on a parabolic format that can be solved more efficiently with a finite element method. The final results are converted back to the Eulerian coordinate system through Eq. (4.210).

In other numerical schemes related to the moving coordinate method, "adaptive grid method" (Hu and Schiesser 1981), "adaptive characteristics method" (Ouyang and Elsworth 1989) and "moving grid method" (Gottardi and Venutelli 1994) are all similar methods to refine the mesh at the steep concentration front so as to minimize dispersion and oscillation, in comparison with traditional schemes.

Function Transformation Method: Functional transformation is an effective way to convert a parabolic-hyperbolic type equation into a more stable parabolic-type equation. However, the trade-off of this transformation is the result of solving a more challenging time-dependent boundary problem.

- *Method of Ogata and Banks*

  Ogata and Banks (1961) derived an analytical solution of the classic dispersion and convection model of one-dimensional transport using function transformation. This technique has been adopted in solving convection-dominated transport equations.

  Assuming that the concentration is a function of an exponential function, as (Ogata and Banks 1961)

$$c = \Lambda_o(x,t)\exp\left(\frac{vx}{2D} - \frac{v^2t}{4D}\right) \tag{4.213}$$

  where $\Lambda_o(x,t)$ is a transformation function, then substituting Eq. (4.213) into Eq. (4.32), gives a parabolic-type equation as

$$D\frac{\partial^2\Lambda_o}{\partial x^2} = \frac{\partial \Lambda_o}{\partial t} \tag{4.214}$$

  with the following initial and boundary conditions:

$$\begin{cases} \Lambda_o = 0 \quad t = 0 \\ \Lambda_o = c_0 \exp\left(\frac{v^2t}{4D}\right) \quad x = 0 \\ \Lambda_o = 0 \quad x = \infty. \end{cases} \tag{4.215}$$

  It may be noted that the concentration at the inlet boundary becomes time-dependent. Applying Duhamel's theorem (Carslaw and Jaeger 1959) and Laplace transforms to Eqs. (4.214) and (4.215), the following well-known analytical solution is obtained after a variety of clever but tedious analytical maneuvers:

$$c = \frac{c_0}{2}\left[\text{erfc}\left(\frac{x-vt}{2\sqrt{Dt}}\right) + \exp(\frac{vx}{D})\text{erfc}\left(\frac{x+vt}{2\sqrt{Dt}}\right)\right] \tag{4.216}$$

  where "erfc" is the complementary error function, which is related to the error function "erf" through $\text{erfc}(x) = 1 - \text{erf}(x)$.

- *Method of Guymon*

  In solving the convection-dominated transport equation, the presence of numerical oscillations was first recognized by Price et al. (1968). To circumvent this difficulty, Guymon (1970) applied the following functional transformation:

$$c = \Lambda_g(x,t)\exp\left(-\frac{vx}{2D}\right) \tag{4.217}$$

  where $\Lambda_g(x,t)$ is also a transformation function.

  Substituting Eq. (4.217) into (4.32), yields

$$D\frac{\partial^2 \Lambda_g}{\partial x^2} - \frac{v^2 \Lambda_g}{4D} = \frac{\partial \Lambda_g}{\partial t}. \tag{4.218}$$

  In Eq. (4.218), the first-order derivative term with respect to x, which is problematic in representing convection-dominated transport, is eliminated.

  The initial and boundary conditions are transformed to

$$\begin{cases} \Lambda_g = 0 \quad t = 0 \\ \Lambda_g = c_0 \quad x = 0 \\ \Lambda_g = c_l \exp\left(\frac{v^2 t}{4D}\right) \quad x = L^* \end{cases} \tag{4.219}$$

  where c_l is the concentration value at the outlet.

  For zero concentration at the outlet, the outlet boundary condition is simplified to $\Lambda_g = 0$. As a result, the boundary conditions become time-independent. Under such conditions, Guymon's method is simpler than Ogata and Banks' method. However, Guymon's method is not effective in reducing numerical instability at higher Peclet numbers (Guymon 1970).

- *Method of Bai et al.*

  The problems associated with the convection-dominated transport equation are, in general, attributed to the dominance of convection over dispersion. In general, this problem exists for both steady state and transient behavior. Bai et al. (1994d) proposed the following procedure.

  For steady state transport, Eq. (4.208) reduces to

$$\frac{\partial^2 c}{\partial x_d^2} - \gamma^* \frac{\partial c}{\partial x_d} = 0. \tag{4.220}$$

  Assuming

$$\gamma^* = \frac{\partial \Lambda_b}{\Lambda_b \partial x_d} \tag{4.221}$$

where $\Lambda_b(x)$ is an arbitrary function. Eq. (4.221) enables Λ_b to be defined from

$$\Lambda_b = \Lambda_b^0 \exp(\gamma^* x_d) \tag{4.222}$$

where Λ_b^0 is the function Λ_b at its initial value. Similarly, an analog to Eq. (4.220) may be defined as

$$\frac{\partial}{\Lambda_b \partial x_d}\left(\Lambda_b \frac{\partial \Lambda_b^*}{\partial x_d}\right) = \frac{1}{\Lambda_b}\frac{\partial \Lambda_b}{\partial x_d}\frac{\partial \Lambda_b^*}{\partial x_d} + \frac{\partial^2 \Lambda_b^*}{\partial x_d^2} \tag{4.223}$$

where Λ_b^* is another arbitrary function.

From the development of Eqs. (4.221), (4.222), and (4.223), it is amenable to re-write Eq. (4.220) as

$$\frac{\partial}{\partial x_d}\left\{\exp(-\gamma^* x_d)\frac{\partial c}{\partial x_d}\right\} = 0. \tag{4.224}$$

Eq. (4.224) yields the exact form of Eq. (4.220). Since the boundary and initial conditions are unchanged from Eq. (4.209), analytical solution of (4.224) may be easily obtained for the prescribed outlet concentration boundary condition as

$$c = \frac{1}{1 - \exp(-\gamma^*)} + \frac{\exp(\gamma^* x_d)}{1 - \exp(\gamma^*)}. \tag{4.225}$$

For modeling steady state solute transport, the method developed by Bai et al. (1994d) provided more accurate solution in comparison with the analytical solution (van Genuchten 1982) than the Galerkin finite element method (Fletcher 1984) and the upwind weighting method (Huyakorn and Pinder 1983), even at high Peclet numbers.

Chapter 5

ANALYTICAL SOLUTION

5.1 INTRODUCTION

A mathematical model is a replica of some real-world object or system. It is an attempt to take our understanding of the conceptual process and translate it into mathematical terms. There are many basic ways to solve mathematical equations. Analytical methods represent classical approaches to solve the particular system equations, sometimes in closed form, and sometimes requiring that the final simplified equations are solved numerically. Even though typically constrained to reduced spatial dimensions, simple boundary geometries and initial conditions, analytical methods serve as effective means for preliminary simulation, sensitivity analysis, and benchmark study for numerical validations, due primarily to their convenience and ease of application. In reality, however, closed form analytical solutions are difficult to obtain, and therefore are rare in their application to coupled problems. However, semi-analytical solutions in which numerical inversion is used are useful and popular in the solution of partial differential equations.

Many solutions have already been presented using analytical means, such as function transformations. However, these solutions were provided only as illustrations of some basic or well-known approaches and corresponding results. In contrast, this chapter introduces some popular analytical solution techniques relevant to solving coupled processes. Specifically, three different function transformation techniques, together with a method of differential operators, are presented.

5.2 LAPLACE TRANSFORM

As previously demonstrated, Laplace transform is a powerful and broadly used method for the solution of partial differential equations associated with transient fluid flow and transport phenomena. The technique removes the time derivative through transformation, rendering partial differential equations as ordinary differential equations. The predominant application of Laplace transforms is in the solution of decoupled or partially decoupled systems.

5.2.1 Flow

For simulation using the dual-porosity concept, Warren and Root's (1963) model may be solved by Laplace transforms only because the fluid pressures in the matrix and the fractures are decoupled in Laplace space. The following evaluations are limited to fluid flow in nondeformable fractured porous media.

5.2.1.1 Solution Method

As an alternative, Barenblatt et al. (1960, 1990) proposed a more complete dual-porosity formulation than those expressed in Chapter 3 by Eqs. (3.83) and (3.84) through considering the cross-phase storage interaction

$$\frac{k_1}{\mu}\nabla^2 p_1 = c_1^* n_1 \frac{\partial p_1}{\partial t} - c_{12}^* n_1 \frac{\partial p_2}{\partial t} + \Gamma(p_1 - p_2) \tag{5.1}$$

$$\frac{k_2}{\mu}\nabla^2 p_2 = c_2^* n_2 \frac{\partial p_2}{\partial t} - c_{21}^* n_2 \frac{\partial p_1}{\partial t} - \Gamma(p_1 - p_2) \tag{5.2}$$

where c_{12}^* and c_{21}^* are the cross-coefficients. The second terms on the right-hand side of Eqs. (5.1) and (5.2) were claimed to have an insignificant impact on fluid mass exchange, and therefore were omitted in the final formulation (Barenblatt et al. 1990), as shown in Eqs. (3.83) and (3.84) (Chapter 3). However, it is understood that the formulation of Eqs. (5.1) and (5.2) is based on the phenomenology of the system. A more rigorously derived counterpart to this equation, represented by Eqs. (3.124) and (3.125), can be rewritten in a similar form to Eqs. (5.1) and (5.2) as

$$\frac{k_1}{\mu}\nabla^2 p_1 = g_1 \frac{\partial p_1}{\partial t} - g_2 \frac{\partial p_2}{\partial t} + \Gamma(p_1 - p_2) \tag{5.3}$$

$$\frac{k_2}{\mu}\nabla^2 p_2 = g_3 \frac{\partial p_2}{\partial t} - g_4 \frac{\partial p_1}{\partial t} - \Gamma(p_1 - p_2) \tag{5.4}$$

where g_1, g_2, g_3, and g_4 correspond to c_{11}^*, c_{12}^*, c_{21}^* and c_{22}^*, respectively, defined in Eq. (3.137) (Chapter 3).

For convenience of comparison, the Warren and Root model in Chapter 3 is rewritten using terms represented in the following two equations, assuming the matrix-fracture flow maintains a quasi-steady state, as

$$0 = g_1 \frac{\partial p_1}{\partial t} + \Gamma(p_1 - p_2) \tag{5.5}$$

$$\frac{k_2}{\mu}\nabla^2 p_2 = g_3\frac{\partial p_2}{\partial t} - \Gamma(p_1 - p_2) \tag{5.6}$$

where the variables and constants have been defined previously.

In contrast, for the model represented by Eqs. (5.3) and (5.4) using the concept of quasi-steady matrix flow Eq. (5.3) should reduce to

$$0 = g_1\frac{\partial p_1}{\partial t} - g_2\frac{\partial p_2}{\partial t} + \Gamma(p_1 - p_2). \tag{5.7}$$

Assuming the following dimensionless quantities for the quasi-steady matrix flow:

$$\begin{cases} P_{Di} = \dfrac{2\pi k_2 h_r(p_0 - p_i)}{q\mu} \quad (i = 1, 2) \\ r_D = \dfrac{r}{r_w} \qquad \lambda_w = \dfrac{\Gamma\mu r_w^2}{k_2} \\ t_D = \dfrac{k_2 t}{\mu\sum\limits_{j=1}^{4} g_j r_w^2} \qquad \omega_j = \dfrac{g_j}{\sum\limits_{j=1}^{4} g_j} \quad (j = 1 \text{ to } 4) \end{cases} \tag{5.8}$$

where h_r is the reservoir thickness, p_0 is the initial reservoir pressure, q is the flow rate at the well, and r_w is the wellbore radius. For a block-type matrix, $\Gamma = [60k_1]/[\mu(s^*)^2]$ where s^* is the average fracture spacing.

More specifically,

$$\begin{cases} \sum\limits_{j=1}^{4} g_j = \dfrac{n_1 + n_2}{K_f} \\ \omega_1 = \dfrac{1}{n_1 + n_2}[(\alpha_3 - n_1 - n_2)\dfrac{K_f}{K_s} + n_1] \\ \omega_2 = -\dfrac{\alpha_3 - n_1 - n_2}{n_1 + n_2}\dfrac{K_f}{K_{fr}} \\ \omega_3 = \dfrac{1}{n_1 + n_2}[(\alpha_3 - n_1 - n_2)\dfrac{K_f}{K_{fr}} + n_2] \\ \omega_4 = -\dfrac{\alpha_3 - n_1 - n_2}{n_1 + n_2}\dfrac{K_f}{K_s} \end{cases} \tag{5.9}$$

where K_{fr}, K_s and K_f are the bulk moduli of fractures, solid grain, and fluid, respectively. Substituting the dimensionless terms given in Eq. (5.8) into the governing equations (5.7) and (5.4), the new formulation for the radial flow is

$$\lambda_w(P_{D2} - P_{D1}) = \omega_1\frac{\partial P_{D1}}{\partial t_D} - \omega_2\frac{\partial P_{D2}}{\partial t_D} \tag{5.10}$$

$$\frac{1}{r_D}\frac{\partial}{\partial r_D}(r_D\frac{\partial P_{D2}}{\partial r_D}) + \lambda_w(P_{D1} - P_{D2}) = \omega_3\frac{\partial P_{D2}}{\partial t_D} - \omega_4\frac{\partial P_{D1}}{\partial t_D}. \tag{5.11}$$

Using a Laplace transform to convert the partial differential equations (5.7) and (5.4) to ordinary differential equations, and solving these equations in Laplace space, gives

$$\frac{d^2\overline{P}_{D2}}{dr_D^2} + \frac{1}{r_D}\frac{d\overline{P}_{D2}}{dr_D} - sf(s)\overline{P}_{D2} = 0 \tag{5.12}$$

where s is a Laplace parameter, and

$$f(s) = \frac{(\omega_1 + \omega_3 - \omega_2 - \omega_4)\lambda_w + s(\omega_1\omega_3 - \omega_2\omega_4)}{s\omega_1 + \lambda_w}. \tag{5.13}$$

If the governing equations are substituted by Eqs. (5.5) and (5.6) of the Warren and Root (1963) model, then one has

$$f(s) = \frac{(\omega_1 + \omega_3)\lambda_w + s(\omega_1\omega_3)}{s\omega_1 + \lambda_w}. \tag{5.14}$$

Initially neglecting constants g_2 and g_4 in ω_1 and ω_3 of Eq. (5.14), then it can be easily verified that if in Eq. (5.14),

$$\omega_3 = \omega = \frac{n_2c_2^*}{n_1c_1^* + n_2c_2^*},$$

and $\omega_1 = 1 - \omega_3$, then Eq. (5.14) carries an identical form of $f(s)$ as in Warren and Root (1963).

It is understood that $f(s)$ in Eqs. (5.13) and in (5.14) of the Warren and Root (1963) model is solely responsible for the dual-porosity effect. For convenience of comparison, restoring the dummy constants g_2 and g_4 in ω_1 and ω_3 of Eq. (5.14), Eqs. (5.13) and (5.14) can be compared in a relative fashion as

$$\left|\frac{\Delta f(s)_{[Eq.\ (5.13)-Eq.\ (5.14)]}}{f(s)_{[Eq.\ (5.14)]}}\right| = \frac{(\omega_2 + \omega_4)\lambda_w + s(\omega_2\omega_4)}{(\omega_1 + \omega_3)\lambda_w + s(\omega_1\omega_3)}. \tag{5.15}$$

Eq. (5.15) represents the relative difference between the current and the traditional dual-porosity models, which depend on the magnitudes of ω_i ($i = 1, 2, 3, 4$). In the most extreme case, this difference can reach 100% if $\omega_1 = \omega_4$ and $\omega_2 = \omega_3$. As a result, it may be concluded that neglecting the second terms on the right-hand side of Eqs. (5.3) and (5.4) may not always be justifiable.

For the case of quasi-steady matrix flow, the comparison between the improved model and the Warren and Root (1963) model is made in the following.

The boundary and initial conditions for an infinite reservoir may be described as

$$\begin{cases} P_{D2}(\infty, t_D) = 0 \qquad \text{(outer boundary)} \\ \\ \dfrac{\partial P_{D2}}{\partial r_D}|_{r_D=1} = -1 \qquad \text{(inner boundary)} \\ \\ P_{D1}(r_D, t_D = 0) = P_{D2}(r_D, t_D = 0) = 0 \qquad \text{(initial condition)} \end{cases} \tag{5.16}$$

For quasi-steady flow within the porous matrix blocks, the fracture fluid pressure at the well can be obtained from Eq. (5.12) in Laplace space, using the modified Bessel functions, i.e.,

$$\overline{P}_{D2} = \frac{K_0(r_D\sqrt{sf(s)})}{s\sqrt{sf(s)}K_1(\sqrt{sf(s)})} \tag{5.17}$$

where $f(s)$ is shown in Eq. (5.13) for the improved model, and in Eq. (5.14) for the Warren and Root (1963) model.

For the late time approximation, Bessel functions can be simplified as $K_0(z) \approx \ln(z) - \ln 2 - 0.5772$ and $K_1(z) \approx \frac{1}{z}$. The fluid pressure in the fractures may be determined by inversion of Eq. (5.17) as

$$P_{D2} = \frac{1}{2}[\ln(t_D) + 0.80909 - \ln(\phi_1) + E_i(-\frac{\lambda_w \phi_1}{\phi_2}t_D) - E_i(-\frac{\lambda_w}{\omega_1}t_D)] \tag{5.18}$$

where

$$\begin{cases} \phi_1 = \omega_1 + \omega_3 - \omega_2 - \omega_4 \\ \\ \phi_2 = \omega_1\omega_3 - \omega_2\omega_4 \end{cases} \tag{5.19}$$

and

$$E_i(-x) = -\int_x^\infty \frac{e^{-u}}{u}du = 0.5772 + \ln(x) + \sum_{n=1}^{\infty}\frac{(-1)^n x^n}{n\, n!}. \tag{5.20}$$

In view of the spatial distribution of matrix-fracture interaction, where the fluid compressibility is also considered as one of the primary factors of influence, ϕ_1 and ϕ_2 may be given explicitly as

$$\phi_1 = 1 + \frac{2K_f(\alpha_3 - n_1 - n_2)}{n_1 + n_2}\left(\frac{1}{K_s} + \frac{1}{K_{fr}}\right) \tag{5.21}$$

and

$$\phi_2 = \frac{1}{(n_1+n_2)^2}\left[K_f(\alpha_3 - n_1 - n_2)\left(\frac{n_2}{K_s} + \frac{n_1}{K_{fr}}\right) + n_1 n_2\right] \tag{5.22}$$

where K_f is the fluid bulk modulus, K_s is the bulk modulus of the solid grains, and K_{fr} is the fracture bulk modulus.

It is interesting to examine a limiting case. For a slightly compressible fluid, or $K_f \ll K_s$, and $K_f \ll K_{fr}$, then $\phi_1 = 1$ in Eq. (5.21), so $\ln(\phi_1) = 0$;

as a result, P_{D2} in Eq. (5.18) holds a similar form to that suggested by Warren and Root (1963).

5.2.1.2 Illustrative Example

Assuming that $r_1 = K_f/K_{fr}$, $r_2 = K_f/K_s$ then Figure 5-1 is a comparison between Warren and Root's (1963) model and the solution described in Eq. (5.18), when λ_w in Eq. (5.8) is taken as 5×10^{-6}. Two intermediate results between Warren and Root's solutions ($\omega = 1$ and 0.01) and the solution in Eq. (5.18) are also compared for the combined ratios of $r_1 = 1$, $r_2 = 1$ and $r_1 = 1$, $r_2 = 20$, respectively. The straight line solution is approached when the ratio of the compressibility of the solid grains to the fluid is small ($r_2 = 1$). It is also interesting to note that when the ratios of compressibilities of the fluid to the solid grains and to the fractures are equal ($r_1 = 1$, $r_2 = 1$), the pressure-time relationship at large times does not merge with the solution for a homogeneous reservoir. The residual difference implies the inherently greater pressure drop in a fractured reservoir than in a homogeneous one under the similar prescribed conditions.

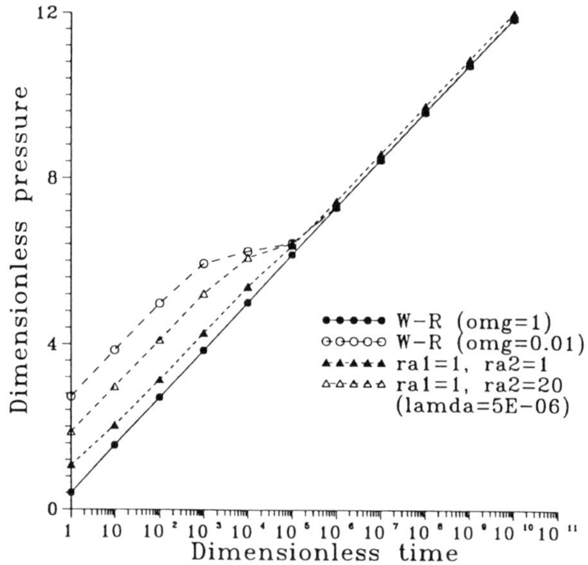

Figure 5-1. Comparison with Warren and Root's (1963) model.

5.2.2 Transport

Although numerous factors related to the capacitance effect have been investigated, analytical solutions are basically confined to the one-dimensional model proposed by Coats and Smith (1964) for a narrow range of parameters that provide an acceptable match to experimental data (Baker 1977). As an extension to the capacitance model, this section presents a semi-analytical solution for solute transport in a two-dimensional region. With respect to time, the Laplace transform is initially used to convert the system of equations defined in real time, into Laplace time space. The Laplace transform with respect to the space dimension is subsequently applied to one of the domain dimensions to transform the partial differential equations into ordinary differential equations. The solution is subsequently numerically inverted twice to achieve a solution in real space and time. These double Laplace transforms are superior to Laplace-Fourier transforms (Johns and Roberts 1991) because of their systematic application and simplicity.

5.2.2.1 Solution Method

As mentioned in the previous section, Coats and Smith's model (1964) assumes that the regions of immobile water are well mixed and that the process of diffusion into and out of these regions is governed by the first-order rate rule. As a result, two coupled equations are required, in which the first equation represents the solute transport in the mobile region while the second equation defines diffusion in the immobile region following the first-order rate rule. Such a model can be written in two dimension as (Bai et al. 1999a)

$$D_{x*}\frac{\partial^2 c_1}{\partial x_*^2} + D_{y*}\frac{\partial^2 c_1}{\partial y_*^2} - v_{x*}\frac{\partial c_1}{\partial x_*} - v_{y*}\frac{\partial c_1}{\partial y_*} = f\frac{\partial c_1}{\partial t_*} + (1-f)\frac{\partial c_2}{\partial t_*} \tag{5.23}$$

$$(1-f)\frac{\partial c_2}{\partial t_*} = \Gamma^{\#}(c_1 - c_2) \tag{5.24}$$

where c_i is the solute concentration, subscripts $i = 1, 2$ represent macropores and micropores, respectively; t_* is the time after the inception of transport; x_* and y_* are the coordinates; subscript $*$ indicates real time and space; D is the dispersion coefficient; v is the average flow velocity with respect to the porous medium porosity; f is the fraction of pore space occupied by mobile fluid; and $\Gamma^{\#}$ is the rate of mass exchange between macropores and micropores.

For a step increase in solute infiltration penetrating from one side, no variation of concentration at infinity in the x_* direction, and constant concentrations at the boundary in the y_* direction, the boundary and initial conditions can be expressed as

$$\begin{cases} c_1 = c_2 = 0 & (t_* = 0) \\ c_1 = c_0 & (x_* = 0) \\ c_1 = 0 & (x_* = \infty) \\ c_1 = c_{b1} & (y_* = 0) \\ c_1 = c_{b2} & (y_* = y_b) \end{cases} \tag{5.25}$$

where c_0 is the injected concentration, c_{b1} and c_{b2} are the boundary concentrations, and y_b is the domain length in the y_* direction.

For convenience, the following dimensionless terms are introduced

$$x = \frac{x_*}{y_b}, \quad y = \frac{y_*}{y_b}, \quad t = \frac{v_{x*}t_*}{y_b}, \quad \gamma_x = \frac{v_{x*}y_b}{D_{x*}}, \quad \gamma_y = \frac{v_{x*}y_b}{D_{y*}}, \quad \beta = \frac{v_{x*}}{v_{y*}}, \quad \Theta = \frac{\Gamma^{\#}y_b}{v_{x*}}. \tag{5.26}$$

Incorporating all dimensionless terms, Eqs. (5.23) and (5.24) are rewritten as

$$\frac{1}{\gamma_x}\frac{\partial^2 c_1}{\partial x^2} + \frac{1}{\gamma_y}\frac{\partial^2 c_1}{\partial y^2} - \frac{\partial c_1}{\partial x} - \frac{1}{\beta}\frac{\partial c_1}{\partial y} = f\frac{\partial c_1}{\partial t} + (1-f)\frac{\partial c_2}{\partial t} \tag{5.27}$$

$$(1-f)\frac{\partial c_2}{\partial t} = \Theta(c_1 - c_2). \tag{5.28}$$

Applying Laplace transformation to Eqs. (5.27) and (5.28), yields

$$\frac{1}{\gamma_x}\frac{\partial^2 \bar{c}_1}{\partial x^2} + \frac{1}{\gamma_y}\frac{\partial^2 \bar{c}_1}{\partial y^2} - \frac{\partial \bar{c}_1}{\partial x} - \frac{1}{\beta}\frac{\partial \bar{c}_1}{\partial y} = fs\bar{c}_1 + (1-f)s\bar{c}_2 \tag{5.29}$$

$$(1-f)s\bar{c}_2 = \Theta(\bar{c}_1 - \bar{c}_2) \tag{5.30}$$

where s is the Laplace transform parameter.

Boundary conditions in the Laplace domain are transformed as

$$\begin{cases} \bar{c}_1 = \frac{c_0}{s} & (x = 0) \\ \bar{c}_1 = 0 & (x = \infty) \\ \bar{c}_1 = \frac{c_{b1}}{s} & (y = 0) \\ \bar{c}_1 = \frac{c_{b2}}{s} & (y = 1). \end{cases} \tag{5.31}$$

From Eq. (5.30), one has

$$\bar{c}_2 = \frac{\Theta}{(1-f)s + \Theta}\bar{c}_1. \tag{5.32}$$

Substituting Eq. (5.32) into Eq. (5.29), then

$$\frac{1}{\gamma_x}\frac{\partial^2 \bar{c}_1}{\partial x^2} + \frac{1}{\gamma_y}\frac{\partial^2 \bar{c}_1}{\partial y^2} - \frac{\partial \bar{c}_1}{\partial x} - \frac{1}{\beta}\frac{\partial \bar{c}_1}{\partial y} + \xi\bar{c}_1 = 0, \tag{5.33}$$

where

$$\xi = -s\left[\frac{f(1-f)s + \Theta}{(1-f)s + \Theta}\right]. \tag{5.34}$$

The solution of hyperbolic-type partial differential equations, such as Eq. (5.33), requires the definition of Cauchy-type (or third-type) boundary conditions. In other words, both slope and value are known quantities at the boundary (Lapidus and Pinder 1982). While the condition $\dfrac{\partial \bar{c}_1(0,y)}{\partial x} = 0$ is implied in the present derivation, applying the Laplace transform in Eq. (5.33) with respect to x and incorporating the boundary conditions in Eq. (5.31) results in the following ordinary differential equation

$$\frac{d^2\hat{c}_1}{dy^2} - \phi_3^{\#}\frac{d\hat{c}_1}{dy} + \phi_1^{\#}\hat{c}_1 = \phi_2^{\#} \tag{5.35}$$

where

$$\begin{cases} \phi_1^{\#} = \gamma_y[w^*(\frac{w^*}{\gamma_x} - 1) + \xi] \\ \\ \phi_2^{\#} = \frac{\gamma_y c_0}{s}(\frac{w^*}{\gamma_x} - 1), \quad \phi_3^{\#} = \frac{\gamma_y}{\beta} \end{cases} \tag{5.36}$$

and w^* is another Laplace parameter.

The boundary conditions are modified again, as

$$\begin{cases} \hat{c}_1 = \frac{c_{b1}}{w^* s} & (y = 0) \\ \\ \hat{c}_1 = \frac{c_{b2}}{w^* s} & (y = 1). \end{cases} \tag{5.37}$$

The inhomogeneous differential equation (5.35) can be solved using the superposition principle, then

$$\hat{c}_1 = (\hat{c}_1)_h + (\hat{c}_1)_i \tag{5.38}$$

where $(\hat{c}_1)_h$ and $(\hat{c}_1)_i$ are the homogeneous and inhomogeneous solutions, respectively.

After satisfying the boundary conditions of Eq. (5.37), the homogeneous and inhomogeneous solutions are separately derived. The final solution can be expressed as

$$\hat{c}_1 = \frac{1}{\psi_3 - \psi_2}[(\psi_1\psi_3 - \psi_4)e^{\tau_1 y} + (\psi_4 - \psi_1\psi_2)e^{\tau_2 y}] + \frac{\phi_2^{\#}}{\phi_1^{\#}} \tag{5.39}$$

where

$$
\begin{cases}
\psi_1 = -\dfrac{\phi_2^\#}{\phi_1^\#} + \dfrac{c_{b1}}{w^* s}, \quad \psi_2 = e^{\tau_1} \\[2ex]
\psi_3 = e^{\tau_2}, \quad \psi_4 = -\dfrac{\phi_2^\#}{\phi_1^\#} + \dfrac{c_{b2}}{w^* s} \\[2ex]
\tau_1 = \dfrac{\phi_3^\#}{2} + \sqrt{\dfrac{(\phi_3^\#)^2}{4} - \phi_1^\#} \\[2ex]
\tau_2 = \dfrac{\phi_3^\#}{2} - \sqrt{\dfrac{(\phi_3^\#)^2}{4} - \phi_1^\#}.
\end{cases}
\tag{5.40}
$$

$\hat{c}_1$ can be inverted twice back to c_1 using the numerical inversion technique of Stehfest (1970).

5.2.2.2 Illustrative Example

A comparison of the two-dimensional solution, presented previously, and the one-dimensional model of Coats and Smith (1964), provides an immediate justification of the need for the 2-D solution. For $\gamma_y = 100$, Figure 5-2 reveals the significant difference between the 1-D and the 2-D models in terms of spatial spread of contaminant. For the examined spatial range, the 2-D model shows more restricted down-gradient solute spreading, as a result of the effect of vertical transport, which cannot be revealed by the 1-D model. However, comparison of breakthrough curves at a specific point between the 1-D and the 2-D models under the same conditions reveals a less significant contrast, as illustrated in Figure 5-3.

5.3 FOURIER TRANSFORM

Even though the Laplace transform provides a straightforward solution procedure, the method is most effective in transforming a single variable such as time, and a single equation frequently is preferred. This preference requires that the solution procedure is decoupled. As a result, for fully coupled formulations, the Fourier transform is often a better choice in lieu of the Laplace transform.

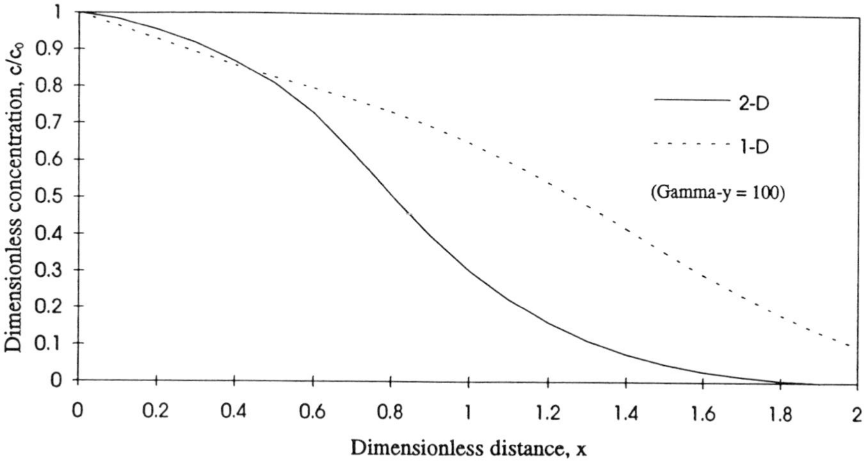

Figure 5-2. Spatial comparison between 1-D and 2-D models. (Reprinted from Bai et al. 1999a, with permission from Elsevier Science.)

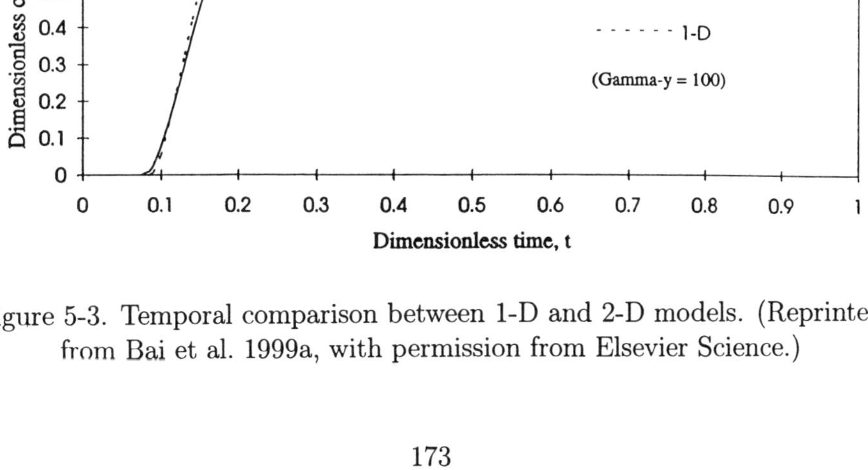

Figure 5-3. Temporal comparison between 1-D and 2-D models. (Reprinted from Bai et al. 1999a, with permission from Elsevier Science.)

5.3.1 Flow

The fully coupled dual-porosity formulation shown in Eqs. (5.3) and (5.4) can be tackled effectively by the Fourier transform.

5.3.1.1 Solution Method

The finite Fourier transform may be defined as

$$\bar{p}_i(m^*, t) = \frac{2}{L}\int_0^L p_i^*(x,t)\cos(\xi_p x)dx \tag{5.41}$$

where

$$p_i^*(x,t) = \sum_{m^*=0}^{\infty} \bar{p}_i(m^*,t)\cos(\xi_p x) \tag{5.42}$$

and $i = 1$ and 2; also

$$\xi_p = \frac{(2m^*+1)\pi}{2L} \tag{5.43}$$

where m^* is a Fourier parameter and L is the linear length of the simulated 1-D domain.

For the case of a constant flow rate at the pumping location ($x = 0$) and a constant pressure outer boundary ($x = L$), the general solution may be expressed as

$$p_i = p_i^0 + Q(L - x) + p_i^* \tag{5.44}$$

where $Q = \dfrac{q\mu}{k_2 A_c}$ and A_c is the flow cross-sectional area. The initial condition is assumed as $p_i = p_i^0$ ($t = 0$).

Applying the Fourier transform to Eqs. (5.3) and (5.4) gives

$$-\frac{k_1}{\mu}\xi_p^2\bar{p}_1^* = g_1\frac{d\bar{p}_1^*}{dt} - g_2\frac{d\bar{p}_2^*}{dt} + \Gamma(\bar{p}_1^* - \bar{p}_2^*) \tag{5.45}$$

$$-\frac{k_2}{\mu}\xi_p^2\bar{p}_2^* = g_3\frac{d\bar{p}_2^*}{dt} - g_4\frac{d\bar{p}_1^*}{dt} - \Gamma(\bar{p}_1^* - \bar{p}_2^*). \tag{5.46}$$

Solving the ordinary differential equations (5.45) and (5.46) yields

$$\bar{p}_1^* = D_1^{\#}\frac{b_{22}^*}{G\#}e^{-b_{11}^*t} - D_2^{\#}\frac{b_{12}^*}{G\#}e^{-b_{22}^*t} \tag{5.47}$$

$$\bar{p}_2^* = D_2^{\#}\frac{b_{11}^*}{G\#}e^{-b_{22}^*t} - D_1^{\#}\frac{b_{21}^*}{G\#}e^{-b_{11}^*t} \tag{5.48}$$

where $D_1^{\#}$ and $D_2^{\#}$ are the integral constants, and

$$\begin{cases} b_{11}^* = [g_3(a_1+\Gamma) - g_2\Gamma]/G^*, & b_{21}^* = [g_4(a_1+\Gamma) - g_1\Gamma]/G^* \\ b_{12}^* = [g_2(a_2+\Gamma) - g_3\Gamma]/G^*, & b_{22}^* = [g_1(a_2+\Gamma) - g_4\Gamma]/G^* \end{cases} \tag{5.49}$$

$$\begin{cases} a_1 = \frac{k_1}{\mu}\xi_p^2 \quad a_2 = \frac{k_2}{\mu}\xi_p^2 \\ G^{\#} = b_{11}^* b_{22}^* - b_{12}^* b_{21}^*, \quad G^* = g_1 g_3 - g_2 g_4. \end{cases} \tag{5.50}$$

Constants $D_1^{\#}$ and $D_2^{\#}$ in the fractional form $D_{1j}^{\#}$ and $D_{2j}^{\#}$ may be obtained by satisfying the initial conditions and using orthogonal properties. Correspondingly, one has

$$\begin{cases} D_{1j}^{\#} = -\frac{8QL(b_{11}^* + b_{12}^*)}{[(2m^*+1)\pi]^2 G^{\#}} \\ D_{2j}^{\#} = -\frac{8QL(b_{21}^* + b_{22}^*)}{[(2m^*+1)\pi]^2 G^{\#}}. \end{cases} \tag{5.51}$$

The solutions for fluid pressures in the matrix and in the fractures are then derived as

$$p_1 = p_1^0 + Q(L-x) + \sum_{m=0}^{\infty} (D_{1j}^{\#} b_{22}^* e^{-b_{11}^* t} - D_{2j}^{\#} b_{12}^* e^{-b_{22}^* t}) \cos(\xi_p x) \tag{5.52}$$

$$p_2 = p_2^0 + Q(L-x) + \sum_{m=0}^{\infty} (-D_{1j}^{\#} b_{21}^* e^{-b_{11}^* t} + D_{2j}^{\#} b_{11}^* e^{-b_{22}^* t}) \cos(\xi_p x). \tag{5.53}$$

The solutions in Eqs. (5.52) and (5.53) satisfy the limiting boundary and initial conditions. In addition, the solution of quasi-steady flow within the matrix can be simply achieved by letting $a_1 = 0$ in Eq. (5.49).

5.3.1.2 Illustrative Example

A simple case study is considered to quantify the proposed dual-porosity solutions given by Eqs. (5.52) and (5.53), and more specifically to examine the temporal pressure variation at the pumping location in a naturally-fractured reservoir. The basic reservoir conditions are listed in Table 5.1 (Craft and Hawkins 1959; Bai et al. 1994c) for the case of one-dimensional linear flow with constant flow rate applied at the perforation points and constant pressure at the outer boundary.

Table 5.1. Basic parameters.

Parameter	Symbol	Value	Unit
Fluid-flow rate	q	0.01	m^3/s
Dynamic viscosity	μ	2×10^{-4}	Pa·s
Reservoir length	L	1000	m
Flow area	A_c	100	m
Fluid compressibility	c_f	10^{-5}	$(Pa)^{-1}$

Using the parameters given in Table 5.2 (Streltsova 1988; Chilingarian et al. 1992), Figure 5-4 illustrates the temporal pressure variation in fractures as a result of changing fracture permeabilities. The magnitudes of fluid pressure are inversely proportional to the permeability magnitudes. As expected, the dual-porosity behavior is characterized by a dampening of pressure change amplitude as a result of the matrix-fracture interaction. This behavior appears to be more obvious for lower fracture permeabilities (or smaller permeability ratio between fractures and matrix), at earlier times.

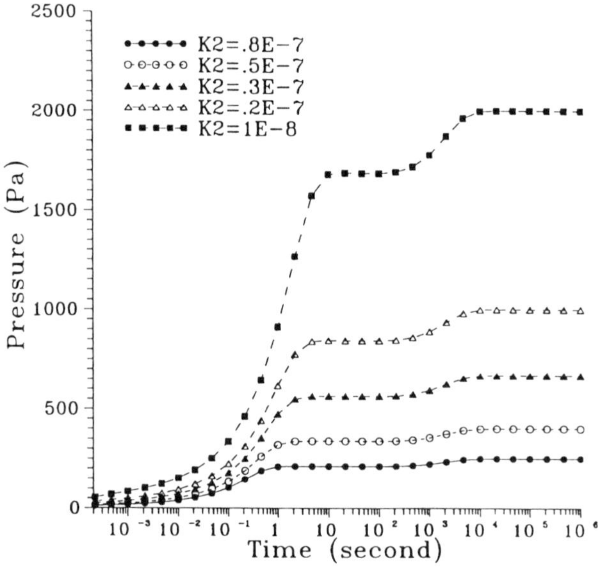

Figure 5-4. Pressure for various permeabilities.

Table 5.2. Parameters for studying dual-porosity effects.

Parameter	Symbol	Matrix	Fracture	Unit
Initial pressure	p_1^0, p_2^0	0	0	Pa
Permeability	k_1, k_2	10^{-14}	$10^{-8} \sim .8 \times 10^{-7}$	m^2
Porosity	n_1, n_2	0.1	0.05	-
Compressibility	c_1, c_2	10^{-10}	2×10^{-10}	$(Pa)^{-1}$

5.3.2 Nonisothermal Flow and Deformation

Strategies to implement either full coupling or partial decoupling may be demonstrated by examining a case of nonisothermal one-dimensional consolidation, frequently referred to as a "column problem." A laterally confined porous column of height h_c is situated over a rigid, impermeable, and adiabatic base. A load of magnitude F_0 is applied instantaneously at the top of the column ($x = 0$), forcing the column to consolidate while allowing the fluid to escape and heat to dissipate from the top.

5.3.2.1 Solution Method

By neglecting forced thermal convection as well as fluid and heat sources, the governing equations for 1-D nonisothermal consolidation due to external force can be written as (Bai and Abousleiman 1997)

$$(\lambda + 2G)\frac{\partial^2 u}{\partial x^2} = \alpha\frac{\partial p}{\partial x} + \beta_h\frac{\partial T}{\partial x} \tag{5.54}$$

$$\frac{k}{\mu}\frac{\partial^2 p}{\partial x^2} = c^*\frac{\partial p}{\partial t} + \alpha\frac{\partial^2 u}{\partial x \partial t} - \alpha_h\frac{\partial T}{\partial t} \tag{5.55}$$

$$K_h^*\frac{\partial^2 T}{\partial x^2} = \beta_h T_0\frac{\partial^2 u}{\partial x \partial t} + s_h^*\frac{\partial T}{\partial t} \tag{5.56}$$

where u is the displacement, p is the pressure, T is the temperature, λ and G are *Lamé*'s constants, α is Biot's coefficient, β_h is the thermal expansion factor, k is the fluid permeability, μ is the fluid dynamic viscosity, c^* is the lumped compressibility, α_h is the thermal expansion coefficient, K_h^* is the thermal conductivity, and s_h^* is the lumped intrinsic heat capacity.

More specific definitions for the above parameters in terms of heat flow include

$$\beta_h = \alpha_h(3\lambda + 2G) \tag{5.57}$$

$$\alpha_h = n\alpha_h^f + (1 - n)\alpha_h^s \tag{5.58}$$

where α_h^f and α_h^s are the thermal expansion coefficients for fluid and solid;

$$K_h^* = nK_h^{*f} + (1 - n)K_h^{*s} \tag{5.59}$$

where K_h^{*f} and K_h^{*s} are the thermal conductivities for fluid and solid;

$$s_h^* = \rho_h^f c_h^f n + \rho_h^s c_h^s (1 - n) \tag{5.60}$$

where ρ_h^f and c_h^f are the density and heat capacity of the fluid, and ρ_h^s and c_h^s are the density and heat capacity of the solid.

The boundary and initial conditions can be given as

$$\begin{cases} \dfrac{\partial u(0,t)}{\partial x} = E_0 = -\dfrac{F_0}{\lambda + 2G}, \quad p(0,t) = T(0,t) = 0; \\ u(h,t) = 0, \quad \dfrac{\partial p(h,t)}{\partial x} = \dfrac{\partial T(h,t)}{\partial x} = 0; \\ p(x,0) = p_0, \quad T(x,0) = T_0 \end{cases} \tag{5.61}$$

where F_0 is the external load, p_0 and T_0 are the initial pressure and temperature, respectively.

Integration with respect to x in Eq. (5.54), yields

$$\eta \frac{\partial u}{\partial x} = \alpha p + \beta_h T + f_1^{\#} \tag{5.62}$$

where

$$\eta = \lambda + 2G \tag{5.63}$$

and the integration constant $f_1^{\#}$ can be determined by applying the conditions in Eq. (5.61), which gives

$$f_1^{\#} = \eta\ E_0 = -F_0. \tag{5.64}$$

The spatial derivative of u can be obtained from Eq. (5.62). Differentiating this derivative with respect to t, substituting the result into Eqs. (5.55) and (5.56), yields

$$\frac{\partial^2 p}{\partial x^2} = g_{11}^* \frac{\partial p}{\partial t} + g_{12}^* \frac{\partial T}{\partial t} \tag{5.65}$$

$$\frac{\partial^2 T}{\partial x^2} = g_{21}^* \frac{\partial p}{\partial t} + g_{22}^* \frac{\partial T}{\partial t} \tag{5.66}$$

where

$$\begin{cases} g_{11}^* = \frac{\mu}{k}\left(c^* + \frac{\alpha^2}{\eta}\right), \quad g_{12}^* = \frac{\mu}{k}\left(\frac{\alpha\beta_h}{\eta} - \alpha_h\right) \\ g_{21}^* = \dfrac{\alpha\beta_h T_0}{K_h^*\eta}, \quad g_{22}^* = \dfrac{1}{K_h^*}\left(s_h^* + \frac{\beta_h^2 T_0}{\eta}\right). \end{cases} \tag{5.67}$$

The finite Fourier transform can be used, which is defined by

$$\begin{cases} \bar{p}(m,t) = \int_0^{h_c} p(x,t) \sin \xi_p x\ dx, \quad p(x,t) = \dfrac{2}{h_c} \displaystyle\sum_{m=0}^{\infty} \bar{p}(m,t) \sin \xi_p x \\ \bar{T}(m,t) = \int_0^{h_c} T(x,t) \sin \xi_p x\ dx, \quad T(x,t) = \dfrac{2}{h_c} \displaystyle\sum_{m=0}^{\infty} \bar{T}(m,t) \sin \xi_p x \end{cases} \tag{5.68}$$

$$\xi_p = \frac{(2m+1)\pi}{2h_c}. \tag{5.69}$$

The transformation defined in Eq. (5.68) automatically satisfies the boundary conditions described in Eq. (5.61). The initial conditions in Eq. (5.61) in the Fourier domain can be expressed as

$$\begin{cases} \bar{p}(m,0) = p_0 \int_0^{h_c} \sin \xi_p x \; dx = \frac{p_0}{\xi_p} \\ \\ \bar{T}(m,0) = T_0 \int_0^{h_c} \sin \xi_p x \; dx = \frac{T_0}{\xi_p}. \end{cases} \tag{5.70}$$

Using the Fourier transform, Eqs. (5.65) and (5.66) can be rewritten as

$$-\xi_p^2 \bar{p} = g_{11}^* \frac{d\bar{p}}{dt} + g_{12}^* \frac{d\bar{T}}{dt} \tag{5.71}$$

$$-\xi_p^2 \bar{T} = g_{21}^* \frac{d\bar{p}}{dt} + g_{22}^* \frac{d\bar{T}}{dt}. \tag{5.72}$$

Solving Eqs. (5.71) and (5.72) simultaneously, and applying the initial conditions in Eq. (5.70), gives

$$\bar{p} = \frac{1}{\xi_p} \left(\phi_{12}^{\#} e^{-\gamma_{11}\xi_p^2 t} + \phi_{11}^{\#} e^{-\gamma_{22}\xi_p^2 t} \right) \tag{5.73}$$

$$\bar{T} = \frac{1}{\xi_p} \left(\phi_{22}^{\#} e^{-\gamma_{11}\xi_p^2 t} + \phi_{21}^{\#} e^{-\gamma_{22}\xi_p^2 t} \right) \tag{5.74}$$

where

$$\phi_{ij}^{\#} = -\frac{\gamma_{ij}\gamma_j^{\#}}{\gamma^{\#}} \qquad (i = 1,2; j = 1,2) \tag{5.75}$$

and where

$$\begin{cases} \gamma_{ij} = \frac{g_{ij}^*}{g^*} \qquad (i = 1,2; j = 1,2) \\ \\ g^* = g_{11}^* g_{22}^* - g_{12}^* g_{21}^*, \quad \gamma_1^{\#} = \gamma_{12} T_0 - \gamma_{22} p_0 \\ \\ \gamma_2^{\#} = \gamma_{21} p_0 - \gamma_{11} T_0, \quad \gamma^{\#} = \gamma_{11}\gamma_{22} - \gamma_{12}\gamma_{21}. \end{cases} \tag{5.76}$$

The pressure and temperature can be finally determined by substituting Eqs. (5.73) and (5.74) into Eq. (5.68), which yields

$$p(x,t) = \frac{2}{h_c} \sum_{m=0}^{\infty} \frac{1}{\xi_p} \left(\phi_{12}^{\#} e^{-\gamma_{11}\xi_p^2 t} + \phi_{11}^{\#} e^{-\gamma_{22}\xi_p^2 t} \right) \sin \xi_p x \tag{5.77}$$

$$T(x,t) = \frac{2}{h_c} \sum_{m=0}^{\infty} \frac{1}{\xi_p} \left(\phi_{22}^{\#} e^{-\gamma_{11}\xi_p^2 t} + \phi_{21}^{\#} e^{-\gamma_{22}\xi_p^2 t} \right) \sin \xi_p x. \tag{5.78}$$

The total displacement can be obtained through integration with respect to the derivative of u in Eq. (5.62), then

$$\begin{aligned} u &= \int_0^{h_c} \left[\frac{1}{\eta}(\alpha p + \beta_h T) + E_0 \right] dx \\ &= E_0 h_c + \frac{2}{h_c \eta} \sum_{m=0}^{\infty} \frac{1}{\xi_p^2} \left(\psi_1^* e^{-\gamma_{11}\xi_p^2 t} + \psi_2^* e^{-\gamma_{22}\xi_p^2 t} \right) \end{aligned} \tag{5.79}$$

where

$$\begin{cases} \psi_1^* = \alpha \phi_{12}^{\#} + \beta_h \phi_{22}^{\#} \\ \psi_2^* = \alpha \phi_{11}^{\#} + \beta_h \phi_{21}^{\#}. \end{cases} \tag{5.80}$$

The general solution of displacement u, however, can be expressed as

$$u(x,t) = E_0(x - h_c) - \frac{2}{h_c \eta} \sum_{m=0}^{\infty} \frac{1}{\xi_p^2} \left(\psi_1^* e^{-\gamma_{11}\xi_p^2 t} + \psi_2^* e^{-\gamma_{22}\xi_p^2 t} \right) \cos \xi x. \tag{5.81}$$

Eqs. (5.77), (5.78), and (5.81) represent the solutions of one-dimensional nonisothermal consolidation.

5.3.2.2 Illustrative Example

This comparative analysis is designed to identify the relevant influence of each coupling term described in Eqs. (5.54), (5.55) and (5.56). The selected geometric, hydraulic, thermal, and mechanical parameters are listed in Table 5.3. For generality, the pressure p, temperature T, and displacement u are expressed in a normalized fashion with respect to their maximum values, as is the spatial dimension. Similarly, dimensionless time is defined as

$$\tau = \frac{c_v t}{h_c^2} \tag{5.82}$$

where h is the height of the consolidating column and c_v is the coefficient of consolidation and is described as

$$c_v = \frac{k\eta}{\mu} \tag{5.83}$$

where η is described in Eq. (5.63).

Using these parameters, the results shown in Figure 5-5 represent an interactive response of dissipating pressure and temperature to the increase of displacement due to consolidation of the column. In comparison with the diffusive temperature dissipation, pressure change occurs earlier with a slight change in slope, as a result of the thermoporoelastic response.

Table 5.3. Selected thermoporoelastic parameters.

Category	Symbol	Value	Unit
Hydraulic	k	10^{-9}	m^2
	μ	20	kg/(m,h)
	n	0.2	-
	p_0	10^4	Pa
Thermal	α_h^f	10^{-5}	$1/^oC$
	α_h^s	10^{-6}	$1/^oC$
	K_h^f	1000	$J/(m,h,^oC)$
	K_h^s	100	$J/(m,h,^oC)$
	ρ_h^f	1000	kg/m^3
	ρ_h^s	2000	kg/m^3
	c_h^f	500	$J/(kg,^oC)$
	c_h^s	200	$J/(kg,^oC)$
	T_0	100	^oC
Mechanical	α	0.9	-
	ν	0.25	-
	E	5×10^9	Pa
	K_f	5×10^9	Pa
	K	2×10^9	Pa
	K_s	2×10^{10}	Pa
	F_0	10^8	Pa
Geometric	h_c	100	m
	x	5	m

5.4 HANKEL TRANSFORM

Hankel transform is an appropriate method for domains defined in polar or cylindrical coordinate systems.

5.4.1 Flow

Hankel transforms may be applied in the analysis of fluid flow, typically for the simulation of well pumping in radial or cylindrical coordinate systems.

5.4.1.1 Solution Method

The following dimensionless parameters are introduced for the governing equations of a dual-porosity model [i.e., Eqs. (5.3) and (5.4)], in the radial flow

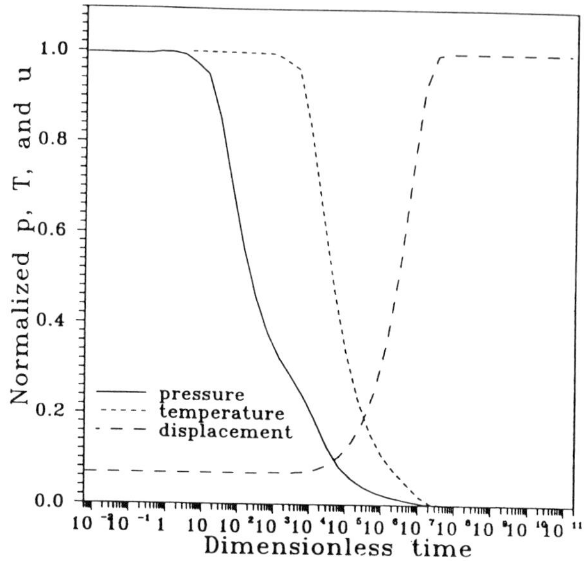

Figure 5-5. Thermoporoelastic evolution. (Reprinted from Bai and Abousleiman 1997, with permission from John Wiley and Sons.)

case:

$$P_{Di} = \frac{2\pi h_r(k_1 + k_2)(p_0 - p)}{q\mu} \qquad (i = 1, 2) \tag{5.84}$$

where h_r is the reservoir thickness, p_0 is the initial reservoir pressure, and q is the flow rate at the well. Dimensionless radius is defined as

$$r_D = \frac{r}{r_w} \tag{5.85}$$

where r_w is the wellbore radius and dimensionless time is

$$t_D = \frac{(k_1 + k_2)t}{\mu r_w^2 \sum_{i=1}^{4} g_i} \tag{5.86}$$

where g_1, g_2, g_3, and g_4 are defined as c_{11}^*, c_{12}^*, c_{21}^* and c_{22}^*, respectively, in Eq. (3.137) of Chapter 3.

$$\omega_i^* = \frac{(k_1 + k_2)g_i}{k_j \sum_{i=1}^{4} g_i} \qquad (i = 1, 2, 3, 4) \tag{5.87}$$

where in k_j, $i = 1$ and 2, $j = 1$; $i = 3$ and 4, $j = 2$. Assuming

$$\lambda^* = (\Gamma)\frac{\mu}{k_1}(r_w^2) = \left(\frac{k_1}{\mu}\alpha_g\right)\frac{\mu}{k_1}(r_w^2) = \alpha_g(r_w^2) \tag{5.88}$$

where α_g is a geometric constant related to the matrix-fracture geometric pattern. For a block-type matrix dissected by orthogonally arranged planar fractures, $\alpha_g = 60/(s^*)^2$ where s^* is the fracture spacing (Warren and Root 1963).

ω_i^* in Eq. (5.87) can be explicitly expressed as

$$\begin{cases} \omega_1^* = \dfrac{1+R_k^{-1}}{n_1+n_2}\left[(\alpha_3 - n_1 - n_2)\dfrac{K_f}{K_s} + n_1\right] \\[2ex] \omega_2^* = -\dfrac{(1+R_k^{-1})(\alpha_3 - n_1 - n_2)}{n_1+n_2}\dfrac{K_f}{K^*} \\[2ex] \omega_3^* = \dfrac{1+R_k}{n_1+n_2}\left[(\alpha_3 - n_1 - n_2)\dfrac{K_f}{K^*} + n_2\right] \\[2ex] \omega_4^* = -\dfrac{(1+R_k)(\alpha_3 - n_1 - n_2)}{n_1+n_2}\dfrac{K_f}{K_s} \end{cases} \tag{5.89}$$

where R_k is the ratio between matrix and fracture permeabilities (i.e., $R_k = k_1/k_2$).

Substituting all the above dimensionless terms into governing equations (5.3) and (5.4), yields the following equations in radial coordinates:

$$\frac{\partial^2 P_{D1}}{\partial r_D^2} + \frac{1}{r_D}\frac{\partial P_{D1}}{\partial r_D} = \omega_1^*\frac{\partial P_{D1}}{\partial t_D} - \omega_2^*\frac{\partial P_{D2}}{\partial t_D} + \lambda^*(P_{D1} - P_{D2}) \tag{5.90}$$

$$\frac{\partial^2 P_{D2}}{\partial r_D^2} + \frac{1}{r_D}\frac{\partial P_{D2}}{\partial r_D} = \omega_3^*\frac{\partial P_{D2}}{\partial t_D} - \omega_4^*\frac{\partial P_{D1}}{\partial t_D} - \lambda^* R_k(P_{D1} - P_{D2}). \tag{5.91}$$

Reservoir pressure measurements are usually recorded in the production well while maintaining a constant flow rate, which is also a normal operation requirement in well testing. It is difficult to determine the proportion of fluid released to the well from either the matrix or fractures. As a result, it is reasonable to assume that an average flow between fractures and matrix occurs at the well. Therefore, the boundary conditions for a finite reservoir with a constant outer boundary and a constant pumping rate are

$$\left.\frac{\partial P_{D1}}{\partial r_D} = \frac{\partial P_{D2}}{\partial r_D}\right|_{r_D=1} = -1 \quad \text{(inner boundary)} \tag{5.92}$$

$$P_{D1}(r_{De}, t_D) = P_{D2}(r_{De}, t_D) = 0 \quad \text{(outer boundary)}. \tag{5.93}$$

The initial condition is assumed as

$$P_{D1}(r_D, t_D = 0) = P_{D2}(r_D, t_D = 0) = 0. \tag{5.94}$$

Also assuming a linear system, the solution of Eqs. (5.90) and (5.91) with nonhomogeneous boundary conditions can be solved using a decoupled mechanism, and is given by (Bai et al. 1994b)

$$\Delta P_{Di}(r_D, t_D) = \Delta V_{Di}(r_D) + \Delta U_{Di}(r_D, t_D) \quad (i = 1, 2) \tag{5.95}$$

where $i = 1$ and 2 represent matrix and fractures, respectively.

It is therefore necessary to solve two systems with reference to V_{Di} and U_{Di}, respectively, as shown below.

System 1: This is an ordinary differential equation with nonhomogeneous boundary conditions but homogeneous initial condition, such that

$$\frac{d^2V_{D1}}{dr_D^2} + \frac{1}{r_D}\frac{dV_{D1}}{dr_D} = 0 \tag{5.96}$$

$$\frac{d^2V_{D2}}{dr_D^2} + \frac{1}{r_D}\frac{dV_{D2}}{dr_D} = 0. \tag{5.97}$$

The boundary conditions are

$$\frac{dV_{D1}}{dr_D} = \left.\frac{dV_{D2}}{dr_D}\right|_{r_D=1} = -1 \tag{5.98}$$

$$V_{D1} = V_{D2}|_{r_D=r_{De}} = 0. \tag{5.99}$$

System 1 has a simple solution. Indeed, Eqs. (5.96) and (5.97) may be written as

$$\frac{1}{r_D}\frac{d}{dr_D}\left(r_D\frac{dV_{Di}}{dr_D}\right) = 0. \tag{5.100}$$

Solving Eq. (5.100) with boundary conditions (5.98) and (5.99) yields the solution

$$V_{Di} = \ln(r_{De}) - \ln(r_D). \tag{5.101}$$

System 2: Alternatively, System 2 is a partial differential equation with homogeneous boundary conditions but nonhomogeneous initial condition such as

$$\frac{\partial^2U_{D1}}{\partial r_D^2} + \frac{1}{r_D}\frac{\partial U_{D1}}{\partial r_D} = \omega_1^*\frac{\partial U_{D1}}{\partial t_D} - \omega_2^*\frac{\partial U_{D2}}{\partial t_D} + \lambda^*(U_{D1} - U_{D2}) \tag{5.102}$$

$$\frac{\partial^2 U_{D2}}{\partial r_D^2} + \frac{1}{r_D}\frac{\partial U_{D2}}{\partial r_D} = \omega_3^* \frac{\partial U_{D2}}{\partial t_D} - \omega_4^* \frac{\partial U_{D1}}{\partial t_D} - \lambda^* R_k (U_{D1} - U_{D2}). \qquad (5.103)$$

The boundary conditions are

$$\frac{\partial U_{D1}}{\partial r_D} = \left.\frac{\partial U_{D2}}{\partial r_D}\right|_{r_D=1} = 0 \qquad (5.104)$$

$$U_{D1} = U_{D2}|_{r_D=r_{De}} = 0. \qquad (5.105)$$

The initial condition is

$$U_{D1} = U_{D2}|_{t_D=0} = -V_{Di}(r_D). \qquad (5.106)$$

System 2 may be solved using a general finite Hankel transform defined as

$$\bar{U}_{Di}(j, t_D) = \int_1^{r_{De}} U_{Di}(r_D, t_D)\phi_j(r_D) r_D dr_D \qquad (i = 1, 2) \qquad (5.107)$$

where

$$U_{Di}(r_D, t_D) = \sum_{j=1}^{\infty} G_j^c \bar{U}_{Di}(j, t_D)\phi_j(r_D) \qquad (5.108)$$

where G_j^c is a coefficient, which can be determined from the initial condition. $\phi_j(r_D)$ can be obtained by solving the following eigenvalue problem

$$\frac{d^2\phi_j}{dr_D^2} + \frac{1}{r_D}\frac{d\phi_j}{dr_D} + \lambda_j^2 \phi_j = 0 \qquad (5.109)$$

with boundary conditions

$$\left.\frac{d\phi_j}{dr_D}\right|_{r_D=1} = 0 \qquad (5.110)$$

$$\phi_j|_{r_D=r_{De}} = 0. \qquad (5.111)$$

The solution of Eq. (5.109) exists if λ_j satisfies the following equation

$$J_1(\lambda_j)Y_0(\lambda_j r_{De}) - J_0(\lambda_j r_{De})Y_1(\lambda_j) = 0. \qquad (5.112)$$

The solution of the eigenvalue system (5.109) to (5.111) is given by

$$\phi_j(r_D) = Y_0(\lambda_j r_{De})J_0(\lambda_j r_D) - J_0(\lambda_j r_{De})Y_0(\lambda_j r_D). \qquad (5.113)$$

Applying the Hankel transform, governing equations (5.102) and (5.103) may be written as

$$-a_1\bar{U}_{D1} + \lambda^* \bar{U}_{D2} = \omega_1^* \frac{d^*\bar{U}_{D1}}{dt_D} - \omega_2^* \frac{d^*\bar{U}_{D2}}{dt_D} \qquad (5.114)$$

$$-a_2\bar{U}_{D2} + \lambda^* R_k \bar{U}_{D1} = \omega_3^* \frac{d^* \bar{U}_{D2}}{d^* t_D} - \omega_4^* \frac{d^* \bar{U}_{D1}}{d^* t_D} \tag{5.115}$$

where

$$a_1 = (\lambda_j^2 + \lambda^*) \tag{5.116}$$

$$a_2 = (\lambda_j^2 + \lambda^* R_k). \tag{5.117}$$

Solving Eqs. (5.114) and (5.115) simultaneously, and integrating the result with respect to time t_D, gives

$$\bar{U}_{D1} = \frac{c_{22}^@ D^c}{d^@} e^{-c_{11}^@ t_D} - \frac{c_{12}^@ E^c}{d^@} e^{-c_{22}^@ t_D} \tag{5.118}$$

$$\bar{U}_{D2} = -\frac{c_{21}^@ D^c}{d^@} e^{-c_{11}^@ t_D} + \frac{c_{11}^@ E^c}{d^@} e^{-c_{22}^@ t_D} \tag{5.119}$$

where

$$\begin{cases} d^@ = c_{11}^@ c_{22}^@ - c_{12}^@ c_{21}^@, & c_{11}^@ = (a_1\omega_3^* - \lambda^*\omega_2^* R_k)(b^\#)^{-1} \\ c_{12}^@ = (a_2\omega_2^* - \lambda^*\omega_3^*)(b^\#)^{-1}, & c_{21}^@ = (a_1\omega_4^* - \lambda^*\omega_1^* R_k)(b^\#)^{-1} \\ a_{22}^@ = (a_2\omega_1^* - \lambda^*\omega_4^*)(b^\#)^{-1}, & b^\# = \omega_1^*\omega_3^* - \omega_2^*\omega_4^* \end{cases} \tag{5.120}$$

D^c and E^c are constants that will be determined later.

Substituting Eqs. (5.118) and (5.119) into Eq. (5.108), gives

$$U_{D1} = \sum_{j=1}^{\infty} \phi_j(r_D) \left(c_{22}^@ D_j^c e^{-c_{11}^@ t_D} - c_{12}^@ E_j^c e^{-c_{22}^@ t_D} \right) \tag{5.121}$$

$$U_{D2} = \sum_{j=1}^{\infty} \phi_j(r_D) \left(-c_{21}^@ D_j^c e^{-c_{11}^@ t_D} + c_{11}^@ E_j^c e^{-c_{22}^@ t_D} \right) \tag{5.122}$$

where $D_j^c = D^c G_j^c$ and $E_j^c = E^c G_j^c$; coefficients D_j^c and E_j^c can be determined from the initial condition (5.106). Using orthogonal properties, D_j^c and E_j^c can be obtained by solving the following equations:

$$c_{22}^@ D_j^c - c_{12}^@ E_j^c = -[\ln(r_{De}) - \ln(r_D)] Q^\# \tag{5.123}$$

$$-c_{21}^@ D_j^c + c_{11}^@ E_j^c = -[\ln(r_{De}) - \ln(r_D)] Q^\# \tag{5.124}$$

where

$$Q^\# = \frac{I_1}{I_2}, \quad I_1 = \int_1^{r_{De}} \phi_j(r_D) r_D dr_D, \quad I_2 = \int_1^{r_{De}} \phi_j^2(r_D) r_D dr_D. \tag{5.125}$$

The solution of the integrals in Eq. (5.125) is determined in conjunction with the corresponding boundary conditions. For the present particular case,

$$I_1 = \frac{1}{\lambda_j} [\phi_1(\lambda_j) - r_{De}\phi_1(\lambda_j r_{De})] \tag{5.126}$$

where

$$\phi_1(\lambda_j r_D) = J_1(\lambda_j r_D)Y_0(\lambda_j r_{De}) - Y_1(\lambda_j r_D)J_0(\lambda_j r_{De}). \tag{5.127}$$

Solving Eq. (5.126), one obtains

$$I_1 = \frac{2}{\lambda_j^2 \pi}. \tag{5.128}$$

The integral I_2 can be solved in a similar fashion:

$$I_2 = \frac{1}{2\lambda_j^2}\left[\left(r_D \frac{d\phi_j(r_D)}{dr_D}\right)^2\right]_1^{r_{De}} = \frac{2}{\lambda_j^2 \pi^2}. \tag{5.129}$$

$Q^{\#}$ is readily known from Eqs. (5.125), (5.128), and (5.129) as $Q = \pi$. D_j^c and E_j^c are determined from Eqs. (5.123) and (5.124), as follows:

$$D_j^c = -\frac{(c_{11}^{@} + c_{12}^{@})[\ln(r_{De}) - \ln(r_D)]\pi}{d^{@}} \tag{5.130}$$

$$E_j^c = -\frac{(c_{21}^{@} + c_{22}^{@})[\ln(r_{De}) - \ln(r_D)]\pi}{d^{@}}. \tag{5.131}$$

Incorporating all solutions, the final governing equations may be expressed as

$$P_{D1} = \ln(r_{De}) - \ln(r_D) + \sum_{j=1}^{\infty} \phi_j(r_D)\left(c_{22}^{@} D_j^c e^{-c_{11}^{@} t_D} - c_{12}^{@} E_j^c e^{-c_{22}^{@} t_D}\right) \tag{5.132}$$

$$P_{D2} = \ln(r_{De}) - \ln(r_D) + \sum_{j=1}^{\infty} \phi_j(r_D)\left(-c_{21} D_j^c e^{-c_{11} t_D} + c_{11} E_j^c e^{-c_{22} t_D}\right). \tag{5.133}$$

The governing equations (5.132) and (5.133) satisfy the boundary and initial conditions expounded in Eqs. (5.92) to (5.94). In addition, as t_D approaches infinity, fluid pressures in the matrix blocks and in the fractures approach a constant value $[\ln(r_{De}) - \ln(r_D)]$. If the system falls into a single-porosity or equivalent single-porosity category, one of the exponential terms in Eqs. (5.132) and (5.133) drops out. The governing equations then conform to those normally obtained in a single-porosity formulation.

The solution for the fluid pressures given by Eqs. (5.132) to (5.133) may be accurately obtained by implementing an optimized solution technique. As an example, the traditional method to evaluate the cross-product of the Bessel functions in Eq. (5.113) may be represented for $r_D = 1$ by

$$U(r_D) = U(1) = \sum_{j=1}^{J} \Phi_j(1). \tag{5.134}$$

Due to the infinite summation involved in the determination of the Bessel function, the solution of Eq. (5.134) is slow and inaccurate. The error incurred in using Eq. (5.134) may be assessed through the following equation:

$$E_{rr} = \frac{2\sqrt{2}}{\pi} \sum_{j=J+1}^{\infty} (-1)^j \frac{1}{\sqrt{4j-1}}. \tag{5.135}$$

However, if the following expression is applied

$$U(r_D) = U(1) = \sum_{j=1}^{J} \Phi_j(1) + \frac{1}{2}\Phi_{J+1}(1) \tag{5.136}$$

then convergence can be rapidly achieved. For instance, to achieve 2% accuracy (or 98%), 1,000 iterations are required by the conventional method, and only 17 iterations are needed via the optimized method. Furthermore, the exact solution can be obtained after about 60 iterations, which is practically unachievable if the traditional technique is adopted. The error resulting from the optimized method may be expressed as

$$Err^* = \frac{2\sqrt{2}}{\pi} \sum_{j=J+1}^{\infty} (-1)^j \frac{1}{2\sqrt{j(4j-1)(4j+3)}}. \tag{5.137}$$

Comparing the errors given by Eqs. (5.135) and (5.137), it is apparent that the optimized method produces at least one order lower error than the conventional method, thus explaining the significant improvement in convergence rate.

5.4.1.2 Illustrative Example

For $r_0 = k_1/k_2$, $r_1 = K_f/K_{fr}$, $r_2 = K_f/K_s$ (defined as permeability ratio, fracture compressibility ratio, and grain compressibility ratio, respectively). The dual-porosity behavior is explicitly demonstrated in Figure 5-6 where the dimensionless time is small. Clearly, the transition period of fluid flow between the matrix blocks and the fractures depends primarily on the relative compressibility ratios. This transition period will be extended if r_1 is reduced and r_2 increases. Alternatively, dual-porosity behavior may not be observed if this ratio approaches unity. Therefore, the magnitude of fluid compressibility is also critical in the identification of dual-porosity behavior.

5.4.2 Flow and Deformation

In general, conservation laws hold for momentum, mass, and energy. For an isothermal case, total stress equilibrium must be maintained for the load-deformation behavior of the reservoir, while fluid mass is concurrently conserved. Reservoir deformation, as a result of production, is coupled with induced pore pressure change. This effect is most pronounced in areas of large

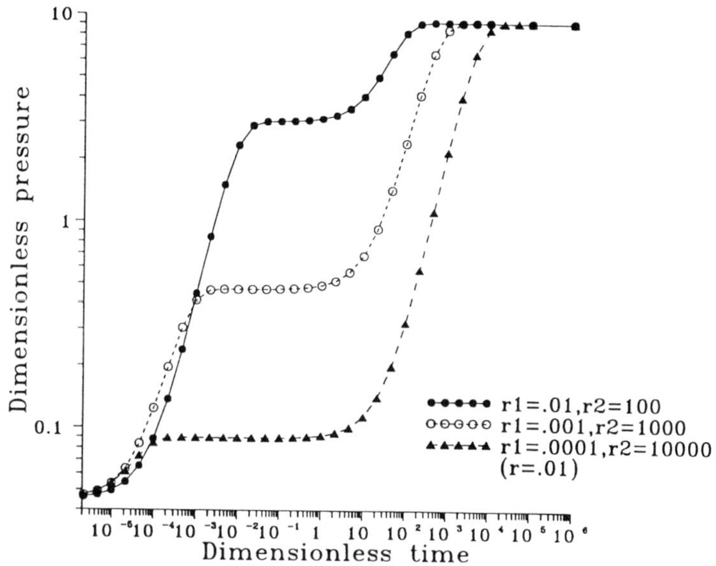

Figure 5-6. Temporal pressure for various compressibilities. (Reprinted from Bai et al. 1994b, with permission from John Wiley and Sons.)

change in total stress, such as in the vicinity of producing wells, with the magnitude of this impact diminishing with distance.

5.4.2.1 Solution Method

In the dual-porosity poromechanical formulation, the governing equations for the solid and fluid phases can be written as (Wilson and Aifantis 1982; Bai et al. 1995a)

$$Gu_{i,jj} + (\lambda + G)u_{k,ki} - \sum_{m=1}^{2} \alpha_m p_{m,i} = 0 \tag{5.138}$$

$$\frac{k_m}{\mu} p_{m,kk} - \alpha_m \dot{\varepsilon}_{kk} - c_m^* \dot{p}_m \pm \Gamma(\Delta p) = 0 \tag{5.139}$$

where m=1 and 2, represent the matrix and fractures, respectively; λ and G are the *Lamé* constants; α is the fluid pressure ratio factor or Biot coefficient in single-porosity cases (Biot 1941); c^* is the total compressibility representing the lumped deformability of the fluid and the fractured or intact medium; k is the permeability; μ is the fluid dynamic viscosity; Γ is the interporosity flow coefficient; u is the solid displacement; p is the fluid pressure; ε_{kk} is the total body strain; and Δp is the pressure difference between fractures and matrix blocks. The formulation in Eqs. (5.138) and (5.139) does not provide characteristic deformation for the fractures. This drawback can be eliminated by using the conversion introduced in Chapter 2 on deformation.

The stress-strain relationship in a poroelastic medium may be expressed as

$$\sigma_{ij} = 2G\varepsilon_{ij} + \delta_{ij}\lambda\varepsilon_{kk} - \delta_{ij}\sum_{m=1}^{2}\alpha_m p_m. \tag{5.140}$$

Biot (1956) first identified the complete analogy between poroelasticity and thermoelasticity. Rice and Cleary (1976) pointed out that the analogy holds only when the coupling between the fluid pressure (or temperature) and stresses is rigorously retained. In other words, the pressure should be derived from the governing equations simultaneously with either stresses or temperature (Cleary 1977). In poroelasticity, complete decoupling may cause a significant error, however, only in close proximity to the application of load (Cryer 1963). In an infinite medium, partial decoupling using the concept of displacement potential offers an acceptable approximation to that of rigorous coupling, even adjacent to a fluid source (Curran and Carvalho 1987). As a result, a simple sequential solution procedure is then permissible. For an infinite medium with isotropic and homogeneous properties, the solid displacement can be expressed in terms of a displacement potential proposed by Goodier (1936), assuming the displacement field is irrotational as

$$u_i = \Phi_{,i}. \tag{5.141}$$

Substituting Eq. (5.141) into Eq. (5.138), gives

$$G\Phi_{,ikk} + (\lambda + G)\Phi_{,kki} - \sum_{m=1}^{2}\alpha_m p_{m,i} = 0. \tag{5.142}$$

Integrating Eq. (5.142) with respect to x_i, results in a Poisson equation

$$\Phi_{,kk} = \varepsilon_{kk} = \beta_e\sum_{m=1}^{2}\alpha_m p_m \tag{5.143}$$

where

$$\beta_e = \frac{1}{2G+\lambda} = \frac{(1+\nu)(1-2\nu)}{E(1-\nu)} \tag{5.144}$$

where E is the elastic modulus and ν is the Poisson's ratio. Substituting Eq. (5.143) into Eq. (5.139) yields

$$\frac{k_m}{\mu}p_{m,kk} - \beta_e\alpha_m\sum_{m=1}^{2}\alpha_m\dot{p}_m - c_m^*\dot{p}_m \pm \Gamma(\Delta p) = 0. \tag{5.145}$$

Eq. (5.145) is based on a phenomenological postulation. For a more rigorous formulation, Eqs. (3.138) and (3.139) and terms in Eq. (3.137) of Chapter

3 should be incorporated. Expressing the total compressibilities in more explicit forms and incorporating Eq. (3.137), Eq. (5.145) can be reformulated for the matrix and the fractures, respectively, as

$$\frac{k_1}{\mu}\nabla^2 p_1 = c_1^{\#}\frac{\partial p_1}{\partial t} - c_2^{\#}\frac{\partial p_2}{\partial t} + \Gamma(p_1 - p_2) \tag{5.146}$$

$$\frac{k_2}{\mu}\nabla^2 p_2 = c_3^{\#}\frac{\partial p_2}{\partial t} - c_4^{\#}\frac{\partial p_1}{\partial t} - \Gamma(p_1 - p_2) \tag{5.147}$$

where

$$\begin{cases} c_1^{\#} = \dfrac{\alpha_3 - n_1 - n_2}{K_s} + \dfrac{n_1}{K_f} + \alpha_1^2\beta_e \\ c_2^{\#} = -\dfrac{\alpha_3 - n_1 - n_2}{K_{fr}} - \alpha_1\alpha_2\beta_e \\ c_3^{\#} = \dfrac{\alpha_3 - n_1 - n_2}{K_{fr}} + \dfrac{n_2}{K_f} + \alpha_2^2\beta_e \\ c_4^{\#} = -\dfrac{\alpha_3 - n_1 - n_2}{K_s} - \alpha_1\alpha_2\beta_e. \end{cases} \tag{5.148}$$

To represent a long producing well zone, stress and pressure changes are assumed plane radially symmetric and therefore independent of the circumferential and vertical orientations. Strains are therefore given as

$$\begin{cases} \varepsilon_{rr} = \dfrac{\partial u}{\partial r}, \quad \varepsilon_{\theta\theta} = \dfrac{u}{r} \\ \varepsilon_{r\theta} = \varepsilon_{zz} = 0. \end{cases} \tag{5.149}$$

The total strain is then

$$\Phi_{,kk} = \varepsilon_{kk} = \varepsilon_{rr} + \varepsilon_{\theta\theta} = \frac{\partial u}{\partial r} + \frac{u}{r} \tag{5.150}$$

and stresses can be evaluated from Eq. (5.140) as

$$\begin{aligned} \sigma_{rr} &= 2G\varepsilon_{rr} + \lambda\varepsilon_{kk} - \sum_{m=1}^{2}\alpha_m p_m \\ &= \beta_e^{-1}\frac{\partial u}{\partial r} + \lambda\frac{u}{r} - \sum_{m=1}^{2}\alpha_m p_m \\ \sigma_{\theta\theta} &= 2G\varepsilon_{\theta\theta} + \lambda\varepsilon_{kk} - \sum_{m=1}^{2}\alpha_m p_m \end{aligned} \tag{5.151}$$

$$= \beta_e^{-1}\frac{u}{r} + \lambda\frac{\partial u}{\partial r} - \sum_{m=1}^{2} \alpha_m p_m, \tag{5.152}$$

and,

$$\sigma_{zz} = \nu(\sigma_{rr} + \sigma_{\theta\theta}) = \lambda\varepsilon_{kk} - 2\nu \sum_{m=1}^{2} \alpha_m p_m. \tag{5.153}$$

In a radial system, Eq. (5.138) can be reformulated as

$$\beta_e^{-1}\frac{\partial}{\partial r}\left[\frac{1}{r}\frac{\partial}{\partial r}(ru)\right] - \sum_{m=1}^{2} \alpha_m \frac{\partial p_m}{\partial r} = 0. \tag{5.154}$$

Integrating Eq. (5.154) with respect to radius r, and assuming $r \to \infty$, Δu and $\Delta p_m \to 0$ (Δ indicates the change over initial values), yields

$$\frac{\partial u}{\partial r} + \frac{u}{r} = \beta_e \sum_{m=1}^{2} \alpha_m p_m. \tag{5.155}$$

Note that Eq. (5.155) is identical to the combination of Eqs. (5.143) and (5.150).

The dimensionless forms of pressure P_{Di}, radius r_D, and interporosity flow coefficient λ^* are identical to those defined in Eqs. (5.84), (5.85), and (5.88). Other dimensionless parameters are defined as

- Dimensionless time:

$$t_D = \frac{(k_1 + k_2)t}{\mu r_w^2 \sum_{i=1}^{4} c_i^{\#}} \tag{5.156}$$

  where $c_i^{\#}$ are described in Eq. (5.148).

- Dimensionless displacement:

$$U_D^* = \frac{u_0 - u}{r_w} \tag{5.157}$$

  where u_0 is the initial displacement.

- Elastic coefficient:

$$\beta_e^* = \beta_e \frac{q\mu}{2\pi h_r(k_1 + k_2)} \tag{5.158}$$

- Operational coefficient:

$$\omega_i^{@} = \frac{(k_1 + k_2)c_i^{\#}}{k_j \sum_{i=1}^{4} c_i^{\#}} \qquad (i = 1, 2, 3, 4) \tag{5.159}$$

where

$$k_j = \begin{cases} k_1 & \text{when } i = 1 \text{ and } 2 \\ k_2 & \text{when } i = 3 \text{ and } 4 \end{cases}$$

- Permeability ratio:

$$R_k = \frac{k_1}{k_2}. \tag{5.160}$$

The governing equations (5.155), (5.146), and (5.147) can be rewritten in dimensionless form as

$$\frac{\partial U_D^*}{\partial r_D} + \frac{U_D^*}{r_D} = \beta_e^*(\alpha_1 P_{D1} + \alpha_2 P_{D2}) \tag{5.161}$$

$$\frac{\partial^2 P_{D1}}{\partial r_D^2} + \frac{1}{r_D}\frac{\partial P_{D1}}{\partial r_D} = \omega_1^{@}\frac{\partial P_{D1}}{\partial t_D} - \omega_2^{@}\frac{\partial P_{D2}}{\partial t_D} + \lambda^*(P_{D1} - P_{D2}) \tag{5.162}$$

$$\frac{\partial^2 P_{D2}}{\partial r_D^2} + \frac{1}{r_D}\frac{\partial P_{D2}}{\partial r_D} = \omega_3^{@}\frac{\partial P_{D2}}{\partial t_D} - \omega_4^{@}\frac{\partial P_{D1}}{\partial t_D} - \lambda^* R_k(P_{D1} - P_{D2}). \tag{5.163}$$

Eqs. (5.162) and (5.163) have almost identical forms to the flow equations of (5.90) and (5.91) in the previous section. As a result, solutions of Eqs. (5.162) and (5.163) follow the same procedure described in that section. Only the solution of Eq. (5.161) needs to be developed.

It is understood that P_{D1} and P_{D2} can be obtained from Eqs. (5.162) and (5.163) independently from (5.161). Substituting P_{D1} and P_{D2} into Eq. (5.161) and integrating with respect to r_D, yields

$$U_D^* = \frac{1}{r_D}[\beta_e^* \int_1^{r_D} (\alpha_1 P_{D1} + \alpha_2 P_{D2}) r_D dr_D + g_h(t)] \tag{5.164}$$

where $g_h(t)$ can be determined by satisfying the initial and boundary conditions defined in Eqs. (5.92), (5.93), and (5.94), along with the displacement boundary condition,

$$U_D^*|_{r_D = r_{De}} = 0. \tag{5.165}$$

Therefore,

$$U_D^* = -\frac{\beta_e^*}{r_D}\left\{\frac{(\alpha_1 + \alpha_2)}{2}\left[\frac{1}{2}(r_{De}^2 - r_D^2) + r_D^2[\ln(r_D) - \ln(r_{De})]\right] - \right.$$

$$\sum_{j=1}^{\infty}\frac{2H^{\#}}{\pi\xi_j^2}\left[\frac{\ln(r_{De})}{J_0(\xi_j)}[J_0(\xi_j r_{De})-J_0(\xi_j r_D)]+\ln(r_{De})-\ln(r_D)\right]\Bigg\} \tag{5.166}$$

where

$$H^{\#}=\frac{-\pi}{d_h^*}$$

$$\left[(\zeta_{11}+\zeta_{12})(\alpha_1\zeta_{22}-\alpha_2\zeta_{21})e^{-\zeta_{11}t_D}+(\zeta_{21}+\zeta_{22})(\alpha_2\zeta_{11}-\alpha_1\zeta_{12})e^{-\zeta_{22}t_D}\right] \tag{5.167}$$

and

$$\begin{cases}\zeta_{11}=(a_1^{\#}\omega_3^{@}-\lambda^*\omega_2^{@}R_k)\omega_b^{-1}, & \zeta_{12}=(a_2^{\#}\omega_2^{@}-\lambda^*\omega_3^{@})\omega_b^{-1}\\ \zeta_{21}=(a_1^{\#}\omega_4^{@}-\lambda^*\omega_1^{@}R_k)\omega_b^{-1}, & \zeta_{22}=(a_2^{\#}\omega_1^{@}-\lambda^*\omega_4^{@})\omega_b^{-1},\\ a_1^{\#}=(\xi^2+\lambda^*), & a_2^{\#}=(\xi^2+\lambda^*R_k)\\ d_h^*=\zeta_{11}\zeta_{22}-\zeta_{12}\zeta_{21}, & \omega_b=\omega_1^{@}\omega_3^{@}-\omega_2^{@}\omega_4^{@}.\end{cases} \tag{5.168}$$

The derivative of the displacement may be derived from Eq. (5.161) as an alternative to direct differentiation from Eq. (5.166) as

$$\frac{\partial U_D^*}{\partial r_D}=-\frac{U_D^*}{r_D}+\beta_e^*(\alpha_1 P_{D1}+\alpha_2 P_{D2}). \tag{5.169}$$

5.4.2.2 Illustrative Example

The ratios r_0, r_1 and r_2 are as defined in the previous section, with the matrix compressibility ratio, r_3, defined as $r_3=\frac{K_f}{E}$ where E is the elastic modulus of the porous medium. Dual-porosity behavior is exhibited if the matrix compressibility ratio of the rock mass is relatively large for fixed ratios of r_0, r_1 and r_2. The effect of the elastic constant of the rock mass diminishes when the ratio r_3 is smaller than 10^{-5} (Figure 5-7). Although the duration is short at early dimensionless times, the pressure change is significant, depending on the magnitudes of r_3. In addition, the period of pressure stabilization during the fluid transfer between matrix and fractures appears to be prolonged as a result of variation in r_3. It is important to note that the depleting pressure magnitude at early periods appears greater than the reservoir pressure as depicted in Figure 5-7. This poromechanical impact cannot be observed if a conventional flow model is used alone.

5.5 DIFFERENTIAL OPERATOR METHOD

As demonstrated previously, the coupled flow and transport equations can be solved using Fourier transform and Hankel transform methods. However, these

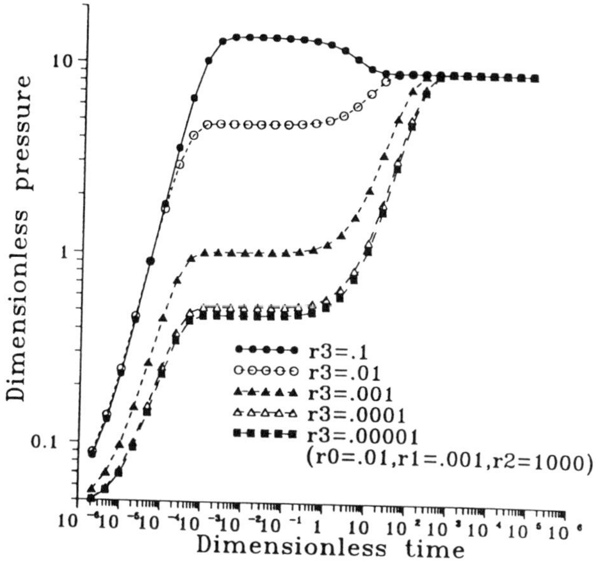

Figure 5-7. Temporal pressure for various elastic constants. (Reprinted from Bai et al. 1995a, with permission from Elsevier Science.)

solutions require special functions with infinite summation series, which are frequently sufficiently tedious that their practical utilization is discouraged. In contrast, the method of Laplace transforms does not require the use of special functions (excluding numerical inversion), but is suitable for only decoupled formulations. An alternative method, incorporating the use of differential operators, is advantageous in avoiding the drawback of the Laplace transforms. The coupled system of equations are solved within the Laplace domain without invoking any special functions.

5.5.1 Flow

The method of differential operators, applied to fluid flow, is evaluated by examining a coupled dual-porosity model.

5.5.1.1 Solution Method

The coupled dual-porosity model proposed by Barenblatt et al. (1960), as shown in Eqs. (3.83) and (3.84) of Chapter 3 can be rewritten as

$$\frac{k_1}{\mu}\nabla^2 p_1^* = c_1^* n_1 \frac{\partial p_1^*}{\partial t^*} + \Gamma(p_1^* - p_2^*) \tag{5.170}$$

$$\frac{k_2}{\mu}\nabla^2 p_2^* = c_2^* n_2 \frac{\partial p_2^*}{\partial t^*} - \Gamma(p_1^* - p_2^*) \tag{5.171}$$

where p^* is the fluid pressure; definitions of all other parameters are as in Eqs. (3.83) and (3.84) of Chapter 3.

For the 1-D radial flow case, the above equations can be written as (Bai 1997)

$$\frac{\partial^2 p_1^*}{\partial (r^*)^2} + \frac{1}{r^*}\frac{\partial p_1^*}{\partial r^*} = \frac{\mu c_1^* n_1}{k_1}\frac{\partial p_1^*}{\partial t^*} + \Gamma^{\#}(p_1^* - p_2^*) \tag{5.172}$$

$$\frac{\partial^2 p_2^*}{\partial (r^*)^2} + \frac{1}{r^*}\frac{\partial p_2^*}{\partial r^*} = \frac{\mu c_2^* n_2}{k_2}\frac{\partial p_2^*}{\partial t^*} - \Gamma^{\#}\frac{k_1}{k_2}(p_1^* - p_2^*) \tag{5.173}$$

where $\Gamma^{\#} = \frac{\Gamma\mu}{k_1}$, and r^* is the radial distance from the well. The radial distance can be converted to a linear distance by using a logarithmic scale.

Conventional application frequently invokes a dimensionless formulation, as described by the following terms:

$$\begin{cases} p_i = \dfrac{2\pi k_2 h_r (p_0 - p_i^*)}{q\mu} \quad (i = 1, 2), \\ r = \dfrac{r^*}{r_w}, \qquad \lambda^* = \Gamma^{\#} r_w^2, \qquad R_k = \dfrac{k_1}{k_2}, \\ t = \dfrac{(k_1 + k_2)t^*}{\mu r_w^2 (c_1^* n_1 + c_2^* n_2)}, \qquad \omega = \dfrac{c_2^* n_2}{c_1^* n_1 + c_2^* n_2} \end{cases} \tag{5.174}$$

where h_r is the reservoir thickness, p_0 is the initial reservoir pressure, q is the flow rate at the well, and r_w is the wellbore radius. Although the rate of flow supply to the well from the matrix is typically undefined as a proportion of the total, it is assumed that fluid flows at the well only through the fractures.

The dimensionless form for fluid flow through a fractured porous formation can be written as

$$\frac{\partial^2 p_1}{\partial r^2} + \frac{1}{r}\frac{\partial p_1}{\partial r} = \left(1 + \frac{1}{R_k}\right)(1 - \omega)\frac{\partial p_1}{\partial t} + \lambda^*(p_1 - p_2) \tag{5.175}$$

$$\frac{\partial^2 p_2}{\partial r^2} + \frac{1}{r}\frac{\partial p_2}{\partial r} = (1 + R_k)\omega\frac{\partial p_2}{\partial t} - \lambda^* R_k(p_1 - p_2) \tag{5.176}$$

where it is indicated that fluid flow is characterized by the equivalent storage ratio for the fractures ω, the permeability ratio R_k, and the interporosity coefficient λ^*.

For a constant flow rate at the well ($r^* = r_w$) and constant pressure at the reservoir boundary ($r^* = r_e^*$), the boundary and initial conditions are expressed in dimensionless form as

$$\left.\frac{\partial p_1}{\partial r}\right|_{r=1} = 0, \quad \left.\frac{\partial p_2}{\partial r}\right|_{r=1} = -1, \quad p_1 = p_2|_{r=r_e} = 0, \quad p_1 = p_2|_{t=0} = 0 \tag{5.177}$$

where r_e is the dimensionless reservoir radius.

Applying a Laplace transform to Eqs. (5.175) and (5.176) results in the following ordinary differential equations:

$$\frac{d^2\bar{p}_1}{dr^2} + \frac{1}{r}\frac{d\bar{p}_1}{dr} = \left(1 + \frac{1}{R_k}\right)(1-\omega)s\bar{p}_1 + \lambda^*(\bar{p}_1 - \bar{p}_2) \tag{5.178}$$

$$\frac{d^2\bar{p}_2}{dr^2} + \frac{1}{r}\frac{d\bar{p}_2}{dr} = (1 + R_k)\omega s\bar{p}_2 - \lambda^* R_k(\bar{p}_1 - \bar{p}_2) \tag{5.179}$$

where s is a Laplace parameter.

Boundary conditions are changed from Eq. (5.177) to

$$\left.\frac{d\bar{p}_1}{dr}\right|_{r=1} = 0, \qquad \left.\frac{d\bar{p}_2}{dr}\right|_{r=1} = -\frac{1}{s}, \qquad \bar{p}_1 = \bar{p}_2|_{r=r_e} = 0. \tag{5.180}$$

Invoking the method of differential operators (Mathematical Handbook 1979), the differential operator, D_{do}^n, can be expressed as

$$D_{do}^n f(x_i) = \frac{d^n f(x_i)}{dx_i^n} \tag{5.181}$$

where i indexes an arbitrary variable and the superscript n is the order of the differential equations.

Applying the differential operator to Eqs. (5.178) and (5.179) gives

$$\left\{D_{do}^2 + \frac{D_{do}}{r} - [\lambda^* + (1 + \frac{1}{R_k})(1-\omega)s]\right\}\bar{p}_1 + \lambda^*\bar{p}_2 = 0 \tag{5.182}$$

$$\left\{D_{do}^2 + \frac{D_{do}}{r} - [\lambda^* R_k + (1 + R_k)\omega s]\right\}\bar{p}_2 + \lambda^* R_k\bar{p}_1 = 0. \tag{5.183}$$

The following relationship is derived from Eq. (5.183):

$$\bar{p}_1 = \frac{-1}{\lambda^* R_k}\left\{D_{do}^2 + \frac{D_{do}}{r} - [\lambda^* R_k + (1 + R_k)\omega s]\right\}\bar{p}_2. \tag{5.184}$$

Substituting Eq. (5.184) into (5.182) results in

$$(D_{do}^4 + B_1^* D_{do}^3 + B_2^* D_{do}^2 + B_3^* D_{do} + B_4^*)\bar{p}_2 = 0 \tag{5.185}$$

where

$$\begin{cases} B_1^* = \frac{2}{r}, \qquad B_2^* = \frac{1}{r^2} - B^*, \qquad B_3^* = -\frac{B^*}{r}, \\ B_4^* = s\left(1 + \frac{1}{R_k}\right)[\lambda^* R_k + \omega s(1 + R_k)(1-\omega)]. \\ B^* = \left(1 + \frac{1}{R_k}\right)\{\lambda^* R_k + s[1 - \omega(1 - R_k)]. \end{cases} \tag{5.186}$$

Eq. (5.185) has four roots from the following equations:

$$D_{do}^2 + \xi_1 D_{do} + \phi_1 = 0 \tag{5.187}$$

$$D_{do}^2 + \xi_2 D_{do} + \phi_2 = 0 \tag{5.188}$$

where

$$\begin{cases} \xi_1 = \frac{1}{2}(B_1^* + \sqrt{\Delta_1}), & \xi_2 = \frac{1}{2}(B_1^* - \sqrt{\Delta_1}), \\ \phi_1 = z_r + \dfrac{B_1^* z_r - B_3^*}{\sqrt{\Delta_1}}, & \phi_2 = z_r - \dfrac{B_1^* z_r - B_3^*}{\sqrt{\Delta_1}}, \\ \Delta_1 = 8z_r + (B_1^*)^2 - 4B_2^* \end{cases} \tag{5.189}$$

where z_r is any real root of the following equation:

$$8z_r^3 + 4E_1^* z_r^2 + E_2^* z_r + E_3^* = 0 \tag{5.190}$$

and where

$$\begin{cases} E_1^* = -B_2^*, \quad E_2^* = 2B_1^* B_3^* - 8B_4^* \\ E_3^* = B_4^*[4B_2^* - (B_1^*)^2] - (B_3^*)^2. \end{cases} \tag{5.191}$$

Further, assume

$$z_r = u^* - \frac{E_1^*}{6} \tag{5.192}$$

where the parameter u^* can be determined from the 3rd order equation

$$(u^*)^3 + \tau_1^* u^* + \tau_2^* = 0 \tag{5.193}$$

and

$$\begin{cases} \tau_1^* = \frac{1}{8}\left[E_2^* - \dfrac{2(E_1^*)^2}{3}\right], \\ \tau_2^* = \frac{1}{8}\left[\dfrac{2(E_1^*)^3}{27} - \dfrac{E_1^* E_2^*}{6} + E_3^*\right]. \end{cases} \tag{5.194}$$

The three roots of u^* may be described as

$$u_1^* = F_1^* + F_2^*, \quad u_2^* = \theta_1 F_1^* + \theta_2 F_2^*, \quad u_3^* = \theta_2 F_1^* + \theta_1 F_2^* \tag{5.195}$$

where

$$\begin{cases} F_1^* = \left(-\frac{\tau_2^*}{2} + \sqrt{\Delta^*}\right)^{\frac{1}{3}}, & F_2^* = \left(-\frac{\tau_2^*}{2} - \sqrt{\Delta^*}\right)^{\frac{1}{3}}, \\ \Delta^* = (\frac{\tau_2^*}{2})^2 + (\frac{\tau_1^*}{3})^3, \quad \theta_1 = -\frac{1}{2} + \frac{\sqrt{3}}{2}i, & \theta_2 = -\frac{1}{2} - \frac{\sqrt{3}}{2}i. \end{cases} \tag{5.196}$$

Three possible solutions for the real root, z_r, exist depending upon the signs of Δ^* in Eq. (5.196), as indicated in the following:

When $\Delta^* > 0$, the real root is

$$z_r = F_1^* + F_2^* - \frac{E_1^*}{6}. \tag{5.197}$$

If $\Delta^* = 0$, then the real root becomes

$$z_r = -\sqrt[3]{4\tau_2^*} - \frac{E_1^*}{6}. \tag{5.198}$$

However, if $\Delta^* < 0$, the real root is recovered in trigonometric form as

$$z_r = 2\sqrt[3]{\bar{r}}\cos(\frac{\alpha_{do}}{3}) - \frac{E_1^*}{6} \tag{5.199}$$

where

$$\bar{r} = \sqrt{-(\frac{\tau_1^*}{3})^3}, \quad \alpha_{do} = \arccos(-\frac{\tau_2^*}{2\bar{r}}). \tag{5.200}$$

Once the real root, z_r, is determined for Eq. (5.190), the four roots of Eqs. (5.187) and (5.188) can then be expressed as

$$\begin{cases} \psi_1 = -\frac{\xi_1}{2} + \sqrt{\Delta_2}, & \psi_2 = -\frac{\xi_1}{2} - \sqrt{\Delta_2}, \\ \psi_3 = -\frac{\xi_2}{2} + \sqrt{\Delta_3}, & \psi_4 = -\frac{\xi_2}{2} - \sqrt{\Delta_3}, \\ \Delta_2 = \frac{\xi_1^2}{4} - \phi_1, & \Delta_3 = \frac{\xi_2^2}{4} - \phi_2. \end{cases} \tag{5.201}$$

The solutions in the Laplace domain may become complicated as a result of uncertainty in the signs of Δ_1, Δ_2, and Δ_3. However, for a choice of the limiting physical parameters: $r(1 \to \infty)$, $s(0 \to \infty)$, $\lambda^*(0 \to \infty)$, $\omega(0 \to 1)$, $R_k(0 \to 1)$, Δ_1 in Eq. (5.189) is predominantly positive.

From the procedure previously described, the following system of equations can be derived, after satisfying the boundary conditions, as expounded in Eq. (5.180):

$$\begin{bmatrix} \beta_1^* & \beta_2^* & \beta_3^* & \beta_4^* \\ \beta_1^@ & \beta_2^@ & \beta_3^@ & \beta_4^@ \\ \gamma_1^* & \gamma_2^* & \gamma_3^* & \gamma_4^* \\ \gamma_1^@ & \gamma_2^@ & \gamma_3^@ & \gamma_4^@ \end{bmatrix} \begin{bmatrix} g_1^\# \\ g_2^\# \\ g_3^\# \\ g_4^\# \end{bmatrix} = \begin{bmatrix} 0 \\ 0 \\ 0 \\ \delta^* \end{bmatrix} \tag{5.202}$$

where $\delta^* = -\frac{1}{s}$. It should be noted that the above four equations imply that the four boundary conditions, as defined in Eq. (5.180), are applied and satisfied.

The solutions of Eq. (5.202) can be expressed as

$$g_i^{\#} = \frac{I_i^*}{J^*} \qquad (i = 1, 2, 3, 4) \tag{5.203}$$

where

$$J^* = \sum_{i=1}^{4} (-1)^{i-1} \beta_i^* J_i^* \tag{5.204}$$

and where J_i^* can be described as

$$J_i^* = \beta_j^@(\gamma_k^*\gamma_l^@ - \gamma_k^@\gamma_l^*) - \beta_k^@(\gamma_j^*\gamma_l^@ - \gamma_j^@\gamma_l^*) + \beta_l^@(\gamma_j^*\gamma_k^@ - \gamma_j^@\gamma_k^*) \tag{5.205}$$

following the rotation rule, $i \neq j \neq k \neq l$. More specifically:

$$\begin{cases} i = 1, \ j = 2, \ k = 3, \ l = 4 \\ i = 2, \ j = 1, \ k = 3, \ l = 4 \\ i = 3, \ j = 1, \ k = 2, \ l = 4 \\ i = 4, \ j = 1, \ k = 2, \ l = 3. \end{cases} \tag{5.206}$$

Similarly,

$$I_i^* = (-1)^i \delta^* [\beta_j^*(\beta_k^@\gamma_l^* - \gamma_k^*\beta_l^@) - \beta_k^*(\beta_j^@\gamma_l^* - \gamma_j^*\beta_l^@) + \beta_l^*(\beta_j^@\gamma_k^* - \gamma_j^*\beta_k^@)] \tag{5.207}$$

where the rotation rule also applies.

The solutions in the Laplace domain can be divided into four groups depending on the signs of Δ_2 and Δ_3 in Eq. (5.201) while $\Delta_1 > 0$.

Case 1. $\Delta_2 \geq 0$ *and* $\Delta_3 \geq 0$*:*

In the Laplace domain, the fluid pressure in the fractures can be derived from Eq. (5.185) as

$$\bar{p}_2 = \sum_{i=1}^{4} g_i^{\#} e^{\psi_i r} \tag{5.208}$$

where constants $g_i^{\#}$ are determined by satisfying the boundary conditions of Eq. (5.180).

For cases with finite fluid pressure and since $\psi_3 > 0$ and $\psi_4 > 0$ in Eq. (5.201), it is known from the outer boundary conditions in Eq. (5.180) that $g_3 = g_4 = 0$. As a result,

$$\bar{p}_2 = \sum_{i=1}^{2} g_i^{\#} e^{\psi_i r}. \tag{5.209}$$

The fluid pressure of the matrix in the Laplace domain can be determined from Eq. (5.184) as

$$\bar{p}_1 = -\frac{1}{\lambda^* R_k} \sum_{i=1}^{2} g_i^{\#} \eta_i^{@} e^{\psi_i r} \tag{5.210}$$

where

$$\eta_i^{@} = \psi_i^2 + \frac{\psi_i}{r} - [\lambda^* R_k + \omega s(1 + R_k)]. \tag{5.211}$$

The constants $g_1^{\#}$ and $g_2^{\#}$ can be determined from the inner boundary conditions in Eq. (5.180), which yield

$$\begin{bmatrix} A_1^* & A_2^* \\ A_1^{@} & A_2^{@} \end{bmatrix} \begin{bmatrix} g_1^{\#} \\ g_2^{\#} \end{bmatrix} = \begin{bmatrix} 0 \\ \delta^* \end{bmatrix} \tag{5.212}$$

where

$$A_i^* = \eta_i^{@} A_i^{@}, \qquad A_i^{@} = \psi_i e^{\psi_i}, \qquad \delta^* = -\frac{1}{s}. \tag{5.213}$$

Solving Eq. (5.212) gives

$$g_1^{\#} = -\frac{\delta^* A_2^*}{A^*}, \qquad g_2^{\#} = \frac{\delta^* A_1^*}{A^*} \tag{5.214}$$

where

$$A^* = A_1^* A_2^{@} - A_1^{@} A_2^*. \tag{5.215}$$

The corresponding parameters required in Eq. (5.202) are

$$\beta_i^* = \eta_i^{@} \beta_i^{@}, \qquad \beta_i^{@} = e^{\psi_i r_e}, \qquad \gamma_i^* = \eta_i^{@} \gamma_i^{@}, \qquad \gamma_i^{@} = \psi_i e^{\psi_i} \tag{5.216}$$

where, again, $i = 1, 2, 3, 4$.

Case 2. $\Delta_2 \geq 0$ *and* $\Delta_3 < 0$*:*

The solution is expressed as

$$\bar{p}_2 = \sum_{i=1}^{2} g_i^{\#} e^{\psi_i r} + e^{W_3 r}[g_3^{\#} \cos(W_4 r) + g_4^{\#} \sin(W_4 r)] \tag{5.217}$$

where

$$W_3 = -\frac{\xi_2}{2}, \qquad W_4 = \sqrt{-\Delta_3} \tag{5.218}$$

$$\bar{p}_1 = -\frac{1}{\lambda^* R_k} \left\{ \sum_{i=1}^{2} g_i^{\#} \eta_i^{\#} e^{\psi_i r} + e^{W_3 r} \{ g_3^{\#} [\lambda_1^{\#} \cos(W_4 r) - \lambda_2^{\#} \sin(W_4 r)] \right.$$

$$+g_4^{\#}[\lambda_2^{\#}\cos(W_4 r) + \lambda_1^{\#}\sin(W_4 r)]\}\Big\} \tag{5.219}$$

where

$$\begin{cases} \lambda_1^{\#} = W_3^2 - W_4^2 + \Omega_1 W_3 + \Omega_2, \quad \lambda_2^{\#} = W_4(2W_3 + \Omega_1), \\ \Omega_1 = \frac{1}{r}, \qquad \Omega_2 = -[\lambda^* R_k + (1 + R_k)\omega s]. \end{cases} \tag{5.220}$$

For $i = 1, 2$, β_i^*, $\beta_i^@$, γ_i^* and $\gamma_i^@$ are identical to those in Case 1. However, for $i = 3, 4$:

$$\begin{cases} \beta_3^* = \eta_3^@ e^{W_3 r_e} \qquad \beta_4^* = \eta_4^@ e^{W_3 r_e} \\ \beta_3^@ = e^{W_3 r_e}\cos(W_4 r_e) \qquad \beta_4^@ = e^{W_3 r_e}\sin(W_4 r_e) \end{cases} \tag{5.221}$$

$$\begin{cases} \gamma_3^* = e^{W_3}[\lambda_3^*\cos(W_4) - \lambda_4^*\sin(W_4)], \\ \gamma_4^* = e^{W_3}[\lambda_4^*\cos(W_4) + \lambda_3^*\sin(W_4)], \\ \gamma_3^@ = e^{W_3}[W_3\cos(W_4) - W_4\sin(W_4)], \\ \gamma_4^@ = e^{W_3}[W_3\sin(W_4) + W_4\cos(W_4)], \end{cases} \tag{5.222}$$

$$\begin{cases} \lambda_3^{\#} = W_3\lambda_1^{\#} - W_4\lambda_2^{\#}, \qquad \lambda_4^{\#} = W_3\lambda_2^{\#} + W_4\lambda_1^{\#}, \\ \eta_3^@ = \lambda_1^{\#}\cos(W_4 r_e) - \lambda_2^{\#}\sin(W_4 r_e), \quad \eta_4^@ = \lambda_2^{\#}\cos(W_4 r_e) + \lambda_1^{\#}\sin(W_4 r_e). \end{cases} \tag{5.223}$$

Case 3. $\Delta_2 < 0$ *and* $\Delta_3 \geq 0$*:*

The solution is written as

$$\bar{p}_2 = e^{W_1 r}[g_1^{\#}\cos(W_2 r) + g_2^{\#}\sin(W_2 r)] + \sum_{i=3}^{4} g_i^{\#} e^{\psi_i r} \tag{5.224}$$

where

$$W_1 = -\frac{\xi_1}{2}, \qquad W_2 = \sqrt{-\Delta_2} \tag{5.225}$$

$$\bar{p}_1 = -\frac{1}{\lambda^* R_k}\{e^{W_1 r}\{g_1^{\#}[\lambda_1^@\cos(W_2 r) - \lambda_2^@\sin(W_2 r)]$$

$$+g_2^{\#}[\lambda_2^@\cos(W_2 r) + \lambda_1^@\sin(W_2 r)]\} + \sum_{i=3}^{4} g_i^{\#}\eta_i^@ e^{\psi_i r}\}, \tag{5.226}$$

and where

$$\begin{cases} \eta_1^@ = \lambda_1^@ \cos(W_2 r_e) - \lambda_2^@ \sin(W_2 r_e), \\ \eta_2^@ = \lambda_2^@ \cos(W_2 r_e) + \lambda_1^@ \sin(W_2 r_e), \\ \lambda_1^@ = W_1^2 - W_2^2 + \Omega_1 W_1 + \Omega_2, \\ \lambda_2^@ = W_2(2W_1 + \Omega_1). \end{cases} \tag{5.227}$$

For $i = 3, 4$, β_i^*, $\beta_i^@$, γ_i^* and $\gamma_i^@$ are identical to those in Case 1. However, for $i = 1, 2$:

$$\begin{cases} \beta_1^* = \eta_1^@ e^{W_1 r_e}, \qquad \beta_2^* = \eta_2^@ e^{W_1 r_e}, \\ \beta_1^@ = e^{W_1 r_e} \cos(W_2 r_e), \qquad \beta_2^@ = e^{W_1 r_e} \sin(W_2 r_e), \end{cases} \tag{5.228}$$

$$\begin{cases} \gamma_1^* = e^{W_1}[\lambda_3^@ \cos(W_2) - \lambda_4^@ \sin(W_2)] \\ \gamma_2^* = e^{W_1}[\lambda_4^@ \cos(W_2) + \lambda_3^@ \sin(W_2)] \\ \gamma_1^@ = e^{W_1}[W_1 \cos(W_2) - W_2 \sin(W_2)] \\ \gamma_2^@ = e^{W_1}[W_1 \sin(W_2) + W_2 \cos(W_2)] \\ \lambda_3^@ = W_1 \lambda_1^@ - W_2 \lambda_2^@, \qquad \lambda_4^@ = W_1 \lambda_2^@ + W_2 \lambda_1^@. \end{cases} \tag{5.229}$$

Case 4. $\Delta_2 < 0$ *and* $\Delta_3 < 0$*:*

The solution is given as follows:

$$\bar{p}_2 = e^{W_1 r}[g_1^\# \cos(W_2 r) + g_2^\# \sin(W_2 r)] + e^{W_3 r}[g_3^\# \cos(W_4 r) + g_4^\# \sin(W_4 r)] \tag{5.230}$$

$$\bar{p}_1 = -\frac{1}{\lambda^* R_k}[\sum_{i=1}^{4} g_i^@] \tag{5.231}$$

where

$$\begin{cases} g_1^@ = g_1^\# e^{W_1 r}[\lambda_1^@ \cos(W_2 r) - \lambda_2^@ \sin(W_2 r)] \\ g_2^@ = g_2^\# e^{W_1 r}[\lambda_2^@ \cos(W_2 r) + \lambda_1^@ \sin(W_2 r)] \\ g_3^@ = g_3^\# e^{W_3 r}[\lambda_1^\# \cos(W_4 r) - \lambda_2^\# \sin(W_4 r)] \\ g_4^@ = g_4^\# e^{W_3 r}[\lambda_2^\# \cos(W_4 r) + \lambda_1^\# \sin(W_4 r)] \end{cases} \tag{5.232}$$

and where

$$\begin{cases} \eta_1^{@} = \lambda_1^{@}\cos(W_2 r_e) - \lambda_2^{@}\sin(W_2 r_e) \\ \eta_2^{@} = \lambda_2^{@}\cos(W_2 r_e) + \lambda_1^{@}\sin(W_2 r_e) \\ \eta_3^{@} = \lambda_1^{\#}\cos(W_4 r_e) - \lambda_2^{\#}\sin(W_4 r_e) \\ \eta_4^{@} = \lambda_2^{\#}\cos(W_4 r_e) + \lambda_1^{\#}\sin(W_4 r_e) \end{cases} \tag{5.233}$$

$$\begin{cases} \lambda_1^{@} = W_1^2 - W_2^2 + \Omega_1 W_1 + \Omega_2 \\ \lambda_2^{@} = W_2(2W_1 + \Omega_1) \\ \lambda_1^{\#} = W_3^2 - W_4^2 + \Omega_1 W_3 + \Omega_2 \\ \lambda_2^{\#} = W_4(2W_3 + \Omega_1) \end{cases} \tag{5.234}$$

$$\begin{cases} \beta_1^{@} = e^{W_1 r_e}\cos(W_2 r_e), \quad \beta_2^{@} = e^{W_1 r_e}\sin(W_2 r_e), \\ \beta_3^{@} = e^{W_3 r_e}\cos(W_4 r_e), \quad \beta_4^{@} = e^{W_3 r_e}\sin(W_4 r_e), \\ \beta_1^{*} = \eta_1^{@} e^{W_1 r_e},\ \beta_2^{*} = \eta_2^{@} e^{W_1 r_e},\ \beta_3^{*} = \eta_3^{@} e^{W_3 r_e},\ \beta_4^{*} = \eta_4^{@} e^{W_3 r_e}, \end{cases} \tag{5.235}$$

$$\begin{cases} \gamma_1^{@} = e^{W_1}[W_1\cos(W_2) - W_2\sin(W_2)] \\ \gamma_2^{@} = e^{W_1}[W_1\sin(W_2) + W_2\cos(W_2)] \\ \gamma_3^{@} = e^{W_3}[W_3\cos(W_4) - W_4\sin(W_4)] \\ \gamma_4^{@} = e^{W_3}[W_3\sin(W_4) + W_4\cos(W_4)] \end{cases} \tag{5.236}$$

$$\begin{cases} \gamma_1^{*} = e^{W_1}[\lambda_3^{@}\cos(W_2) - \lambda_4^{@}\sin(W_2)] \\ \gamma_2^{*} = e^{W_1}[\lambda_4^{@}\cos(W_2) + \lambda_3^{@}\sin(W_2)] \\ \gamma_3^{*} = e^{W_3}[\lambda_3^{\#}\cos(W_4) - \lambda_4^{\#}\sin(W_4)] \\ \gamma_4^{*} = e^{W_3}[\lambda_4^{\#}\cos(W_4) + \lambda_3^{\#}\sin(W_4)] \end{cases} \tag{5.237}$$

$$\begin{cases} \lambda_3^{@} = W_1\lambda_1^{@} - W_2\lambda_2^{@}, \qquad \lambda_4^{@} = W_1\lambda_2^{@} + W_2\lambda_1^{@}, \\ \lambda_3^{\#} = W_3\lambda_1^{\#} - W_4\lambda_2^{\#}, \qquad \lambda_4^{\#} = W_3\lambda_2^{\#} + W_4\lambda_1^{\#}. \end{cases} \tag{5.238}$$

The final p_1 and p_2 can be obtained from numerical inversion (Stehfest 1970).

5.5.1.2 Illustrative Example

The utility of the solution for the previous Case 1 is demonstrated. To compare the solution with that of a finite reservoir, the pumping rate at the well is modified to consider the steady state discharge [proportional to $\ln(r_e/r_w)$] from the reservoir with a finite boundary. The selected parameters are $R_k = 0.1$, $\omega = 0.01 - 0.5$ and $\lambda^* = 1$. The comparison between the present model and that by Bai (1997) is shown in Figure 5-8. Because of the difference in the assumed flow paths at the well, using identical parameters still leads to a time lag in the pressure development for the latter model (curve 2). However, this difference is reduced as the dimensionless storativity ω decreases (curves 3 and 4). The latter curves depict pressure slope changes at later time, evidence of the dual-porosity behavior of the fractured reservoir.

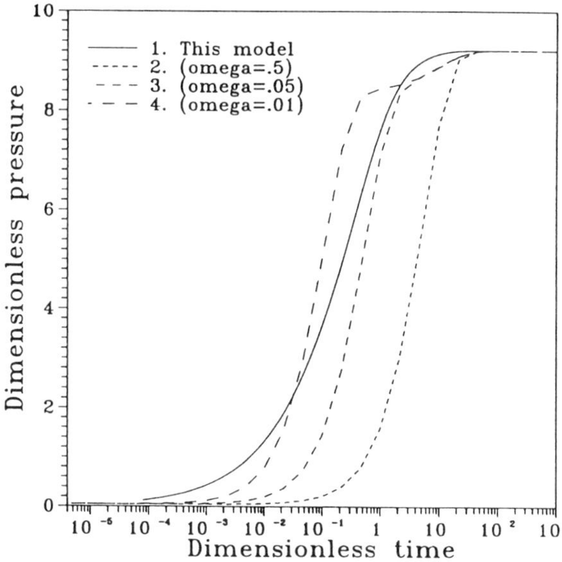

Figure 5-8. Comparison of temporal pressures. (Reprinted from Bai 1997, with permission from Elsevier Science.)

5.5.2 Transport

The method of differential operators is also an efficient method for the solution of the coupled transport equation. The present solution is referred to the formulation of the triple-porosity model with linear sorption, described in Chapter 4.

5.5.2.1 General Solution

In using the method of differential operators to solve for fluid flow, Δ_1 in Eq. (5.189) is assumed to be positive. This restriction can be eliminated, as demonstrated in the following solution where the triple-porosity model with sorption, defined in Eqs. (4.171) and (4.168) of Chapter 4, is employed.

Again, using the differential operator defined in Eq. (5.181), Eqs. (4.171) and (4.168) presented in Chapter 4 on transport can be reformulated as

$$\left(\frac{D_{do}^2}{\gamma_1^*} - D_{do} - \phi_{11}^*\right)\bar{c}_1 + a_{12}^*\bar{c}_2 = \phi_{12}^* \tag{5.239}$$

$$\left(\frac{D_{do}^2}{\gamma_2^*} - D_{do} - \phi_{21}^*\right)\bar{c}_2 + b_2^* a_{21}^*\bar{c}_1 = \phi_{22}^* \tag{5.240}$$

where D_{do}^n is defined in Eq. (5.181).

Solving Eqs. (5.239) and (5.240) simultaneously yields

$$\bar{c}_1 = \frac{1}{b_2^* a_{21}^*}\left[\phi_{22}^* - \left(\frac{D_{do}^2}{\gamma_2^*} - D_{do} - \phi_{21}^*\right)\bar{c}_2\right] \tag{5.241}$$

$$(D_{do}^4 + \theta_1^* D_{do}^3 + \theta_2^* D_{do}^2 + \theta_3^* D_{do} + \theta_4^*)\bar{c}_2 = \theta_5^* \tag{5.242}$$

where

$$\begin{cases} \theta_1^* = -(\gamma_1^* + \gamma_2^*), \quad \theta_2^* = -(\gamma_1^*\phi_{11}^* + \gamma_2^*\phi_{21}^* - \gamma_1^*\gamma_2^*) \\ \theta_3^* = \gamma_1^*\gamma_2^*(\phi_{11}^* + \phi_{21}^*), \quad \theta_4^* = \gamma_1^*\gamma_2^*(\phi_{11}^*\phi_{21}^* - b_2^* a_{12}^* a_{21}^*) \\ \theta_5^* = -\gamma_1^*\gamma_2^*(b_2^* a_{21}^*\phi_{12}^* + \phi_{11}^*\phi_{22}^*). \end{cases} \tag{5.243}$$

The solution of $\bar{c}_2$ from Eq. (5.242) can be expressed as

$$\bar{c}_2 = \bar{c}_2^h + \bar{c}_2^n \tag{5.244}$$

where $\bar{c}_2^h$ and $\bar{c}_2^n$ are the homogeneous and nonhomogeneous solutions, respectively; where $\bar{c}_2^n$ can be derived as $\bar{c}_2^n = (\theta_4^*)^{-1}\theta_5^*$; or

$$\bar{c}_2 = \bar{c}_2^h + (\theta_4^*)^{-1}\theta_5^*. \tag{5.245}$$

For homogeneous solutions $\bar{c}_2^h$, the four roots from Eq. (5.242) can be derived in the following manner. Rewriting Eq. (5.242) as

$$D^2 + \xi_1 D + \phi_1^* = 0, \qquad D^2 + \xi_2 D + \phi_2^* = 0 \tag{5.246}$$

where

$$\begin{cases} \xi_1 = \frac{1}{2}(\theta_1^* + \sqrt{\Delta_1}), & \xi_2 = \frac{1}{2}(\theta_1^* - \sqrt{\Delta_1}), \\ \phi_1^* = z_r + \dfrac{\theta_1^* z_r - \theta_3^*}{\sqrt{\Delta_1}}, & \phi_2^* = z_r - \dfrac{\theta_1^* z_r - \theta_3^*}{\sqrt{\Delta_1}}, \\ \Delta_1 = 8z_r + (\theta_1^*)^2 - 4\theta_2^*, & \end{cases} \tag{5.247}$$

and where z_r is any real root of the following equation:

$$8z_r^3 + 4E_1^* z_r^2 + E_2^* z_r + E_3^* = 0 \tag{5.248}$$

and where

$$E_1^* = -\theta_2^*, \quad E_2^* = 2\theta_1^*\theta_3^* - 8\theta_4^*, \quad E_3^* = \theta_4^*[4\theta_2^* - (\theta_1^*)^2] - (\theta_3^*)^2. \tag{5.249}$$

The solution procedure is identical to the one described by Eqs. (5.192) through (5.200).

Once the real root, z_r, is determined for Eq. (5.248), the four roots of Eq. (5.242) can then be expressed as

$$\begin{cases} \psi_1 = -\frac{\xi_1}{2} + \sqrt{\Delta_2}, & \psi_2 = -\frac{\xi_1}{2} - \sqrt{\Delta_2}, \\ \psi_3 = -\frac{\xi_2}{2} + \sqrt{\Delta_3}, & \psi_4 = -\frac{\xi_2}{2} - \sqrt{\Delta_3}, \\ \Delta_2 = \frac{\xi_1^2}{4} - \phi_1^*, & \Delta_3 = \frac{\xi_2^2}{4} - \phi_2^*. \end{cases} \tag{5.250}$$

The solutions in the Laplace domain also depend on the signs of Δ_1, Δ_2, and Δ_3 in Eqs. (5.247) and (5.250), respectively.

A system of equations can be established after satisfying boundary conditions as expounded in Eq. (5.31):

$$\begin{bmatrix} d_{11} & d_{12} & d_{13} & d_{14} \\ d_{21} & d_{22} & d_{23} & d_{24} \\ d_{31} & d_{32} & d_{33} & d_{34} \\ d_{41} & d_{42} & d_{43} & d_{44} \end{bmatrix} \begin{Bmatrix} g_1 \\ g_2 \\ g_3 \\ g_4 \end{Bmatrix} = \begin{Bmatrix} c_1^@ \\ c_2^@ \\ 0 \\ 0 \end{Bmatrix} \tag{5.251}$$

where

$$c_1^@ = \phi_{22} + \phi_{21}(\theta_4^*)^{-1}\theta_5^* - b_2^* a_{21}^* \frac{c^0}{s}, \qquad c_2^@ = \frac{c^0}{s} - (\theta_4^*)^{-1}\theta_5^*. \tag{5.252}$$

Solutions of Eq. (5.251) can be expressed as

$$g_i^{\#} = \frac{I_i^*}{J^*} \qquad (i = 1, 2, 3, 4) \tag{5.253}$$

where

$$J^* = \sum_{i=1}^{4}(-1)^{i-1} d_{1i} J_i \tag{5.254}$$

and where J_i^* can be described as

$$J_i^* = d_{2j}(d_{3k}d_{4l} - d_{4k}d_{3l}) - d_{2k}(d_{3j}d_{4l} - d_{4j}d_{3l}) + d_{2l}(d_{3j}d_{4k} - d_{4j}d_{3k}) \tag{5.255}$$

and follows the rotation rule, $i \neq j \neq k \neq l$. More specifically:

$$\begin{cases} i = 1, \ j = 2, \ k = 3, \ l = 4 \\ i = 2, \ j = 3, \ k = 4, \ l = 1 \\ i = 3, \ j = 4, \ k = 1, \ l = 2 \\ i = 4, \ j = 1, \ k = 2, \ l = 3. \end{cases} \tag{5.256}$$

Similarly,

$$I_i = (-1)^i[(c_2^@ d_{1j} - c_1^@ d_{2j})(d_{3k}d_{4l} - d_{3l}d_{4k}) -$$

$$(c_2^@ d_{1k} - c_1^@ d_{2k})(d_{3j}d_{4l} - d_{3l}d_{4j}) + (c_2^@ d_{1l} - c_1^@ d_{2l})(d_{3j}d_{4k} - d_{3k}d_{4j})], \tag{5.257}$$

where the retrogressive rotation rule is applied, i.e.,

$$\begin{cases} i = 1, \ j = 2, \ k = 3, \ l = 4 \\ i = 2, \ j = 1, \ k = 3, \ l = 4 \\ i = 3, \ j = 1, \ k = 2, \ l = 4 \\ i = 4, \ j = 1, \ k = 2, \ l = 3. \end{cases} \tag{5.258}$$

The solution procedure includes the derivation of $\bar{c}_1$ and $\bar{c}_2$ along with the components of the system matrix in Eq. (5.251), $d_{\beta^@ i}$ ($\beta^@$=1,2,3,4 and $i = 1, 2, 3, 4$). The coefficients $g_i^{\#}$ (i=1,2,3,4) in the solutions can then be readily calculated from Eq. (5.253).

5.5.2.2 Solution when $\Delta_1 \geq 0$

For $\Delta_1 \geq 0$ in Eq. (5.247), the solutions in the Laplace domain can be divided into four groups depending on the signs of Δ_2 and Δ_3.

Case 1. $\Delta_2 \geq 0$ *and* $\Delta_3 \geq 0$:

The solution from Eqs. (5.242) and (5.245) can be written as

$$\bar{c}_2 = \sum_{i=1}^{4} g_i^{\#} e^{\psi_i x_d} + (\theta_4^*)^{-1}\theta_5^* \tag{5.259}$$

$$\bar{c}_1 = \frac{1}{b_2^* a_{21}^*}\left[\phi_{22}^* + \phi_{21}^*(\theta_4^*)^{-1}\theta_5^* - \sum_{i=1}^{4} g_i^{\#}\Omega_{1i}e^{\psi_i x_d}\right] \tag{5.260}$$

where

$$\Omega_{1i} = \frac{\psi_i^2}{\gamma_2^*} - \psi_i - \phi_{21}^*. \tag{5.261}$$

After satisfying boundary conditions in Eq. (4.159) presented in Chapter 4, $d_{\beta @ i}$ are derived as

$$d_{1i} = \Omega_{1i}, \quad d_{2i} = 1, \quad d_{3i} = \Omega_{3i}, \quad d_{4i} = \Omega_{2i}; \quad (i = 1, 2, 3, 4) \tag{5.262}$$

where

$$\Omega_{2i} = \psi_i e^{\psi_i}, \quad \Omega_{3i} = \Omega_{1i}\Omega_{2i}. \tag{5.263}$$

Case 2. $\Delta_2 \geq 0$ *and* $\Delta_3 < 0$:

The solution is expressed as

$$\bar{c}_2 = \sum_{i=1}^{2} g_i^{\#}e^{\psi_i x_d} + e^{\omega_3^{\#} x_d}[g_3^{\#}\cos(\omega_4^{\#}x_d) + g_4^{\#}\sin(\omega_4^{\#}x_d)] + (\theta_4^*)^{-1}\theta_5^* \tag{5.264}$$

where

$$\omega_3^{\#} = -\frac{\xi_2}{2}, \qquad \omega_4^{\#} = \sqrt{-\Delta_3} \tag{5.265}$$

$$\bar{c}_1 = \frac{1}{b_2^* a_{21}^*}\{\phi_{22}^* + \phi_{21}^*(\theta_4^*)^{-1}\theta_5^* - \sum_{i=1}^{2} g_i^{\#}\Omega_{1i}e^{\psi_i x_d} + e^{\omega_3^{\#} x_d}\{g_3^{\#}[\lambda_3^{@}\cos(\omega_4^{\#}x_d) - \lambda_4^{@}\sin(\omega_4^{\#}x_d)] + g_4^{\#}[\lambda_3^{@}\sin(\omega_4^{\#}x_d) + \lambda_4^{@}\cos(\omega_4^{\#}x_d)]\}\} \tag{5.266}$$

where

$$\lambda_3^{@} = \frac{1}{\gamma_2^*}[(\omega_3^{\#})^2 - (\omega_4^{\#})^2] - \omega_3^{\#} - \phi_{21}^*, \qquad \lambda_4^{@} = \omega_4^{\#}\left(\frac{2\omega_3^{\#}}{\gamma_2^*} - 1\right). \tag{5.267}$$

For $i = 1, 2$, $d_{\beta @ i}$ are identical to those in Case 1. However, for $i = 3, 4$:

$$\begin{cases} d_{1i} = \lambda_i^{@}, \quad d_{2i} = 1 \;\; (i = 3), \quad d_{2i} = 0 \;\; (i = 4) \\ d_{33} = e^{\omega_3^{\#}}[\lambda_3^*\cos(\omega_4^{\#}) - \lambda_4^*\sin(\omega_4^{\#})], \quad d_{34} = e^{\omega_3^{\#}}[\lambda_3^*\sin(\omega_4^{\#}) + \lambda_4^*\cos(\omega_4^{\#})], \\ d_{43} = e^{\omega_3^{\#}}[\omega_3^{\#}\cos(\omega_4^{\#}) - \omega_4^{\#}\sin(\omega_4^{\#})], \quad d_{44} = e^{\omega_3^{\#}}[\omega_3^{\#}\sin(\omega_4^{\#}) + \omega_4^{\#}\cos(\omega_4^{\#})], \end{cases} \tag{5.268}$$

where

$$\lambda_3^* = \omega_3^{\#}\lambda_3^{@} - \omega_4^{\#}\lambda_4^{@}, \qquad \lambda_4^* = \omega_3^{\#}\lambda_4^{@} + \omega_4^{\#}\lambda_3^{@}. \tag{5.269}$$

Case 3. $\Delta_2 < 0$ *and* $\Delta_3 \geq 0$*:*

The solution is described as

$$\bar{c}_2 = e^{\omega_1^{\#} x_d}[g_1^{\#} \cos(\omega_2^{\#} x_d) + g_2^{\#} \sin(\omega_2^{\#} x_d)] + \sum_{i=3}^{4} g_i^{\#} e^{\psi_i x_d} + (\theta_4^*)^{-1}\theta_5^* \quad (5.270)$$

where

$$\omega_1^{\#} = -\frac{\xi_1}{2}, \qquad \omega_2^{\#} = \sqrt{-\Delta_2} \quad (5.271)$$

$$\bar{c}_1 = \frac{1}{b_2^* a_{21}^*}\Big\{\phi_{22}^* + \phi_{21}^*(\theta_4^*)^{-1}\theta_5^* - e^{\omega_1^{\#} x_d}\{g_1^{\#}[\lambda_1^{@} \cos(\omega_2^{\#} x_d) - \lambda_2^{@} \sin(\omega_2^{\#} x_d)]$$

$$+ g_2^{\#}[\lambda_1^{@} \sin(\omega_2^{\#} x_d) + \lambda_2^{@} \cos(\omega_2^{\#} x_d)]\}\Big\} + \sum_{i=3}^{4} g_i^{\#} \Omega_{1i} e^{\psi_i x_d} \quad (5.272)$$

where

$$\lambda_1^{@} = \frac{1}{\gamma_2^*}[(\omega_1^{\#})^2 - (\omega_2^{\#})^2] - \omega_1^{\#} - \phi_{21}^*, \qquad \lambda_2^{@} = \omega_2^{\#}\left(\frac{2\omega_1^{\#}}{\gamma_2^*} - 1\right) \quad (5.273)$$

For $i = 3, 4$, $d_{\beta^{@} i}$ are identical to those in Case 1. However, for $i = 1, 2$:

$$\begin{cases} d_{1i} = \lambda_i^{@}, \quad d_{2i} = 1 \ \ (i = 1), \quad d_{2i} = 0 \ \ (i = 2) \\ d_{31} = e^{\omega_1^{\#}}[\lambda_1^* \cos(\omega_2^{\#}) - \lambda_2^* \sin(\omega_2^{\#})], \quad d_{32} = e^{\omega_1^{\#}}[\lambda_1^* \sin(\omega_2^{\#}) + \lambda_2^* \cos(\omega_2^{\#})], \\ d_{41} = e^{\omega_1^{\#}}[\omega_1^{\#} \cos(\omega_2^{\#}) - \omega_2^{\#} \sin(\omega_2^{\#})], \quad d_{42} = e^{\omega_1^{\#}}[\omega_1^{\#} \sin(\omega_2^{\#}) + \omega_2^{\#} \cos(\omega_2^{\#})], \end{cases} \quad (5.274)$$

where

$$\lambda_1^* = \omega_1^{\#}\lambda_1^{@} - \omega_2^{\#}\lambda_2^{@}, \qquad \lambda_2^* = \omega_1^{\#}\lambda_2^{@} + \omega_2^{\#}\lambda_1^{@}. \quad (5.275)$$

Case 4. $\Delta_2 < 0$ *and* $\Delta_3 < 0$*:*

The solution is written as

$$\bar{c}_2 = e^{\omega_1^{\#} x_d}[g_1^{\#} \cos(\omega_2^{\#} x_d) + g_2^{\#} \sin(\omega_2^{\#} x_d)] +$$

$$e^{\omega_3^{\#} x_d}[g_3^{\#} \cos(\omega_4^{\#} x_d) + g_4^{\#} \sin(\omega_4^{\#} x_d)] + (\theta_4^*)^{-1}\theta_5^* \quad (5.276)$$

$$\bar{c}_1 = \frac{1}{b_2^* a_{21}^*}\{\phi_{22}^* + \phi_{21}^*(\theta_4^*)^{-1}\theta_5^* - e^{\omega_1^{\#} x_d}\{g_1^{\#}[\lambda_1^{@} \cos(\omega_2^{\#} x_d) - \lambda_2^{@} \sin(\omega_2^{\#} x_d)]$$

$$+g_2^{\#}[\lambda_1^{@}\sin(\omega_2^{\#}x_d)+\lambda_2^{@}\cos(\omega_2^{\#}x_d)]\}+e^{\omega_3^{\#}x_d}\{g_3^{\#}[\lambda_3^{@}\cos(\omega_4^{\#}x_d)-\lambda_4^{@}\sin(\omega_4^{\#}x_d)]$$
$$+g_4^{\#}[\lambda_3^{@}\sin(\omega_4^{\#}x_d)+\lambda_4^{@}\cos(\omega_4^{\#}x_d)]\}\}. \tag{5.277}$$

$d_{\beta @ i}$ can be expressed as

$$\begin{cases} d_{1i}=\lambda_i^{@}, \quad (i=1,2,3,4),\ d_{2i}=1, \quad (i=1,3); \qquad d_{2i}=0, \quad (i=2,4) \\ d_{31}=e^{\omega_1^{\#}}[\lambda_1^*\cos(\omega_2^{\#})-\lambda_2^*\sin(\omega_2^{\#})], \quad d_{32}=e^{\omega_1^{\#}}[\lambda_1^*\sin(\omega_2^{\#})+\lambda_2^*\cos(\omega_2^{\#})], \\ d_{33}=e^{\omega_3^{\#}}[\lambda_3^*\cos(\omega_4^{\#})-\lambda_4^*\sin(\omega_4^{\#})], \quad d_{34}=e^{\omega_3^{\#}}[\lambda_3^*\sin(\omega_4^{\#})+\lambda_4^*\cos(\omega_4^{\#})], \\ d_{41}=e^{\omega_1^{\#}}[\omega_1^{\#}\cos(\omega_2^{\#})-\omega_2^{\#}\sin(\omega_2^{\#})], \quad d_{42}=e^{\omega_1^{\#}}[\omega_1^{\#}\sin(\omega_2^{\#})+\omega_2^{\#}\cos(\omega_2^{\#})], \\ d_{43}=e^{\omega_3^{\#}}[\omega_3^{\#}\cos(\omega_4^{\#})-\omega_4^{\#}\sin(\omega_4^{\#})], \quad d_{44}=e^{\omega_3^{\#}}[\omega_3^{\#}\sin(\omega_4^{\#})+\omega_4^{\#}\cos(\omega_4^{\#})]. \end{cases} \tag{5.278}$$

All the parameters have been defined in the previous cases.

5.5.2.3 Solution when $\Delta_1 < 0$

If $\Delta_1 < 0$ in Eq. (5.247), ξ_1, ξ_2, ϕ_1^* and ϕ_2^* in the equation become complex variables. It is not necessary, as in the previous cases, to determine the signs for Δ_2 and Δ_3 in Eq. (5.250), which are redefined as

$$\Delta_2=\Omega_1+i\Omega_2, \qquad \Delta_3=\Omega_1-i\Omega_2 \tag{5.279}$$

where

$$\Omega_1=\frac{1}{16}(\theta_1^2+\Delta_1)-z_r, \quad \Omega_2=\frac{\theta_1^*}{8}\sqrt{-\Delta_1}+\frac{\theta_1^*z_r-\theta_3^*}{\sqrt{-\Delta_1}}. \tag{5.280}$$

Δ_2 and Δ_3 need to be expressed in trigonometric forms, which imply a dual set of solutions. For the first set of solutions, the corresponding four roots in Eq. (5.250) can be written as

$$\begin{cases} \psi_1=\omega_1^*+i\omega_2^*, \qquad \psi_2=\omega_3^*-i\omega_4^* \\ \psi_3=\omega_1^*-i\omega_2^*, \qquad \psi_4=\omega_3^*+i\omega_4^* \end{cases} \tag{5.281}$$

where

$$\begin{cases} \omega_1^*=\sqrt{r_d^*}\cos(\frac{\theta^{\#}}{2})-\frac{1}{4}\theta_1^*, \quad \omega_2^*=\sqrt{r_d^*}\sin(\frac{\theta^{\#}}{2})-\frac{1}{4}\sqrt{-\Delta_1} \\ \omega_3^*=-[\sqrt{r_d^*}\cos(\frac{\theta^{\#}}{2})+\frac{1}{4}\theta_1^*], \quad \omega_4^*=\sqrt{r_d^*}\sin(\frac{\theta^{\#}}{2})+\frac{1}{4}\sqrt{-\Delta_1} \\ \sqrt{r_d^*}=\sqrt{\Omega_1^2+\Omega_2^2}, \qquad \theta^{\#}=\arctan(\frac{\Omega_2}{\Omega_1}). \end{cases} \tag{5.282}$$

The solutions can be derived as

$$\bar{c}_2 = e^{\omega_1^* x_d}[g_1^{\#}\cos(\omega_2^* x_d) + g_3^{\#}\sin(\omega_2^* x_d)] + e^{\omega_3^* x_d}[g_2^{\#}\sin(\omega_4^* x_d) + g_4^{\#}\cos(\omega_4^* x_d)] + (\theta_4^*)^{-1}\theta_5^* \tag{5.283}$$

$$\bar{c}_1 = \frac{1}{b_2^* a_{21}^*}\{\phi_{22}^* + \phi_{21}^*(\theta_4^*)^{-1}\theta_5^* - g_1^{\#} e^{\omega_1^* x_d}[\alpha_{11}^{@}\cos(\omega_2^* x_d) + \alpha_{12}^{@}\sin(\omega_2^* x_d)] + g_2^{\#} e^{\omega_3^* x_d}[\alpha_{21}^{@}\sin(\omega_4^* x_d) + \alpha_{22}^{@}\cos(\omega_4^* x_d)] + g_3^{\#} e^{\omega_1^* x_d}[\alpha_{31}^{@}\sin(\omega_2^* x_d) + \alpha_{32}^{@}\cos(\omega_2^* x_d)] + g_4^{\#} e^{\omega_3^* x_d}[\alpha_{41}^{@}\cos(\omega_4^* x_d) + \alpha_{42}^{@}\sin(\omega_4^* x_d)]\} \tag{5.284}$$

where

$$\begin{cases} \alpha_{11}^{@} = \frac{1}{\gamma_2^*}[(\omega_1^*)^2 - (\omega_2^*)^2)] - \omega_1^* - \phi_{21}^*, \quad \alpha_{12}^{@} = \omega_2^*\left(1 - \frac{2\omega_1^*}{\gamma_2^*}\right) \\ \alpha_{21}^{@} = \frac{1}{\gamma_2^*}[(\omega_3^*)^2 - (\omega_4^*)^2)] - \omega_3^* - \phi_{21}^*, \quad \alpha_{22}^{@} = -\omega_4^*\left(1 - \frac{2\omega_3^*}{\gamma_2^*}\right) \\ \alpha_{31}^{@} = \alpha_{11}^{@}, \quad \alpha_{32}^{@} = -\alpha_{12}^{@}, \quad \alpha_{41}^{@} = \alpha_{21}^{@}, \quad \alpha_{42}^{@} = -\alpha_{22}^{@}. \end{cases} \tag{5.285}$$

From Eq. (5.251), $d_{\beta @_i}$ can be expressed as

$$\begin{cases} d_{1i} = \alpha_{11}^{@}, \alpha_{22}^{@}, \alpha_{32}^{@}, \alpha_{41}^{@}, \quad (i = 1, 2, 3, 4) \\ d_{2i} = 1, \quad (i = 1, 4), \qquad d_{2i} = 0, \quad (i = 2, 3) \\ d_{31} = e^{\omega_1^*}[\alpha_{11}^*\cos(\omega_2^*) + \alpha_{12}^*\sin(\omega_2^*)], \quad d_{32} = e^{\omega_3^*}[\alpha_{21}^*\sin(\omega_4^*) + \alpha_{22}^*\cos(\omega_4^*)], \\ d_{33} = e^{\omega_1^*}[\alpha_{31}^*\sin(\omega_2^*) + \alpha_{32}^*\cos(\omega_2^*)], \quad d_{34} = e^{\omega_3^*}[\alpha_{41}^*\cos(\omega_4^*) + \alpha_{42}^*\sin(\omega_4^*)], \\ d_{41} = e^{\omega_1^*}[\omega_1^*\cos(\omega_2^*) - \omega_2^*\sin(\omega_2^*)], \quad d_{42} = e^{\omega_3^*}[\omega_3^*\sin(\omega_4^*) + \omega_4^*\cos(\omega_4^*)], \\ d_{43} = e^{\omega_1^*}[\omega_1^*\sin(\omega_2^*) + \omega_2^*\cos(\omega_2^*)], \quad d_{44} = e^{\omega_3^*}[\omega_3^*\cos(\omega_4^*) - \omega_4^*\sin(\omega_4^*)], \end{cases} \tag{5.286}$$

where

$$\begin{cases} \alpha_{11}^* = \omega_1^*\alpha_{11}^{@} + \omega_2^*\alpha_{12}^{@}, \quad \alpha_{12}^* = \omega_1^*\alpha_{12}^{@} - \omega_2^*\alpha_{11}^{@} \\ \alpha_{21}^* = \omega_3^*\alpha_{21}^{@} - \omega_4^*\alpha_{22}^{@}, \quad \alpha_{22}^* = \omega_3^*\alpha_{22}^{@} + \omega_4^*\alpha_{21}^{@} \\ \alpha_{31}^* = \omega_1^*\alpha_{31}^{@} - \omega_2^*\alpha_{32}^{@}, \quad \alpha_{32}^* = \omega_1^*\alpha_{32}^{@} + \omega_2^*\alpha_{31}^{@} \\ \alpha_{41}^* = \omega_3^*\alpha_{41}^{@} + \omega_4^*\alpha_{42}^{@}, \quad \alpha_{42}^* = \omega_3^*\alpha_{42}^{@} - \omega_4^*\alpha_{41}^{@}. \end{cases} \tag{5.287}$$

For the second set of solutions when $\Delta_1 < 0$, simply substitute $\cos(\frac{\theta^{\#}}{2})$ with $\cos(\frac{\theta^{\#}}{2} + \pi)$, and substitute $\sin(\frac{\theta^{\#}}{2})$ with $\sin(\frac{\theta^{\#}}{2} + \pi)$.

5.5.2.4 Illustrative Example

Comparing the triple-porosity model with the traditional single- and dual-porosity models is a means to demonstrate the flexibility of using a higher-order porosity modeling approach. Due to the difficulties associated with maintaining compatible parameters and compatible boundary and initial conditions, between various models of similar conceptualizations, both single- and dual-porosity models are derived from the triple-porosity model by independently assuming "$a^*_{12} = 0$" for the former and "$a^*_{23} = a^*_{32} = 0$" for the latter case [refer to Eqs. (4.146) to (4.148) in Chapter 4]. The subsequent comparison is shown in Figure 5-9. Early breakthrough (slight in this case) and extended tailing are documented effects that index the level of transport heterogeneity and result from the interaction between preferential flow channels and less permeable regions. This behavior can be identified in both dual-porosity and triple-porosity models.

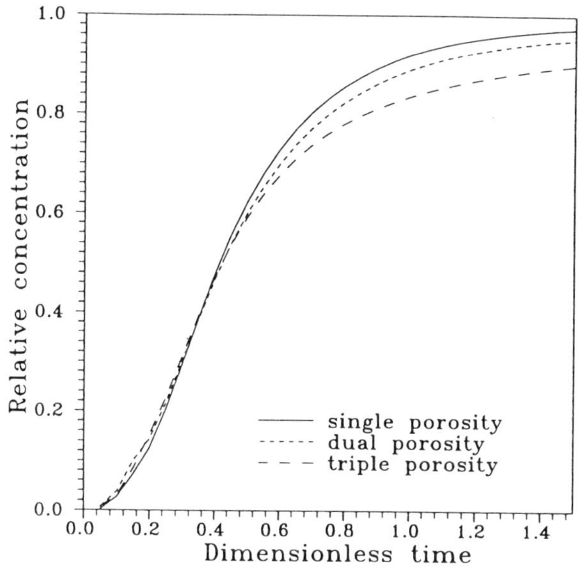

Figure 5-9. Comparison for various models in transport. (Reprinted from Bai et al. 1997a, with permission from ASCE.)

Chapter 6

NUMERICAL SOLUTION

6.1 INTRODUCTION

With the rapid development of computer methods, numerical methods have been actively exploited as an approximation to analytical solutions. In terms of their usefulness in solving practical problems, numerical approaches are often superior to analytical methods since the controlling parameters may vary in either space or time. In particular, numerical methods allow the replication of practical problems with the complex geometrical, geological and hydrological conditions common in nature.

Nearly all the numerical procedures involve replacing the continuous form of the governing differential equation by a finite number of algebraic equations. To develop these equations, it is necessary to subdivide the region. Three of the most widely used numerical methods are the finite difference method, the finite element method, and the boundary element method. The finite difference method provides a direct discretization of the governing differential equations. The finite difference method has been popular in simulating practical problems typically involving relatively simple geometries. The boundary element method invokes the direct integration of a defined fundamental solution via integral techniques. This method has been effective in evaluating linear problems in unbounded domains since the fundamental solutions are readily available for simple uncoupled phenomena. Although applicable to nonlinear or inhomogeneous problems, this method has been restricted from widespread application due to the necessity of applying volume discretization in these cases; the method loses its innate advantage of boundary-only discretization. The finite element method provides a direct discretization of the physical domain of the simulated problems using either the variational principle or the weighted residual method.

Among these three solution methods, the finite element method is perhaps the most versatile technique, despite its conceptual complexity. The finite dif ference method cannot generally compete with the finite element method with regard to the handling of irregular mesh and boundary conditions, nor does the boundary element method serve well in representing the nonlinearities or heterogeneities of the medium. For this reason, the present analysis focuses on finite element techniques. Tensor notation and matrix notation in this chapter are used interchangeably, such as in the definition of the shape function, $N_i = \mathbf{N}$.

6.2 FINITE ELEMENT PRELIMINARIES

Because the finite element method (FEM) is based on certain advanced mathematics, a substantial number of users consider the FEM a "black box." The following section provides basic details on the finite element formulation.

6.2.1 Numerical Integration

In general, the solutions of partial differential equations for a continuous domain are represented in an integral form. In the finite element method, the integration is most effectively approximated by Gauss-Legendre (G-L) quadrature (numerical integration), as illustrated in a simple one-dimensional (1-D) case.

In 1-D, the integration of a function f in the region $(-1 \ll \bar{r}^c \ll 1)$ can be expressed by the G-L quadrature as

$$\int_{\bar{r}^c=-1}^{\bar{r}^c=1} f(\bar{r}^c) d\bar{r}^c \approx \sum_{i=1}^{I} w_i^c f(\bar{r}_i^c) \tag{6.1}$$

where subscript i indexes the integration point, I is the total number of integration points (or order of integration), w_i^c is the weight at point i, and $\bar{r}_i^c$ is the coordinate at point i.

6.2.2 Shape Functions

Numerical discretization requires the sectioning of a continuous domain into a number of segments. Specifically, the domain of interest can be divided into a finite number of elements in which a continuous function can be represented by the summation of that function at the element level. This task can be best accomplished through the application of shape functions.

To use the G-L quadrature described in the previous section, the shape function must be compatible with the system where the coordinates are normalized. The isoparametric element complies with this requirement, which can be generally expressed as

$$\Xi_i^* = \sum_{i=1}^{i=I} N_i \Xi_i \tag{6.2}$$

where subscript i now indexes nodal points of the element instead of the integration point, Ξ is a dummy variable that can be a coordinate or an unknown, Ξ_i is a dummy variable at node i, Ξ_i^* represents the approximate value of Ξ_i (frequently, the difference between Ξ_i^* and Ξ_i is omitted in the notation for convenience), and N_i is the shape function at node i, which satisfies the following conditions:

$$N_i = \delta_{ij}, \qquad \sum_{i=1}^{i=I} N_i = 1 \tag{6.3}$$

and the summation is completed over I nodes.

For a 1-D case, the linear or bar element shape function can be described as

$$N_i = \frac{1}{2}(1 + \bar{r}^c \bar{r}_i^c) \tag{6.4}$$

where subscripts i = 1 and 2 index the nodal point, and where $\bar{r}^c$ is the local coordinate. When $\bar{r}^c$ = -1 and 1, i = 1 and 2, $\bar{r}^c = \bar{r}_i^c$, and $N_i = 1$. This is the simplest shape function.

For the 2-D four-node quadrilateral element, the shape function is written as

$$N_i = \frac{1}{4}(1 + \bar{r}_i^c \bar{r}^c)(1 + \bar{s}_i^c \bar{s}^c) \tag{6.5}$$

where subscripts i = 1, 2, 3, and 4, $\bar{s}^c$ is the second local coordinate orthogonal to $\bar{r}^c$. The shape function depicts the local coordinates of the bi-unit quadrilateral with the origin at the centroid, and are confined by the magnitudes of -1 and 1.

For the 3-D eight-node brick element, the shape function is described as

$$N_i = \frac{1}{8}(1 + \bar{r}_i^c \bar{r}^c)(1 + \bar{s}_i^c \bar{s}^c)(1 + \bar{t}_i^c \bar{t}^c) \tag{6.6}$$

where subscripts range from 1 to 8 and $\bar{t}^c$ is the third local coordinate orthogonal to $\bar{r}^c$ and $\bar{s}^c$. The shape function represents the local coordinates of the bi-unit cube with the origin at the centroid and are limited also by the magnitudes of -1 and 1. Since two points define a straight line, all the above shape functions are therefore linear.

In the coupled deformation-flow finite element method, the shape functions are used to map the element displacements and fluid pressures at the nodal points. For the 3-D eight-node brick element, solid displacement and fluid pressure can be approximated as

$$\mathbf{u}^* = \mathbf{N}\mathbf{u}, \qquad \mathbf{p}^* = \mathbf{M}\mathbf{p} \tag{6.7}$$

where **N** and **M** are the shape functions for displacement and pressure, respectively; the variables with the superscript "*" represents the approximate value of the variable at any point within the element. For simplicity, this distinction between the dependent variable and its approximation value is frequently omitted. In contrast, the variable without the superscript denotes the nodal value.

In tensor form, the relations of Eq. (6.7) are

$$u_i^* = \sum_{i=1}^{8} N_i u_i, \qquad p_i^* = \sum_{i=1}^{8} M_i p_i.$$

Shape functions defined in Eq. (6.6) may be applied to both **N** and **M** for a 3-D element. However, it may be inappropriate to assume that the interpolation functions for solid displacements **N** and for fluid pressures **M** are identical. The component of the partial stress tensor is continuously differentiable to the first order. As a result, the polynomial interpolation functions for the pore pressure distribution must be one order lower than that chosen for the displacement field. Sandhu and Wilson (1969) were among the first to apply quadratic displacement and linear pressure expressions to evaluate finite element functions using triangular elements.

In this work, a quadratic displacement field and a linear pressure field are chosen. For the choice of a 3-D element, a higher order representation can be accommodated by adding a central node to the element. This internal nodal variable is designed only to achieve a higher order interpolation for the displacement than for the pressure. For the nine-node element (eight corner nodes and one central node), the expressions for the approximation in mapping nodal displacements may be described as

$$\mathbf{u}^* = \mathbf{N}\mathbf{u} \tag{6.8}$$

or in a tensor form,

$$(u_{\bar{r}^c})_i^* = \sum_{i=1}^{9} N_i (u_{\bar{r}^c})_i, \qquad (u_{\bar{s}^c})_i^* = \sum_{i=1}^{9} N_i (u_{\bar{s}^c})_i, \qquad (u_{\bar{t}^c})_i^* = \sum_{i=1}^{9} N_i (u_{\bar{t}^c})_i. \tag{6.9}$$

For the first eight nodes in the brick element, **N** are chosen to be identical to **M** defined in Eq. (6.6). For the central node, however, **N** is given as

$$N_9 = [1 - (\bar{r}^c)^2][1 - (\bar{s}^c)^2][1 - (\bar{t}^c)^2]. \tag{6.10}$$

6.2.3 Global and Local Coordinate Mapping

Only local coordinates have been discussed in the previous sections. Since both the input data and output solutions are expressed in the global coordinate systems, global and local coordinates must be mapped. This task is usually termed "Jacobian mapping."

In general, the infinitesimal coordinates of global and local systems are related by the following relations:

$$dxdydz = \det \mathbf{J} \; d\bar{r}^c d\bar{s}^c d\bar{t}^c \tag{6.11}$$

where "det **J**" is the determinant of the Jacobian matrix (in 1-D, it is a scalar); **J** represents the derivative relation between global and local coordinates via the chain rule.

6.2.4 Construction of a System of Equations

The system of equations is the central component of the finite element formulation representing the transformation of the analytical governing partial differential equations into an equivalent numerical form. Also, it represents entire nodal connectivities at the element level. As a result, constructing a system of equations is a primary task of a finite element algorithm, as demonstrated in the following section.

6.3 FINITE ELEMENT FORMULATION

The traditional finite element method has been standardized using structured coding approaches. Typically, the programmer only needs to focus on the element subroutine, occasionally paying attention on the modification of shape functions. In most circumstances, construction of a global system, solution methodologies, and structured input/output are quite similar among the majority of finite element codes; therefore, no additional effort is required. Because the matrix may represent a vector or tensor of any form, matrix notation and operations dominate in the finite element formulation due to their flexibility. At the element level, the major task is to establish the system of equations, as described in the following sections.

6.3.1 Deformation

The governing equation for solid equilibrium can be derived by minimizing the potential energy of a forced system. The energy form may be written as

$$\Psi_e = \frac{1}{2} \int_V \boldsymbol{\varepsilon}^T \boldsymbol{\sigma} dV \tag{6.12}$$

where $\boldsymbol{\varepsilon}$ and $\boldsymbol{\sigma}$ denote strain and stress nodal vectors, and V is the volume of integration. For a 2-D case, V should be substituted with the area of integration, A_e, since a unit thickness is applied.

In two dimensions, the strain vector can be related to the displacement vector using the following expression:

$$\begin{Bmatrix} \varepsilon_x \\ \varepsilon_y \\ \gamma_{xy} \end{Bmatrix} = \begin{bmatrix} \dfrac{\partial}{\partial x} & 0 \\ 0 & \dfrac{\partial}{\partial y} \\ \dfrac{\partial}{\partial y} & \dfrac{\partial}{\partial x} \end{bmatrix} \begin{Bmatrix} u_x \\ u_y \end{Bmatrix} \tag{6.13}$$

where the matrix on the right-hand side of the equation is labeled as a strain-displacement matrix, **B**.

In the finite element formulation, Eq. (6.13) can be written as

$$\boldsymbol{\varepsilon} = \mathbf{B}^T \mathbf{u} \tag{6.14}$$

where $\boldsymbol{\varepsilon}$ and $\mathbf{u}$ denote strain displacement nodal vectors, $\mathbf{B}$ is the strain-displacement matrix, i.e.,

$$\mathbf{B} = \begin{bmatrix} \dfrac{\partial N_i}{\partial x} & 0 \\ 0 & \dfrac{\partial N_i}{\partial y} \\ \dfrac{\partial N_i}{\partial y} & \dfrac{\partial N_i}{\partial x} \end{bmatrix} \tag{6.15}$$

where N_i is the shape function, and the displacement at the element level is mapped by the shape functions, i.e.,

$$u_x = \sum_{i=1}^{I} N_i u_{xi}; \quad u_y = \sum_{i=1}^{I} N_i u_{yi}. \tag{6.16}$$

Stresses can be derived from Hooke's law as

$$\boldsymbol{\sigma} = \mathbf{D}\boldsymbol{\varepsilon} \tag{6.17}$$

where $\mathbf{D}$ is the elastic modulus matrix. Substituting Eq. (6.14) into Eq. (6.17), yields

$$\boldsymbol{\sigma} = \mathbf{D}\mathbf{B}^T \mathbf{u}. \tag{6.18}$$

Substituting both Eqs. (6.14) and (6.18) into Eq. (6.12), for a 2-D element, yields

$$\Psi_e = \frac{1}{2} \int_{A_e} \mathbf{B}^T \mathbf{D} \mathbf{B} \mathbf{u}^2 dA_e. \tag{6.19}$$

Minimizing total potential energy with respect to displacement, $\mathbf{u}$, leads to

$$\frac{\partial \Psi_e}{\partial \mathbf{u}} = \int_{A_e} \mathbf{B}^T \mathbf{D} \mathbf{B} \mathbf{u} dA_e = 0. \tag{6.20}$$

Since $\mathbf{u}$ represents the nodal value in the element, it can be taken outside of the integration; therefore, Eq. (6.20) can be expressed as

$$\frac{\partial \Psi_e}{\partial \mathbf{u}} = \mathbf{K}_e \mathbf{u} = 0 \tag{6.21}$$

where $\mathbf{K}_e$ is the stiffness matrix, i.e.,

$$\mathbf{K}_e = \int_{A_e} \mathbf{B}^T \mathbf{D} \mathbf{B} dA_e. \tag{6.22}$$

$\mathbf{K}_e$ can be written in a local coordinate system for a 2-D case as

$$\mathbf{K}_e = \int_{-1}^{1} \int_{-1}^{1} \mathbf{B}^T \mathbf{D} \mathbf{B} \det \mathbf{J} \; dr_c ds_c. \tag{6.23}$$

The evaluation of the stiffness matrix is the primary task in evaluating solid deformation.

For the system, subject to external load, Eq. (6.21) may be modified as

$$\mathbf{K}_e \mathbf{u} = \mathbf{F} \tag{6.24}$$

where $\mathbf{F}$ is the boundary traction.

For the 1-D two-node linear bar element subjected to boundary tension at both nodes, the system of equations can be expanded from Eq. (6.24) to the following matrix form:

$$\begin{bmatrix} (K_e)_{11} & (K_e)_{12} \\ (K_e)_{21} & (K_e)_{22} \end{bmatrix} \begin{Bmatrix} u_1 \\ u_2 \end{Bmatrix} = \begin{Bmatrix} F_1 \\ F_2 \end{Bmatrix}. \tag{6.25}$$

For a one-dimensional problem, the shear strain in Eq. (6.13) is eliminated. It can be easily verified that the stiffness matrix in Eq. (6.25) has the following form:

$$(K_e)_i = \frac{(A_e^*)_i E_i}{L_e} \begin{bmatrix} 1 & -1 \\ -1 & 1 \end{bmatrix} \tag{6.26}$$

where subscript i indicates the nodal number, and $i = 1$ and 2 for the present case, A_e^* is the uniform cross-sectional area of the bar element, E is the elastic modulus, and L_e is the element length.

6.3.2 Flow

As demonstrated by Zienkiewicz (1983), the primary task in defining fluid flow using the finite element method is to construct the conductance matrix by reducing the higher order governing equations to a linear form. Unlike the formulation for deformation, where the system is assumed to be quasi-steady state, when the inertial forces are neglected, both steady and transient forms are relevant. A 1-D flow is used for the following example.

6.3.2.1 Steady State

Neglecting flow sources, the governing flow equation can be written in its finite element form as (Zienkiewicz 1983)

$$-A_e^* \int_{L_e} \frac{k}{\mu} \frac{\partial p}{\partial x} \frac{\partial M_i}{\partial x} dL_e = 0 \tag{6.27}$$

where L_e is the length of the flow domain and M_i is a shape function for the flow, which may not be identical to N_i for the deformation. This is especially true in the coupled finite element formulation (refer to Section 6.2.2 on Shape Functions) in which a higher order shape function is required for the deformation than that for the flow.

Substituting the discretized pressure p into Eq. (6.27), omitting the summation sign and the superscript $*$ for simplicity, one has

$$\mathbf{K}_c \mathbf{p} = 0 \tag{6.28}$$

where $\mathbf{K}_c$ is the conductance matrix, i.e.,

$$\mathbf{K}_c = -\frac{k}{\mu} \int_{L_e} \frac{\partial M_i}{\partial x}^T \frac{\partial M_i}{\partial x} dL_e \tag{6.29}$$

where the negative sign on the right-hand side of the equation is due to the application of Green's first identity (Zienkiewicz 1983).

If the permeability is anisotropic, the conductance matrix may be expressed as

$$\mathbf{K}_c = -\frac{1}{\mu} \int_{L_e} \frac{\partial M_i}{\partial x}^T \mathbf{k} \frac{\partial M_i}{\partial x} dL_e \tag{6.30}$$

where $\mathbf{k}$ is the permeability tensor.

Considering the impact of external flow sources, Eq. (6.28) may be modified as

$$\mathbf{K}_c \mathbf{p} = \mathbf{q} \tag{6.31}$$

where $\mathbf{q}$ is the vector of prescribed nodal discharge.

Again, for the 1-D two-node linear bar element, subjected to nodal discharge at the boundaries, the system of equations can be derived from Eq. (6.31) as

$$-\frac{(A_e^*)_i\kappa_i}{L_e}\begin{bmatrix} 1 & -1 \\ -1 & 1 \end{bmatrix}\begin{Bmatrix} p_1 \\ p_2 \end{Bmatrix} = \begin{Bmatrix} q_1 \\ q_2 \end{Bmatrix} \tag{6.32}$$

where κ is the mobility (k/μ) and A_e^* is the uniform cross-sectional flow area.

6.3.2.2 Transient State

In the transient state, the fluid pressure is a time-dependent variable. In addition, the variation of storage as a result of compressibilities of the porous medium and fluid should be considered. Extending from Eq. (6.31), one has

$$\mathbf{K}_c\mathbf{p}_i = \mathbf{c}^*\dot{\mathbf{p}} + \mathbf{q} \tag{6.33}$$

where the *dot* denotes the time derivative, and $\mathbf{c}^*$ is the lumped compressibility matrix (sometimes called the storativity matrix). For the 1-D two-node linear bar element, the storativity matrix may be expressed as

$$\mathbf{c}^* = \frac{(A_e^*)_i L_e c_i^*}{6}\begin{bmatrix} 2 & 1 \\ 1 & 2 \end{bmatrix}. \tag{6.34}$$

The storativity matrix in Eq. (6.34) is usually defined in consistent form, as shown. Although the consistent matrix is theoretically rigorous, its utilization in the finite element modeling of engineering problems has been hindered due to its difficulty of application in some time integration schemes, which is primarily attributed to the existence of the off-diagonal terms in the matrix. To circumvent this problem, a lumped form of the matrix may be used, which can be modified from Eq. (6.34) as

$$\mathbf{c}^* = \frac{(A_e^*)_i L_e c_i^*}{2}\begin{bmatrix} 1 & 0 \\ 0 & 1 \end{bmatrix}. \tag{6.35}$$

The off-diagonal terms in Eq. (6.34) are eliminated. The technique is to eliminate these terms in Eq. (6.34) by horizontally moving the off-diagonal terms in the matrix to the locations of the diagonal terms, then summing them together. This leaves a lumped mass matrix that may be readily applied to solving flow and transport problems, sometimes with better results than using the full consistent matrix.

In the finite element method (FEM), time t, represents a fourth dimension, which theoretically can be discretized using the shape functions otherwise employed in FEM. However, since the time discretization, using the finite difference method (FDM) may not seriously affect the computational accuracy (for non-inertial applications), the FDM is the preferred tool due to its simplicity and efficiency. The majority of current finite element models adopt this combined FEM-FDM approach for non-inertial behavior. A general FDM to discretize the continuous time derivative of p_i in Eq. (6.33) can be given as

$$\mathbf{p}_{\tau^*} = \mathbf{p}_t + \psi_{fd}(\mathbf{p}_{t+\Delta t} - \mathbf{p}_t) \tag{6.36}$$

where subscript τ^* represents the desired time level, subscripts t and $t + \Delta t$ denote the previous and current time levels, respectively, ψ_{fd} is the scaling constant that may determine a full spectrum of different schemes:

- For explicit time discretization:

$$\psi_{fd} = 0, \quad \mathbf{p}_{\tau^*} = \mathbf{p}_t \tag{6.37}$$

- For implicit time discretization:

$$\psi_{fd} = 1, \quad \mathbf{p}_{\tau^*} = \mathbf{p}_{t+\Delta t} \tag{6.38}$$

- For central difference time discretization (Crank-Nicolson scheme):

$$\psi_{fd} = 0.5, \quad \mathbf{p}_{\tau^*} = \frac{1}{2}(\mathbf{p}_t + \mathbf{p}_{t+\Delta t}). \tag{6.39}$$

Based on the theory of the FDM, $\tau^* \geq 0.5$ is a precondition for an unconditionally stable solution. Therefore, the fully explicit scheme may result in an unstable solution, even though it is the most efficient method. Stability is unconditional for the fully implicit method. The Crank-Nicolson method is at the limit of stability, but provides the highest computational accuracy among the three schemes.

Using an implicit time discretization scheme, Eq. (6.33) can be rewritten as

$$\mathbf{K}_c(\mathbf{p})_{t+\Delta t} = \mathbf{c}^*(\dot{\mathbf{p}})_{t+\Delta t} + \mathbf{q}_{t+\Delta t}. \tag{6.40}$$

$\dot{\mathbf{p}}$ can be explicitly expressed using the FDM as

$$\dot{\mathbf{p}} = \frac{1}{\Delta t}(\mathbf{p}_{t+\Delta t} - \mathbf{p}_t). \tag{6.41}$$

Substituting Eq. (6.41) into Eq. (6.40), gives

$$\left(\mathbf{K}_c - \frac{\mathbf{c}^*}{\Delta t}\right)\mathbf{p}_{t+\Delta t} = \mathbf{q}_{t+\Delta t} - \frac{\mathbf{c}^*}{\Delta t}\dot{\mathbf{p}}_t. \tag{6.42}$$

From Eq. (6.42), it is apparent that the current prescribed nodal discharge must be known before the system of equations can be sequentially solved.

6.3.3 Coupled Deformation and Flow

Numerical modeling of fluid flow through a deformable porous medium requires that the equations describing deformation and flow be solved simultaneously. As a result, a coupled formulation must be established.

6.3.3.1 Steady State

A common perception is that coupled poroelastic effects are manifest in porous media only during the transient period of fluid flow. This is true for fully coupled poroelastic phenomena where the compressibilities of the porous medium and interstitial fluid are controlled by the effective volumetric strain during transient changes in mass storage. This behavior retroactively affects changes in solid deformation such as grain expansion and contraction via fluid pressures in a time-dependent fashion. However, poroelastic phenomena also remain in effect during the subsequent steady state, however, in a less direct and nonretroactive manner. Because the flow rate is independent of time, the pore pressure variation no longer affects the time-dependent deformation of the solid skeleton. As a result, the fluid flow in the saturated rock may be influenced by the geometric variations of void spaces and/or grain sizes, characterized through the changes in original permeability, as a result of application of external load. In particular, this coupling occurs in a sequential rather than interactive fashion. Applying external load in a quasi-steady manner, which allows the rearrangement of the structure of the porous media, may result in changes in pressure and stress distributions.

Finite element modeling may be applied to these systems to represent the coupling between rock deformation and fluid flow in the governing poroelastic formulation, as described in Chapter 3.

Indirect Coupling

In this situation, the rock deformation and fluid pressure do not interact with each other. The coupling is only partially recovered through the strain-permeability relationship as described in Chapter 3.

Combining the previous formulations describing deformation and flow, the matrix form of the coupled formulation, where permeability is stress-dependent but otherwise without direct interactive unknowns, can be written as

$$\begin{bmatrix} \mathbf{K}_e & 0 \\ 0 & \mathbf{K}_c \end{bmatrix} \begin{Bmatrix} \mathrm{u} \\ \mathrm{p} \end{Bmatrix} = \begin{Bmatrix} \mathbf{F} \\ \mathbf{Q} \end{Bmatrix} \tag{6.43}$$

where

$$\mathbf{K}_e = \int_V \mathbf{B}^T \mathbf{D} \mathbf{B} dV \tag{6.44}$$

$$\mathbf{K}_c = -\frac{1}{\mu} \int_V \nabla \mathbf{M}^T \mathbf{k} \nabla \mathbf{M} dV \tag{6.45}$$

$$\mathbf{F} = \int_{A_e} \mathbf{N}\mathbf{f}dA_e \tag{6.46}$$

$$\mathbf{Q} = \int_{A_e} \mathbf{M}\mathbf{q}dA_e \tag{6.47}$$

and where $\mathbf{N}$ and $\mathbf{M}$ are the shape functions for the displacement $\mathbf{u}$ and pressure $\mathbf{p}$ (refer to Section 6.6.2 on shape functions), $\mathbf{B}$ is the strain-displacement matrix in Eq. (6.15), $\mathbf{K}_e$ and $\mathbf{K}_c$ are stiffness and conductance matrices in Eqs. (6.22) and (6.29), $\mathbf{F}$ and $\mathbf{Q}$ are the boundary tractions and discharges, $\mathbf{f}$ and $\mathbf{q}$ are vectors of applied nodal boundary tractions and prescribed nodal fluxes, V and A_e are the volume and the surface of the calculated domain, and $\mathbf{D}$ and $\mathbf{k}$ are the elastic modulus and the permeability matrices, respectively.

The numerical procedure can be itemized as

- Derive $\Delta\mathbf{u}$ and $\Delta\mathbf{p}$ from Eq. (6.43)
- Calculate strain $\Delta\varepsilon$ from the derived displacement $\Delta\mathbf{u}$
- Substitute the strain into the permeability-strain relationships given in Chapter 3 to assess the permeability change Δk
- Determine the change of flow rate using Darcy's law and the derived Δk and Δp, e.g.,

$$\Delta\mathbf{Q} = -A^* \frac{\Delta k}{\mu} \frac{\Delta \mathbf{p}}{L} \tag{6.48}$$

  where μ is the fluid dynamic viscosity, L is the flow length, and A^* is the effective flow cross-sectional area.

Direct Coupling

To consider the principle of effective stress and interaction of rock stress and pore pressure, an explicitly coupled numerical analysis can be formulated. Because steady state conditions prevail, this stress-pressure coupling is not retroactive, or only partial decoupling can be achieved. The coupling is maintained in the force equilibrium between change of solid deformation and spatial variation of pore pressure. The matrix form of the finite element formulation can be written as

$$\begin{bmatrix} \mathbf{K}_e & \mathbf{R}_d \\ \mathbf{0} & \mathbf{K}_c \end{bmatrix} \begin{Bmatrix} \mathbf{u} \\ \mathbf{p} \end{Bmatrix} = \begin{Bmatrix} \mathbf{F} \\ \mathbf{Q} \end{Bmatrix} \tag{6.49}$$

where the coupling matrix $\mathbf{R}_d$ can be described as

$$\mathbf{R}_d = -\alpha \int_V \mathbf{B}^T \mathbf{m} \mathbf{M} dV \tag{6.50}$$

and where α is Biot's coefficient [Biot 1941; Eq. (2.61) presented in Chapter 2], $\mathbf{m}$ is a 1-D vector and can be expressed as unity for 1-D; $(1 \quad 1 \quad 0)^T$ for 2-D; and $(1 \quad 1 \quad 1 \quad 0 \quad 0 \quad 0)^T$ for 3-D formulations, respectively. The vector $\mathbf{m}$ ensures that solid shear strain does not impose any impacts on the volumetric strain and correspondingly has a null influence on the fluid deformation.

The numerical procedure is similar to that used in the indirect coupled approach. In general, the direct coupled procedure provides a more accurate description of the interactive response between fluid flow and rock deformation. Strictly speaking, the solution for both indirect and direct coupled approaches is nonlinear because the permeability contains the strain component which is also a primary unknown. As a result, iterative solution is required and solution convergence must be tested at each step of calculation. As an approximation, the solution can be considered as linear if the permeability change is assumed to be ultimate and the strain variation is envisioned at a terminal stage. In other words, the permeability change occurs only after the permanent strain has been mobilized.

6.3.3.2 Transient State

Transient state fluid flow is nonnegligible if the fluid and the porous medium are both compressible or if time-dependent sources are imposed. Transient analysis is especially important for coupled processes where fluid-solid interaction is manifest within the early stages of transient loading.

The process of laminar flow can be characterized by Darcy's law as

$$q_i = \frac{-k_i}{\mu} \nabla (p_i + \gamma_f z) \tag{6.51}$$

where γ_f is the unit weight of the fluid and z is the elevation of the control volume.

The effective stress law for a poroelastic medium may be expressed as

$$\boldsymbol{\sigma} = \boldsymbol{\sigma}^e - \mathbf{m}\alpha\partial p \tag{6.52}$$

where $\boldsymbol{\sigma}$ is the total stress, $\boldsymbol{\sigma}^e$ is the effective stress, α is Biot's coefficient [refer to Eq. (2.61) in Chapter 2], and $\mathbf{m}$ is a 1-D vector introduced in the previous section.

In tensor notation, Eq. (6.52) can be reformulated as

$$\sigma_{ij} = \sigma_{ij}^e - \alpha\delta_{ij}p. \tag{6.53}$$

The linear constitutive relationship and its inverse can be defined as

$$\sigma_{ij}^e = D_{ijkl}\varepsilon_{kl}, \qquad \varepsilon_{ij} = C_{ijkl}\sigma_{kl}^e \tag{6.54}$$

where D_{ijkl} is the elastic modulus tensor and C_{ijkl} is the compliance tensor.

Substituting Eq. (6.53) into Eq. (6.54), gives

$$\varepsilon_{ij} = C_{ijkl}\sigma_{kl} + C_{ijkl}\alpha\delta_{kl}p \tag{6.55}$$

or

$$\sigma_{ij} = D_{ijkl}(\varepsilon_{kl} - C_{klmn}\alpha\delta_{mn}p). \tag{6.56}$$

For an isotropic medium, the elasticity tensor (D_{ijkl}) is equal to the elastic modulus (E) for a 1-D case. For a 2-D geometry subject to plane strain, it becomes

$$D_{ijkl} = \frac{1}{\mid D_{ijkl} \mid}\begin{bmatrix} 1-\nu & \nu & 0 \\ \nu & 1-\nu & 0 \\ 0 & 0 & \frac{1-2\nu}{2} \end{bmatrix} \tag{6.57}$$

where

$$\mid D_{ijkl} \mid = \frac{(1+\nu)(1-2\nu)}{E} \tag{6.58}$$

and where ν is Poisson's ratio.

For a 3-D configuration, one has

$$D_{ijkl} \; = (C_{ijkl})^{-1} =$$

$$\frac{1}{\mid D_{ijkl} \mid}\begin{bmatrix} d_{11} & d_{12} & d_{13} & 0 & 0 & 0 \\ d_{21} & d_{22} & d_{23} & 0 & 0 & 0 \\ d_{31} & d_{32} & d_{33} & 0 & 0 & 0 \\ 0 & 0 & 0 & d_{44} & 0 & 0 \\ 0 & 0 & 0 & 0 & d_{55} & 0 \\ 0 & 0 & 0 & 0 & 0 & d_{66} \end{bmatrix} \tag{6.59}$$

where

$$\mid D_{ijkl} \mid = \frac{8}{E^6}(1+\nu)^3(1-3\nu^2-2\nu^3) \tag{6.60}$$

$$d_{11} = d_{22} = d_{33} = \frac{8}{E^5}(1+\nu)^3(1-\nu^2) \tag{6.61}$$

$$d_{12} = d_{21} = d_{13} = d_{31} = d_{23} = d_{32} = \frac{8\nu(1+\nu)^4}{E^5} \tag{6.62}$$

$$d_{44} = d_{55} = d_{66} = \frac{4}{E^5}(1+\nu)^2(1-3\nu^2-2\nu^3) \tag{6.63}$$

and where E is the elastic modulus and ν is Poisson's ratio.

Substituting Eq. (6.56) into the equilibrium equation of motion for a solid, i.e., $\sigma_{ij,j} = F_i$ where F_i is the vector of body tractions, one has

$$(D_{ijkl}\varepsilon_{kl} - D_{ijkl}C_{klmn}\alpha\delta_{mn}p)_{,j} = F_i \tag{6.64}$$

which can be converted to the displacement based equation with the substitution of the following strain-displacement relationship:

$$\varepsilon_{ij} = \frac{1}{2}\left(u_{i,j} + u_{j,i}\right) \tag{6.65}$$

where u_i is the vector of displacement of the solid.

Expressions for the finite element approximation of fluid pressure and solid displacement can be given as

$$\mathbf{p}^* = \mathbf{M}\mathbf{p}, \qquad \mathbf{u}^* = \mathbf{N}\mathbf{u} \tag{6.66}$$

where shape functions $\mathbf{M}$ and $\mathbf{N}$ are discussed in Section 6.6.2 for the coupled finite element formulation and the superscript "*" represents the finite element approximation, which will be omitted in the following, for convenience.

Strains within a single element may be related to nodal displacements through the derivatives of the shape functions $\mathbf{B}$. The 2-D case is given in Eq. (6.15). For the 3-D case, the relationship can be derived from the following expression:

$$\boldsymbol{\varepsilon} = \mathbf{B}\mathbf{u} \tag{6.67}$$

where

$$\mathbf{B} = \begin{bmatrix} N_{i,x} & 0 & 0 \\ 0 & N_{i,y} & 0 \\ 0 & 0 & N_{i,z} \\ N_{i,y} & N_{i,x} & 0 \\ 0 & N_{i,z} & N_{i,y} \\ N_{i,z} & 0 & N_{i,x} \end{bmatrix} \tag{6.68}$$

where a comma represents differentiation in the standard manner with respect to the global x, y, and z coordinates. For isoparametric elements, the conversion between local and global coordinates must be performed through Jacobian mapping since the shape functions are defined only at the local level.

Applying the variational principle or Galerkin's method to the equilibrium equation (6.64), and differentiating the resulting expression with respect to time ($\partial/\partial t$), the momentum balance in finite element form can be expressed in the matrix form as

$$\int_V \mathbf{B}^T\mathbf{D}\mathbf{B}dV\dot{\mathbf{u}} - \alpha\int_V \mathbf{B}^T\mathbf{D}\mathbf{m}\mathbf{M}dV\dot{\mathbf{p}} = \int_{A_e} \mathbf{N}\dot{\mathbf{f}}dA_e \tag{6.69}$$

where $\mathbf{f}$ is a vector of applied boundary tractions and a superscript "·" denotes time derivative in the usual manner.

Equating the divergence of Darcy's velocity in Eq. (6.51) to the rate of fluid accumulation due to all sources such as the effect of temporal variation of volumetric strain (Biot 1941), applying Galerkin's principle and pressure mapping functions, and neglecting the impact of fluid body force, the mass balance in finite element form may be given as

$$-\frac{1}{\mu}\int_V \nabla\mathbf{M}^T\mathbf{k}\nabla\mathbf{M}dV\mathbf{p} = \alpha\int_V \mathbf{N}^T\mathbf{m}\mathbf{D}\mathbf{B}dV\dot{\mathbf{u}}+$$

$$c^*\int_V \mathbf{M}^T\mathbf{M}dV\dot{\mathbf{p}} + \int_V \mathbf{M}^T\mathbf{M}dV\mathbf{q} \tag{6.70}$$

where $\mathbf{k}$ is the permeability tensor, $\mathbf{q}$ is the vector of boundary nodal discharge, and c^* is the lumped compressibility.

The poroelastic finite element formulation given in Eqs. (6.69) and (6.70) represents a pair of coupled equations. The solutions are, therefore, required to be obtained simultaneously from the system of equations. The governing equations for an arbitrary representative time level of $\tau^* = t + \psi_{fd}\Delta t$ $(0 \leq \psi_{fd} \leq 1)$ may be established. When adopting a fully implicit finite difference scheme $(\psi_{fd} = 1)$ in the time discretization domain, the final matrix form of the equations may be written as

$$\frac{1}{\Delta t^{\tau^*}}\begin{bmatrix} \mathbf{K}_e & \mathbf{R}_d \\ \mathbf{R}_d^T & \mathbf{K}_c \end{bmatrix}\begin{Bmatrix} \mathbf{u} \\ \mathbf{p} \end{Bmatrix}^{\tau^*} = \frac{1}{\Delta t^{\tau^*}}\begin{bmatrix} \mathbf{K}_e & \mathbf{R}_d \\ \mathbf{R}_d^T & \mathbf{C}^* \end{bmatrix}\begin{bmatrix} \mathbf{u} \\ \mathbf{p} \end{bmatrix}^{t} + \begin{Bmatrix} \dot{\mathbf{F}} \\ \mathbf{Q} \end{Bmatrix}^{\tau^*} \tag{6.71}$$

where superscript T represents the matrix transposition, $\tau^* = t + \Delta t$, and individual matrices are

$$\mathbf{K}_e = \int_V \mathbf{B}^T\mathbf{D}\mathbf{B}dV \tag{6.72}$$

$$\mathbf{R}_d = -\alpha\int_V \mathbf{B}^T\mathbf{D}\mathbf{m}\mathbf{M}dV \tag{6.73}$$

$$\mathbf{K}_c = -\frac{\Delta t^{\phi}}{\mu}\int_V \nabla\mathbf{M}^T\mathbf{k}\nabla\mathbf{M}dV + \mathbf{C}^* \tag{6.74}$$

$$\mathbf{C}^* = -c^*\int_V \mathbf{M}^T\mathbf{M}dV \tag{6.75}$$

$$\dot{\mathbf{F}} = \int_{A_e} \mathbf{N}\dot{\mathbf{f}}dA_e \tag{6.76}$$

$$\mathbf{Q} = \int_{A_e} \mathbf{M}\mathbf{q}dA_e \tag{6.77}$$

where $\mathbf{f}$ and $\mathbf{q}$ are the vectors of nodal boundary traction and discharge, respectively.

It is generally assumed that the solid displacement $\mathbf{u}$ and pore pressure $\mathbf{p}$ are continuous functions in time. However, this assumption cannot be made when the load is instantaneously applied, since the Skempton effect (Skempton 1954) requires that the initial pore pressure and stress are prescribed in the undrained state. As a result, initial pore pressure needs to be calculated for the prescribed boundary loading conditions. Normally, the initial displacement field and fluid pressure distributions can be evaluated by solving the following static undrained governing equations:

$$\begin{bmatrix} \mathbf{K}_e & \mathbf{R}_d \\ \mathbf{R}_d^T & \mathbf{K}_c \end{bmatrix} \begin{Bmatrix} \mathbf{u} \\ \mathbf{p} \end{Bmatrix} = \begin{Bmatrix} \mathbf{F} \\ 0 \end{Bmatrix}. \tag{6.78}$$

However, for slightly compressible fluids, the static governing equations in Eq. (6.78) may be ill-conditioned, especially in the numerical solution. An alternative method is to use a ramp loading to approximate the initial conditions where a variable function, assumed to be linear in the first time step, is used to avoid violent oscillation that may result from the instantaneous step loading.

6.4 FINITE ELEMENT MODEL

Based on the previous finite element formulation, the present analysis focuses on developing certain special finite element models such as those suitable for: (a) cylindrical coordinates (including axisymmetric cases), (b) generalized plane strain, (c) dual-porosity media, and (d) two-phase fluid flow.

6.4.1 Cylindrical Model

For a situation with centralized pumping in a cylindrical domain, it is convenient to use a model defined in cylindrical coordinates in which the radius (r), angle of rotation (θ), and depth (z) substitute the Cartesian coordinates (x, y, z). The finite element formulation in the cylindrical coordinates is similar to that defined in Cartesian coordinates, except that certain modifications must be implemented, such as in the strain-displacement relations and in numerical integration.

6.4.1.1 General Cylindrical Case

The permeability tensor in cylindrical coordinates may be written as

$$\mathbf{k} = \begin{bmatrix} (k_{rr}) & (k_{r\theta}) & (k_{rz}) \\ (k_{\theta r}) & (k_{\theta\theta}) & (k_{\theta z}) \\ (k_{zr}) & (k_{z\theta}) & (k_{zz}) \end{bmatrix}. \tag{6.79}$$

The region of integration is a 3-D cylindrical domain. For example, at the element level, the stiffness matrix $\mathbf{K}_e$ can be written as

$$\mathbf{K}_e = \int_{-1}^{1}\int_{-1}^{1}\int_{-1}^{1} \mathbf{B}^T\mathbf{D}\mathbf{B} \mid \mathbf{J} \mid r d\bar{r}^c d\bar{s}^c d\bar{t}^c \tag{6.80}$$

where the general expression for $\mathbf{D}$ is identical to the one for 3-D Cartesian coordinates presented in Eq. (6.59), $\mid \mathbf{J} \mid$ is a Jacobian determinant, $\bar{r}^c$, $\bar{s}^c$ and $\bar{t}^c$ represent local coordinates, and r is the radius of the cylindrical domain, which may be approximated as

$$r = \bar{r} = \frac{\sum_{i=1}^{j} r_i}{j} \tag{6.81}$$

where j is the number of nodes per element. In addition, the strain-displacement transformation matrix, $\mathbf{B}$, is given by

$$\mathbf{B} = \begin{bmatrix} \frac{\partial}{\partial r} & 0 & 0 \\ 0 & \frac{\partial}{\partial z} & 0 \\ \frac{1}{r} & 0 & \frac{1}{r}\frac{\partial}{\partial \theta} \\ \frac{\partial}{\partial z} & \frac{\partial}{\partial r} & 0 \\ \frac{1}{r}\frac{\partial}{\partial \theta} & 0 & \frac{\partial}{\partial r} - \frac{1}{r} \\ 0 & \frac{1}{r}\frac{\partial}{\partial \theta} & \frac{\partial}{\partial z} \end{bmatrix} \mathbf{N} \tag{6.82}$$

where r, θ, and z constitute the global coordinate system.

6.4.1.2 Axisymmetric Case

Where the domain is axisymmetric with respect to the z axis, and the loads and material properties are axisymmetric, the 3-D cylindrical problem degenerates to a 2-D system. With regard to the present problem, the displacements u_1 and u_2 occur in the r and z directions in a cylindrical coordinate system, where the displacement u_3 in the θ direction remains unchanged since the changes of stress and pressure are independent of angular coordinate θ. Consequently, shearing strains $\gamma_{r\theta}$ and $\gamma_{z\theta}$ also vanish. The strain-displacement transformation matrix, $\mathbf{B}$, reduces to

$$\mathbf{B} = \begin{bmatrix} \frac{\partial}{\partial r} & 0 \\ 0 & \frac{\partial}{\partial z} \\ \frac{1}{r} & 0 \\ \frac{\partial}{\partial z} & \frac{\partial}{\partial r} \end{bmatrix} \mathbf{N}. \tag{6.83}$$

The elasticity matrix $\mathbf{D}$ and the related compliance matrix $\mathbf{C}$ can be expressed as

$$\left[\ \mathbf{D}\ \right] = (\mathbf{C})^{-1} = \frac{1}{\mid D \mid}\begin{bmatrix} d_{11} & d_{12} & d_{13} & 0 \\ d_{21} & d_{22} & d_{23} & 0 \\ d_{31} & d_{32} & d_{33} & 0 \\ 0 & 0 & 0 & d_{44} \end{bmatrix} \tag{6.84}$$

where

$$\mid D \mid = \frac{2}{E^4}(1+\nu)(1-3\nu^2-2\nu^3) \tag{6.85}$$

$$d_{11} = d_{22} = d_{33} = \frac{2}{E^3}(1+\nu)(1-\nu^2) \tag{6.86}$$

$$d_{12} = d_{21} = d_{13} = d_{31} = d_{23} = d_{32} = \frac{2\nu(1+\nu)^2}{E^3} \tag{6.87}$$

$$d_{44} = \frac{1}{E^3}(1-3\nu^2-2\nu^3). \tag{6.88}$$

The region of integration is a 2-D axisymmetric domain. Again, at the element level, the stiffness matrix $\mathbf{K}_e$ can be written as

$$\mathbf{K}_e = 2\pi \int_{-1}^{1}\int_{-1}^{1} \mathbf{B}^T\mathbf{D}\mathbf{B} \mid \mathbf{J} \mid r d\bar{r}^c d\bar{s}^c \tag{6.89}$$

where r can be determined from Eq. (6.81). The axisymmetric model is advantageous in that a 2-D algorithm is applied for a 3-D domain geometry.

6.4.2 Generalized Plane Strain

For 3-D engineering structures featuring much greater lengths in one dimension than the other two dimensions, mechanical problems can be treated as one of plane strain, where the change in strain in the longer direction vanishes even though the stress change along that same direction is finite. The plane strain condition is based on the assumption that there is one plane of symmetry perpendicular to the long axis which is parallel to a principal stress. If this assumption cannot be satisfied, then the generalized plane strain concept must be used. The major departure of the generalized plane strain (GPS) from the plane strain (PS) system is that the normal strain and shear stresses along the long axis are induced even though the normal strain and the shear stresses parallel to the principal stress orientation are null. Application of the GPS concept is useful for cases in which the principal stresses are known, but are not aligned parallel to the long axis of the underground structure represented in the analysis (e.g., tunnels, wellbores, storage chambers, etc.). Alternatively, the GPS method is applicable for structural analysis subject to pseudo 3-D geometries. In other words, it is possible to approximate the structural behavior in a 3-D domain with the 2-D formulation. As a result, accuracy is improved in comparison with the classical PS method.

6.4.2.1 Concept

Because the GPS is not a common term in solid mechanics, the concept should be explained using mathematical and mechanical terms.

In a 2-D geometry (e.g., the $x - y$ plane), it is assumed that the out-of-plane direction z is parallel to the long axis of the structures of interest. In PS problems, displacements and shear stresses are restricted along the z direction. In the case of GPS, however, these restrictions are removed. As a result, the number of tensor components for stresses and strains are identical to those of a 3-D setting. In a general GPS formulation, it is assumed that surface tractions, pore pressure, body forces, and material properties do not vary along the z direction. Further assumptions include no torsion, pure bending, and the existence of strain ε_z (Lekhnitskii 1981). As a result, the displacements, stresses, strains, and pore pressure are only functions of x, y, and time t, as

$$\boldsymbol{u}^T(x, y, t) = \{u_x \; u_y \; u_z\} \tag{6.90}$$

$$\boldsymbol{\varepsilon}^T(x, y, t) = \{\varepsilon_{xx} \; \varepsilon_{yy} \; \varepsilon_{zz} \; \gamma_{xy} \; \gamma_{yz} \; \gamma_{xz}\} \tag{6.91}$$

$$\boldsymbol{\sigma}^T(x, y, t) = \{\sigma_{xx} \; \sigma_{yy} \; \sigma_{zz} \; \tau_{xy} \; \tau_{yz} \; \tau_{xz}\} \tag{6.92}$$

$$\boldsymbol{p}^T(x, y, t) = \{p\} \tag{6.93}$$

where the subscript "T" represents vector transposition.

6.4.2.2 Finite Element Formulation

As mentioned previously, GPS solutions maintain primary unknowns that are compatible with the 3-D formulation, but are geometrically not related to the z coordinate, similar to the 2-D cases. With reference to the finite element formulation, the major difference between the GPS and the PS formulations is exhibited in the strain-displacement matrix $\mathbf{B}$. For the GPS formulation, $\mathbf{B}$ should be written as (Cui et al. 1997)

$$\mathbf{B} = \begin{bmatrix} \frac{\partial}{\partial x} & 0 & 0 \\ 0 & \frac{\partial}{\partial y} & 0 \\ 0 & 0 & 0 \\ \frac{\partial}{\partial y} & \frac{\partial}{\partial x} & 0 \\ 0 & 0 & \frac{\partial}{\partial y} \\ 0 & 0 & \frac{\partial}{\partial x} \end{bmatrix} \mathbf{N} \tag{6.94}$$

where **N** is the shape function for the solid part.

For comparison, in the PS formulation, **B** in the 2-D and 3-D geometries are given, respectively, as follows:

$$\mathbf{B} = \begin{bmatrix} \frac{\partial}{\partial x} & 0 \\ 0 & \frac{\partial}{\partial y} \\ \frac{\partial}{\partial y} & \frac{\partial}{\partial x} \end{bmatrix} \mathbf{N} \tag{6.95}$$

and

$$\mathbf{B} = \begin{bmatrix} \frac{\partial}{\partial x} & 0 & 0 \\ 0 & \frac{\partial}{\partial y} & 0 \\ 0 & 0 & \frac{\partial}{\partial z} \\ \frac{\partial}{\partial y} & \frac{\partial}{\partial x} & 0 \\ 0 & \frac{\partial}{\partial z} & \frac{\partial}{\partial y} \\ \frac{\partial}{\partial z} & 0 & \frac{\partial}{\partial x} \end{bmatrix} \mathbf{N}. \tag{6.96}$$

Using an implicit time discretization scheme, the final governing finite element equations for the poroelastic model can be described as

$$\frac{1}{\Delta t^{(t+\Delta t)}} \begin{bmatrix} \mathbf{K}_e & \mathbf{R}_d \\ \mathbf{R}_d^T & \mathbf{K}_c \end{bmatrix} \begin{Bmatrix} \mathbf{u} \\ \mathbf{p} \end{Bmatrix}^{(t+\Delta t)} =$$

$$\frac{1}{\Delta t^{(t+\Delta t)}} \begin{bmatrix} \mathbf{K}_e & \mathbf{R}_d \\ \mathbf{R}_d^T & \mathbf{C}^* \end{bmatrix} \begin{Bmatrix} \mathbf{u} \\ \mathbf{p} \end{Bmatrix}^{t} + \begin{Bmatrix} \dot{\mathbf{F}} \\ \mathbf{Q} \end{Bmatrix}^{(t+\Delta t)} \tag{6.97}$$

where

$$\mathbf{K}_e = \int_{A_e} \mathbf{B}^T \mathbf{D} \mathbf{B} dA_e \tag{6.98}$$

$$\mathbf{R}_d = -\alpha \int_{A_e} \mathbf{B}^T \mathbf{D} \mathbf{m} \mathbf{M} dA_e \tag{6.99}$$

$$\mathbf{K}_c = -\frac{1}{\mu} \int_{A_e} \nabla \mathbf{M}^T \mathbf{k} \nabla \mathbf{M} dA_e \Delta t^{(t+\Delta t)} + C^* \tag{6.100}$$

$$\mathbf{C}^* = -c^* \int_{A_e} \mathbf{M}^T \mathbf{M} dA_e \tag{6.101}$$

$$\dot{\mathbf{F}} = \int_{L_e} \dot{\mathbf{N}}\mathbf{f} dL_e \tag{6.102}$$

$$\mathbf{Q} = \int_{L_e} \mathbf{M}\mathbf{q} dL_e \tag{6.103}$$

where A_e is the area projected from the cylindrical domain, L_e is the contour that bounds the projected area, and $\mathbf{f}$ and $\mathbf{q}$ are the vectors of applied traction and fluid flux, respectively.

For the GPS formulation, the displacements and stresses may be transformed according to the following rules:

$$\begin{bmatrix} u_r \\ u_\theta \\ u_z \end{bmatrix} = \begin{bmatrix} \cos\theta & \sin\theta & 0 \\ -\sin\theta & \cos\theta & 0 \\ 0 & 0 & 1 \end{bmatrix} \begin{bmatrix} u_x \\ u_y \\ u_z \end{bmatrix} \tag{6.104}$$

$$\boldsymbol{\sigma}^* = \mathbf{T}^T \boldsymbol{\sigma} \mathbf{T} \tag{6.105}$$

where

$$\boldsymbol{\sigma}^* = \begin{bmatrix} \sigma_{rr} & \tau_{r\theta} & \tau_{rz} \\ \tau_{\theta r} & \sigma_{r\theta} & \tau_{\theta z} \\ \tau_{zr} & \tau_{z\theta} & \sigma_{zz} \end{bmatrix} \tag{6.106}$$

$$\boldsymbol{\sigma} = \begin{bmatrix} \sigma_{xx} & \tau_{xy} & \tau_{xz} \\ \tau_{yx} & \sigma_{yy} & \tau_{yz} \\ \tau_{zx} & \tau_{zy} & \sigma_{zz} \end{bmatrix} \tag{6.107}$$

$$\mathbf{T} = \begin{bmatrix} \cos\theta & -\sin\theta & 0 \\ \sin\theta & \cos\theta & 0 \\ 0 & 0 & 1 \end{bmatrix} \tag{6.108}$$

where r and θ are the radius and the angle in the measurement of the distance and direction of the location of interest with reference to the center-axis of the coordinate system. This may represent, for example, the center of the well in a polar coordinate system.

6.4.3 Dual-Porosity Media

As discussed in Chapter 2, two approaches are available to evaluate solid equilibrium of the dual-porosity medium, i.e., single and double effective stress laws. The finite element formulation for the former case is identical to that introduced for the single-porosity medium (see the section on coupled deformation-flow using the finite element formulation). For the finite element formulation of flow continuity, there is no difference between the two approaches. The following description focuses on the latter case.

6.4.3.1 Constitutive Relations

The effective stress law for a dual-porosity medium may be expressed as

$$\partial \boldsymbol{\sigma}_i = \partial \boldsymbol{\sigma}_i^e - \mathbf{m}\alpha_i \partial p_i \tag{6.109}$$

where subscript i indexes the matrix blocks and fractures, and α_i is the pressure ratio factor for the matrix blocks or fractures (α_1 is Biot's coefficient).

Further development of the effective stress law is based on the consideration that the solid deformations of the fractures and the matrix are lumped while the stress compatibility condition is maintained, i.e. (Elsworth and Bai 1992),

$$\boldsymbol{\varepsilon} = \sum_{i=1}^{2} \boldsymbol{\varepsilon}_i, \qquad \boldsymbol{\sigma}_1 = \boldsymbol{\sigma}_2 = \boldsymbol{\sigma} \tag{6.110}$$

where $\boldsymbol{\varepsilon}$ are the total strains.

Because the variation of strain in the matrix blocks or fractures is only due to the change in effective stress, then

$$\boldsymbol{\varepsilon}_i = \mathbf{C}_i \boldsymbol{\sigma}_i^e \tag{6.111}$$

where $\mathbf{C}_i$ is the compliance matrix. This relationship can be reversed as

$$\boldsymbol{\sigma}_i^e = \mathbf{D}_i \boldsymbol{\varepsilon}_i \tag{6.112}$$

where $\mathbf{D}_i$ is the elasticity matrix.

Combining Eqs. (6.109) through (6.112) results in a modified effective stress law in a dual-porosity system:

$$\boldsymbol{\sigma} = \mathbf{D}_{12} \left[\boldsymbol{\varepsilon} - \sum_{i=1}^{2} \alpha_i \mathbf{C}_i \mathbf{m} p_i \right] \tag{6.113}$$

where the combined elasticity matrix $\mathbf{D}_{12}$ can be defined in a 3-D geometry explicitly for an isotropic medium as

$$\mathbf{D}_{12} = \left(\sum_{i=1}^{2} \mathbf{C}_i \right)^{-1} = |\mathbf{D}_{12}|^{-1} \begin{bmatrix} d_{11} & d_{12} & d_{13} & 0 & 0 & 0 \\ d_{21} & d_{22} & d_{23} & 0 & 0 & 0 \\ d_{31} & d_{32} & d_{33} & 0 & 0 & 0 \\ 0 & 0 & 0 & d_{44} & 0 & 0 \\ 0 & 0 & 0 & 0 & d_{55} & 0 \\ 0 & 0 & 0 & 0 & 0 & d_{66} \end{bmatrix} \tag{6.114}$$

where

$$|\mathbf{D}_{12}| = \left[\frac{1}{K_{sh}s^*} + \frac{2(1+\nu)}{E}\right]^3 \left[\left(\frac{1}{K_n s^*} + \frac{1}{E}\right)^3 - \frac{3\nu^2}{E^2}\left(\frac{1}{K_n s^*} + \frac{1}{E}\right) - \frac{2\nu^3}{E^3}\right] \tag{6.115}$$

$$d_{11} = d_{22} = d_{33} = \left[\frac{1}{K_{sh}s^*} + \frac{2(1+\nu)}{E}\right]^3 \left[\left(\frac{1}{K_n s^*} + \frac{1}{E}\right)^2 - \frac{\nu^2}{E^2}\right] \tag{6.116}$$

$$d_{12} = d_{21} = d_{13} = d_{31} = d_{23} = d_{32} = \left[\frac{1}{K_{sh}s^*} + \frac{2(1+\nu)}{E}\right]^3 \left[\frac{\nu}{E}\left(\frac{1}{K_n s^*} + \frac{1}{E}\right) + \frac{\nu^2}{E^2}\right] \tag{6.117}$$

$$d_{44} = d_{55} = d_{66} = |\mathbf{D}_{12}| \left[\frac{1}{K_{sh}s^*} + \frac{2(1+\nu)}{E}\right]^{-1} \tag{6.118}$$

and E is the elastic modulus, ν is Poisson ratio, s^* is the fracture spacing, and K_n and K_{sh} are the fracture normal and shear stiffnesses, respectively. For simplicity, it is assumed that orthogonal fractures are aligned parallel to the global axes and are of uniform spacing and stiffness at the element level. These assumptions considerably simplify the form of the constitutive relation, $\mathbf{D}_{12}$, but are not a requirement of the analysis, as more complex relations may be readily incorporated.

In fluid flow, by envisioning two separate flows within the two individual but overlapping media (matrix blocks and fractures), the process of laminar flow can be characterized by Darcy's law as

$$\mathbf{q}_i = \frac{-k_i}{\mu}\boldsymbol{\nabla}(p_i + \gamma_f z) \tag{6.119}$$

where, again, subscripts $i = 1$ and 2 represent the matrix blocks and the fractures, respectively.

6.4.3.2 System of Equations

The solid displacement and fluid pressure can be mapped at the element level through shape functions:

$$\mathbf{p}_i^* = \mathbf{M}\mathbf{p}_i, \qquad \mathbf{u}^* = \mathbf{N}\mathbf{u} \tag{6.120}$$

where subscripts $i = 1$ and 2 again represent matrix and fractures, respectively. Also, for simplicity, the superscript "*" in the above equation is omitted in the following description.

Substituting the modified effective stress law in Eq. (6.113) into the force equilibrium equation $\sigma_{ij,j} = 0$ where inertial effects are neglected, applying the variational principle, and taking the derivative with respect to time, the equilibrium equation in finite element form can be described as

$$\int_V \mathbf{B}^T\mathbf{D}_{12}\mathbf{B}dV\dot{\mathbf{u}} - \sum_{i=1}^{2}\alpha_i \int_V \mathbf{B}^T\mathbf{D}_{12}\mathbf{C}_i\mathbf{m}\mathbf{M}dV\dot{\mathbf{p}}_i = \int_{A_e} \mathbf{N}\dot{\mathbf{f}}dA_e \tag{6.121}$$

where subscript $i = 1$ and 2.

Taking the divergence of the Darcy velocity in Eq. (6.119), incorporating the rate of fluid accumulation due to all sources such as the poroelastic effect (refer to Chapter 3), and applying Galerkin's principle and the pressure mapping functions, the mass balance in finite element form may be given for the matrix blocks and fractures as

$$-\frac{1}{\mu}\int_V \nabla\mathbf{M}^T\mathbf{k}_i\nabla\mathbf{M}dV\mathbf{p}_i = \alpha_i\int_V \mathbf{N}^T\mathbf{m}\mathbf{D}_{12}\mathbf{C}_i\mathbf{B}dV\dot{\mathbf{u}}+$$

$$c_i^*\int_V \mathbf{M}^T\mathbf{M}dV\dot{\mathbf{p}}_i \pm \Gamma\int_V \mathbf{M}^T\mathbf{M}dV\Delta\mathbf{p} + \int_V \mathbf{M}^T\mathbf{M}dV\mathbf{q}_i \tag{6.122}$$

where subscript $i = 1$ and 2, Γ is a geometric factor indexing the rate of interporosity flow due to the pressure difference between fractures and matrix blocks (Warren and Root 1963), the sign $\pm$ reflects two separate flow equations with "+" for the flow in the matrix blocks, and "−" for flow in the fractures, and $\Delta\mathbf{p} = \mathbf{p}_1 - \mathbf{p}_2$ representing the pressure difference between matrix blocks and fractures.

The coupled dual-porosity poroelastic finite element formulation given in Eqs. (6.121) and (6.122) can be defined in matrix form using a fully implicit finite difference scheme for the time discretization as

$$\frac{1}{\Delta t^{\tau^*}}\begin{bmatrix} \mathbf{K}_e & \mathbf{R}_{d1} & \mathbf{R}_{d2} \\ \mathbf{R}_{d1}^T & \mathbf{K}_{c11} & \mathbf{K}_{c12} \\ \mathbf{R}_{d2}^T & \mathbf{K}_{c21} & \mathbf{K}_{c22} \end{bmatrix}\begin{Bmatrix} \mathbf{u} \\ \mathbf{p}_1 \\ \mathbf{p}_2 \end{Bmatrix}^{\tau^*} =$$

$$\frac{1}{\Delta t^{\tau^*}}\begin{bmatrix} \mathbf{K}_e & \mathbf{R}_{d1} & \mathbf{R}_{d2} \\ \mathbf{R}_{d1}^T & \mathbf{C}_1^* & \mathbf{0} \\ \mathbf{R}_2^T & \mathbf{0} & \mathbf{C}_2^* \end{bmatrix}\begin{Bmatrix} \mathbf{u} \\ \mathbf{p}_1 \\ \mathbf{p}_2 \end{Bmatrix}^{t} + \begin{Bmatrix} \dot{\mathbf{F}} \\ \mathbf{Q}_1 \\ \mathbf{Q}_2 \end{Bmatrix}^{\tau^*} \tag{6.123}$$

where $\tau^* = t + \Delta t$, and

$$\mathbf{K}_e = \int_V \mathbf{B}^T\mathbf{D}_{12}\mathbf{B}dV \tag{6.124}$$

$$\mathbf{R}_{d1} = -\alpha_1\int_V \mathbf{B}^T\mathbf{D}_{12}\mathbf{C}_1\mathbf{m}\mathbf{M}dV \tag{6.125}$$

$$\mathbf{R}_{d2} = -\alpha_2\int_V \mathbf{B}^T\mathbf{D}_{12}\mathbf{C}_2\mathbf{m}\mathbf{M}dV \tag{6.126}$$

$$\mathbf{K}_{c11} = -\frac{\Delta t^{\tau^*}}{\mu} \int_V \nabla \mathbf{M}^T \mathbf{k}_1 \nabla \mathbf{M} dV - (c_1^* + \Delta t \Gamma) \int_V \mathbf{M}^T \mathbf{M} dV \tag{6.127}$$

$$\mathbf{K}_{c12} = \Delta t \Gamma \int_V \mathbf{M}^T \mathbf{M} dV, \qquad \mathbf{K}_{c21} = \mathbf{K}_{c12}^T \tag{6.128}$$

$$\mathbf{K}_{c22} = -\frac{\Delta t^{\tau^*}}{\mu} \int_V \nabla \mathbf{M}^T \mathbf{k}_2 \nabla \mathbf{M} dV - (c_2^* + \Delta t \Gamma) \int_V \mathbf{M}^T \mathbf{M} dV \tag{6.129}$$

$$\mathbf{C}_1^* = -c_1^* \int_V \mathbf{M}^T \mathbf{M} dV \tag{6.130}$$

$$\mathbf{C}_2^* = -c_2^* \int_V \mathbf{M}^T \mathbf{M} dV \tag{6.131}$$

$$\dot{\mathbf{F}} = \int_{A_e} \mathbf{N} \dot{\mathbf{f}} dA_e \tag{6.132}$$

$$\mathbf{Q}_1 = \int_{A_e} \mathbf{M} \mathbf{q}_1 dA_e \tag{6.133}$$

$$\mathbf{Q}_2 = \int_{A_e} \mathbf{M} \mathbf{q}_2 dA_e. \tag{6.134}$$

The initial displacement field and fluid pressure distributions can be evaluated by solving the following static undrained governing equations:

$$\begin{bmatrix} \mathbf{K}_e & \mathbf{R}_{d1} & \mathbf{R}_{d2} \\ \mathbf{R}_{d1}^T & \mathbf{K}_{c11} & \mathbf{K}_{c12} \\ \mathbf{R}_{d2}^T & \mathbf{K}_{c21} & \mathbf{K}_{c22} \end{bmatrix} \begin{Bmatrix} \mathbf{u} \\ \mathbf{p}_1 \\ \mathbf{p}_2 \end{Bmatrix} = \begin{Bmatrix} \mathbf{F} \\ 0 \\ 0 \end{Bmatrix}. \tag{6.135}$$

6.4.4 Two-Phase Fluid Flow

Compared with linear analysis of single-phase fluid flow, modeling two-phase fluid flow requires an iterative solution procedure due to the nonlinear nature of the governing equations. In addition, two fluid pressures, one for the wetting fluid and another for the nonwetting fluid, are the primary unknowns (or degrees of freedom at the nodal level) for fluid flow alone. The present analysis focuses on the basic finite element formulation for single-porosity and dual-porosity media, respectively.

6.4.4.1 Single-Porosity Media

The finite element discretization of the equilibrium and flow equations may be expressed in terms of the nodal displacements ($\mathbf{u}$), nodal wetting fluid pressure (p_w), and nodal nonwetting fluid pressure (p_n). Nodal saturations are derived through relationships between capillary pressure and saturation (refer to Chapter 3). Alternatively, nodal saturations can be treated as primary unknowns. In either instance, the global unknowns can be mapped to the nodal unknowns as

$$\begin{cases} \boldsymbol{\varepsilon}^* = \mathbf{Bu} & \\ \mathbf{p}_w^* = \mathbf{Mp}_w, & \mathbf{p}_n^* = \mathbf{Mp}_n, \\ \mathbf{S}_w^* = \mathbf{MS}_w, & \mathbf{S}_n^* = \mathbf{MS}_n \end{cases} \tag{6.136}$$

where the superscript "*" represents the finite element approximation and is omitted for simplicity. The strain-displacement matrix $\mathbf{B}$ is defined previously for the 2-D and 3-D cases, which can be again used in this analysis.

After taking the derivative with respect to time, the equation governing force equilibrium in finite element form can be expressed as

$$\int_V \mathbf{B}^T\mathbf{DB}dV\frac{\partial \mathbf{u}}{\partial t} - \alpha\int_V \mathbf{B}^\mathbf{T}\mathbf{mM}dV\frac{\partial \mathbf{p}}{\partial t} = \frac{\partial \mathbf{F}}{\partial t} \tag{6.137}$$

where the fluid pressure, $\mathbf{p}$, is again the total pressure, appropriately weighted by saturation.

Taking the derivative with respect to time in Eq. (3.41) from Chapter 3, one obtains

$$\frac{\partial \mathbf{p}}{\partial t} = \mathbf{S}_n\frac{\partial \mathbf{p}_n}{\partial t} + \mathbf{p}_n\frac{\partial \mathbf{S}_n}{\partial t} + \mathbf{S}_w\frac{\partial \mathbf{p}_w}{\partial t} + \mathbf{p}_w\frac{\partial \mathbf{S}_w}{\partial t} \tag{6.138}$$

where subscripts w and n denote wetting and nonwetting phases, respectively, and $\mathbf{S}$ is the vector of nodal saturations in which its derivative with respect to time can be related to the capillary pressure, $\mathbf{p}_c$, as

$$\frac{\partial \mathbf{S}_w}{\partial t} = \frac{\partial \mathbf{S}_w}{\partial \mathbf{p}_c}\frac{\partial \mathbf{p}_c}{\partial t}. \tag{6.139}$$

Since $\mathbf{p}_c$ is generally a known coefficient, the purpose of the conversion in Eq. (6.139) is to eliminate one secondary unknown, such as $\mathbf{S}_n$ in the present formulation.

Incorporating Eqs. (6.138) and (6.139) into Eq. (6.137), the equilibrium equation is modified in short form as

$$\mathbf{K}_e\dot{\mathbf{u}} + \mathbf{L}_w\dot{\mathbf{p}}_w + \mathbf{L}_n\dot{\mathbf{p}}_n = \dot{\mathbf{F}} \tag{6.140}$$

where $\mathbf{K}_e$ and $\dot{\mathbf{F}}$ are defined in Eqs. (6.124) and (6.132), and

$$\mathbf{L}_w = -\alpha\int_V \mathbf{B}^\mathbf{T}\mathbf{mM}\left(\mathbf{S}_w + \mathbf{p}_c\frac{\partial \mathbf{S}_w}{\partial \mathbf{p}_c}\right)dV \tag{6.141}$$

$$\mathbf{L}_n = -\alpha\int_V \mathbf{B}^\mathbf{T}\mathbf{mM}\left(1 - \mathbf{S}_w - \mathbf{p}_c\frac{\partial \mathbf{S}_w}{\partial \mathbf{p}_c}\right)dV. \tag{6.142}$$

Using the capillary pressure relationship in Eq. (3.38) from Chapter 3, and relating the capillary pressure to saturation of the wetting fluid by

$$\frac{\partial \mathbf{p}_c}{\partial t} = \frac{\partial \mathbf{p}_c}{\partial \mathbf{S}_w}\frac{\partial \mathbf{S}_w}{\partial t}, \tag{6.143}$$

Eq. (6.140) can be reformulated as

$$\mathbf{K}_e\dot{\mathbf{u}} + (\mathbf{L}_w + \mathbf{L}_n)\dot{\mathbf{p}}_w + \mathbf{L}_n\frac{\partial \mathbf{p}_c}{\partial \mathbf{S}_w}\dot{\mathbf{S}}_w = \dot{\mathbf{F}}. \tag{6.144}$$

This reformulation ensures that unknowns in the fluid flow equation are related to the wetting phase only. The selection of unknowns is primarily based on specific requirements and individual conditions. In the present analysis, the solid displacement and fluid pressures (wetting and nonwetting phases) are chosen as the primary unknowns. As a result, saturation becomes a secondary unknown. The final system of equations, in matrix form, can be expressed as

$$\begin{bmatrix} 0 & 0 & 0 \\ 0 & \mathbf{W}_w & 0 \\ 0 & 0 & \mathbf{W}_n \end{bmatrix}\begin{Bmatrix} \mathbf{u} \\ \mathbf{p}_w \\ \mathbf{p}_n \end{Bmatrix} +$$

$$\begin{bmatrix} \mathbf{K}_e & \mathbf{L}_w & \mathbf{L}_n \\ \mathbf{K}_w & \mathbf{W}_{11} & \mathbf{W}_{12} \\ \mathbf{K}_n & \mathbf{W}_{21} & \mathbf{W}_{22} \end{bmatrix}\frac{d}{dt}\begin{Bmatrix} \mathbf{u} \\ \mathbf{p}_w \\ \mathbf{p}_n \end{Bmatrix} = \begin{Bmatrix} \dot{\mathbf{F}} \\ \mathbf{Q}_w \\ \mathbf{Q}_n \end{Bmatrix} \tag{6.145}$$

where $\mathbf{K}_e$, $\mathbf{L}_w$, $\mathbf{L}_n$ and $\dot{\mathbf{F}}$ are already defined in Eqs. (6.124), (6.141), (6.142), and (6.132), and

$$\mathbf{W}_w = -\int_V \nabla\mathbf{M}^T\frac{\mathbf{k}\mathbf{k}_{rw}}{\mu_w}\rho_w\nabla\mathbf{M}dV \tag{6.146}$$

$$\mathbf{W}_n = -\int_V \nabla\mathbf{M}^T\frac{\mathbf{k}\mathbf{k}_{rn}}{\mu_n}\rho_n\nabla\mathbf{M}dV \tag{6.147}$$

$$\mathbf{K}_w = \alpha\int_V \mathbf{M}^T\mathbf{S}_w\rho_w\mathbf{m}^T\mathbf{B}dV \tag{6.148}$$

$$\mathbf{K}_n = \alpha\int_V \mathbf{M}^T(1-\mathbf{S}_w)\rho_n\mathbf{m}^T\mathbf{B}dV \tag{6.149}$$

$$\mathbf{W}_{11} = \int_V \mathbf{M}^T n\mathbf{S}_w\frac{\rho_w}{K_w}\mathbf{M}dV - \int_V \mathbf{M}^T n\rho_w\frac{\partial \mathbf{S}_w}{\partial \mathbf{p}_c}\mathbf{M}dV + \int_V \mathbf{M}^T$$

$$\mathbf{S}_w\rho_w\frac{(\alpha-n)}{K_s}\left(\mathbf{S}_w + \mathbf{p}_c\frac{\partial \mathbf{S}_w}{\partial \mathbf{p}_c}\right)\mathbf{M}dV +$$

$$\int_V \mathbf{M}^T \mathbf{S}_w \rho_w n \left(\frac{\mathbf{S}_w}{K_w} + \frac{\mathbf{S}_n}{K_{nw}}\right)\left(\mathbf{S}_w + \mathbf{p}_c \frac{\partial \mathbf{S}_w}{\partial \mathbf{p}_c}\right) \mathbf{M} dV \tag{6.150}$$

$$\mathbf{W}_{12} = \int_V \mathbf{M}^T n \frac{\partial \mathbf{S}_w}{\partial \mathbf{p}_c} \rho_w \mathbf{M} dV + \int_V \mathbf{M}^T \mathbf{S}_w \rho_w$$

$$\frac{(\alpha - n)}{K_s}\left(1 - \mathbf{S}_w - \mathbf{p}_c \frac{\partial \mathbf{S}_w}{\partial \mathbf{p}_c}\right) \mathbf{M} dV +$$

$$\int_V \mathbf{M}^T \mathbf{S}_w \rho_w n \left(\frac{\mathbf{S}_w}{K_w} + \frac{\mathbf{S}_n}{K_{nw}}\right)\left(1 - \mathbf{S}_w - \mathbf{p}_c \frac{\partial \mathbf{S}_w}{\partial \mathbf{p}_c}\right) \mathbf{M} dV \tag{6.151}$$

$$\mathbf{W}_{21} = \int_V \mathbf{M}^T n \frac{\partial \mathbf{S}_w}{\partial \mathbf{p}_c} \rho_n \mathbf{M} dV + \int_V \mathbf{M}^T \mathbf{S}_n \rho_n$$

$$\frac{(\alpha - n)}{K_s}\left(\mathbf{S}_w + \mathbf{p}_c \frac{\partial \mathbf{S}_w}{\partial \mathbf{p}_c}\right) \mathbf{M} dV + \int_V \mathbf{M}^T$$

$$\mathbf{S}_n \rho_n n \left(\frac{\mathbf{S}_w}{K_w} + \frac{\mathbf{S}_n}{K_{nw}}\right)\left(\mathbf{S}_w + \mathbf{p}_c \frac{\partial \mathbf{S}_w}{\partial \mathbf{p}_c}\right) \mathbf{M} dV \tag{6.152}$$

$$\mathbf{W}_{22} = \int_V \mathbf{M}^T n \mathbf{S}_n \frac{\rho_n}{K_{nw}} \mathbf{M} dV - \int_V \mathbf{M}^T n \frac{\partial \mathbf{S}_w}{\partial \mathbf{p}_c} \rho_n \mathbf{M} dV +$$

$$\int_V \mathbf{M}^T \mathbf{S}_n \rho_n \frac{(\alpha - n)}{K_s}\left(1 - \mathbf{S}_w - \mathbf{p}_c \frac{\partial \mathbf{S}_w}{\partial \mathbf{p}_c}\right)$$

$$\mathbf{M} dV + \int_V \mathbf{M}^T \mathbf{S}_n \rho_n n \left(\frac{\mathbf{S}_w}{K_w} + \frac{\mathbf{S}_n}{K_{nw}}\right)\left(1 - \mathbf{S}_w - \mathbf{p}_c \frac{\partial \mathbf{S}_w}{\partial \mathbf{p}_c}\right) \mathbf{M} dV \tag{6.153}$$

$$\mathbf{Q}_w = \int_{A_e} \mathbf{M} \mathbf{q}_w dA_e \tag{6.154}$$

$$\mathbf{Q}_n = \int_{A_e} \mathbf{M} \mathbf{q}_n dA_e \tag{6.155}$$

where the subscripts w and n (i.e., nw) represent wetting and nonwetting phases, ρ is the density of the fluid, $\mathbf{k}$ is the absolute permeability, $\mathbf{k}_{rw}$ and $\mathbf{k}_{rn}$ are the relative permeabilities of the wetting and nonwetting phase fluids, K_w and K_{nw} are the bulk moduli of the wetting and nonwetting phase fluids, n is the porosity, α is Biot's coefficient (Biot 1941), and $\mathbf{q}$ is the nodal boundary discharge.

The solution at various times may be obtained via a finite difference algorithm, for example by applying a fully implicit time-stepping scheme such as

$$\begin{cases} \dot{\mathbf{u}}^{t+\Delta t} = \frac{1}{\Delta t}\left(\mathbf{u}^{t+\Delta t} - \mathbf{u}^{t}\right) \\ \dot{\mathbf{p}}_w^{t+\Delta t} = \frac{1}{\Delta t}\left(\mathbf{p}_w^{t+\Delta t} - \mathbf{p}_w^{t}\right) \\ \dot{\mathbf{p}}_n^{t+\Delta t} = \frac{1}{\Delta t}\left(\mathbf{p}_n^{t+\Delta t} - \mathbf{p}_n^{t}\right) \end{cases} . \tag{6.156}$$

Substituting Eq. (6.156) into Eq. (6.145) to replace the continuous time derivative, and reformulating, yields

$$\frac{1}{\Delta t}\begin{bmatrix} \mathbf{K}_e & \mathbf{L}_w & \mathbf{L}_n \\ \mathbf{K}_w & \mathbf{W}_w\Delta t + \mathbf{W}_{11} & \mathbf{W}_{12} \\ \mathbf{K}_n & \mathbf{W}_{21} & \mathbf{W}_n\Delta t + \mathbf{W}_{22} \end{bmatrix}\begin{Bmatrix} \mathbf{u} \\ \mathbf{p}_w \\ \mathbf{p}_n \end{Bmatrix}^{\tau^*}$$

$$= \frac{1}{\Delta t}\begin{bmatrix} \mathbf{K}_e & \mathbf{L}_w & \mathbf{L}_n \\ \mathbf{K}_w & \mathbf{W}_{11} & \mathbf{W}_{12} \\ \mathbf{K}_n & \mathbf{W}_{21} & \mathbf{W}_{22} \end{bmatrix}\begin{Bmatrix} \mathbf{u} \\ \mathbf{p}_w \\ \mathbf{p}_n \end{Bmatrix}^{t} + \begin{Bmatrix} \dot{\mathbf{F}} \\ \mathbf{Q}_w \\ \mathbf{Q}_n \end{Bmatrix}^{\tau^*} \tag{6.157}$$

where $\tau^* = (t + \Delta t)$, and there are five explicit unknowns at the nodal level, i.e., displacements, u_x, u_y, u_z, wetting phase pressure p_w and nonwetting phase pressure p_n. For an eight-node brick 3-D element, with five degrees of freedom per node, there are a total of 40 equations in the system at the element level. An additional node is applied at the center of the element to improve the displacement mode (refer to Section 6.6.2), which does not result in an additional degree of freedom. Extra degrees of freedom result, if additional nodes are arranged at the edges of the element.

Even though the primary unknowns from the governing equation (6.157) are $\mathbf{u}$, $\mathbf{p}_w$, and $\mathbf{p}_n$, the solution may not be straightforward since these unknowns are also embedded in the coefficients through capillary-pressure versus saturation relationships. The solution procedure may be simplified by assuming that these embedded unknowns are defined at the previous time level. From this assumption, and with reference to the present time level, these unknowns become known values. This approximation is justifiable if the time step is sufficiently small.

Two secondary unknowns are the saturations of the wetting and nonwetting phases, $\mathbf{S}_w$ and $\mathbf{S}_n$. They are related to the primary unknowns (i.e., wetting and nonwetting phase pressures) via the relationships between capillary pressure and pressure, or between capillary pressure and saturation (Dagger 1997). For single-porosity media, two auxiliary variables are relative permeability of the wetting and nonwetting phases, k_{rw} and k_{rn}, respectively. They are related to the secondary unknowns via the relative-permeability versus saturation relationships (Mattax and Dalton 1990). The relationships between the capillary pressure and saturation, and between the relative permeability and saturation, are empirically defined.

The final system of equations represents a fully coupled and highly nonlinear system that requires an iterative solution. The nonlinearity is mainly embedded in the coefficients of the partial differential equations, which are functions of the independent variables (primary unknowns), dependent variables (secondary unknowns), as well as variable coefficients (auxiliary variables).

The stability in solving the final system of equations is monitored during each iteration by applying a convergence criterion, which is based on the control of the relative changes of the unknowns, i.e,

$$\left\| \frac{\Upsilon_i^{k+1} - \Upsilon_i^k}{\Upsilon_i^k} \right\| \leq error \tag{6.158}$$

where Υ_i is the dummy unknown at node i; $(k+1)$ and k are the notations for current and previous iteration levels, respectively; and *error* is the convergence limit, chosen according to the requirement of the solution accuracy.

6.4.4.2 Dual-Porosity Media

As was established in the preceding, an additional flow equation is required to characterize the behavior of the fractured porous medium for the dual-porosity model over the single-porosity case.

In the present analysis, solid displacement, wetting and nonwetting fluid pressures are chosen as primary unknowns (i.e., $\mathbf{u}$, p_{1w}, p_{2w}, p_{1n}, and p_{2n}, where subscripts 1 and 2 represent matrix blocks and fractures, subscripts w and n indicate wetting and nonwetting phases, respectively).

All quantities are defined in term of nodal variables as follows:

$$\begin{cases} \varepsilon^* = \mathbf{Bu} \\ \mathbf{p}_{1w}^* = \mathbf{Mp}_{1w}, \quad \mathbf{p}_{2w}^* = \mathbf{Mp}_{2w} \\ \mathbf{p}_{1n}^* = \mathbf{Mp}_{1n}, \quad \mathbf{p}_{2n}^* = \mathbf{Mp}_{2n} \\ \mathbf{S}_{1w}^* = \mathbf{MS}_{1w}, \quad \mathbf{S}_{2w}^* = \mathbf{MS}_{2w}, \\ \mathbf{S}_{1n}^* = \mathbf{MS}_{1n}, \quad \mathbf{S}_{2n}^* = \mathbf{MS}_{2n} \end{cases} \tag{6.159}$$

where the superscript "*" represents the nodal value, and for convenience is omitted in the following description.

Based on a similar procedure to that given previously for single-porosity media, and referring to the dual-porosity poroelastic formulation described as a coupled process in Chapter 3, the governing equations of force equilibrium and flow continuity for the dual-porosity medium can be represented in matrix form as (Meng and Bai 1998)

$$\begin{bmatrix} 0 & 0 & 0 & 0 & 0 \\ 0 & \mathbf{W}_{e1} & 0 & -\mathbf{E}_w & 0 \\ 0 & 0 & \mathbf{W}_{e2} & 0 & -\mathbf{E}_n \\ 0 & -\mathbf{E}_w & 0 & \mathbf{W}_{e3} & 0 \\ 0 & 0 & -\mathbf{E}_n & 0 & \mathbf{W}_{e4} \end{bmatrix} \begin{Bmatrix} \mathbf{u} \\ \mathbf{p}_{1w} \\ \mathbf{p}_{1n} \\ \mathbf{p}_{2w} \\ \mathbf{p}_{2n} \end{Bmatrix} +$$

$$\begin{bmatrix} \mathbf{K}_e & \mathbf{L}_{1w} & \mathbf{L}_{1n} & \mathbf{L}_{2w} & \mathbf{L}_{2n} \\ \mathbf{K}_{1w} & \mathbf{W}_{11} & \mathbf{W}_{12} & \mathbf{W}_{13} & \mathbf{W}_{14} \\ \mathbf{K}_{1n} & \mathbf{W}_{21} & \mathbf{W}_{22} & \mathbf{W}_{23} & \mathbf{W}_{24} \\ \mathbf{K}_{2w} & \mathbf{W}_{31} & \mathbf{W}_{32} & \mathbf{W}_{33} & \mathbf{W}_{34} \\ \mathbf{K}_{2n} & \mathbf{W}_{41} & \mathbf{W}_{42} & \mathbf{W}_{43} & \mathbf{W}_{44} \end{bmatrix} \frac{d}{dt} \begin{Bmatrix} \mathbf{u} \\ \mathbf{p}_{1w} \\ \mathbf{p}_{1n} \\ \mathbf{p}_{2w} \\ \mathbf{p}_{2n} \end{Bmatrix} = \begin{Bmatrix} \dot{\mathbf{F}} \\ \mathbf{Q}_{1w} \\ \mathbf{Q}_{1n} \\ \mathbf{Q}_{2w} \\ \mathbf{Q}_{2n} \end{Bmatrix} \tag{6.160}$$

where $\mathbf{W}_{e1} = \mathbf{W}_{1w} + \mathbf{E}_w$, $\mathbf{W}_{e2} = \mathbf{W}_{1n} + \mathbf{E}_n$, $\mathbf{W}_{e3} = \mathbf{W}_{2w} + \mathbf{E}_w$, $\mathbf{W}_{e4} = \mathbf{W}_{2n} + \mathbf{E}_n$, $\mathbf{K}_e$ and $\dot{\mathbf{F}}$ are identical to Eqs. (6.124) and (6.132), and

$$\mathbf{E}_w = -\int_V \mathbf{M}^T \frac{\mathbf{k}_1 \mathbf{k}_{r1w}}{\mu_{1w}} \rho_{1w} \Gamma \mathbf{M} dV \tag{6.161}$$

$$\mathbf{E}_n = -\int_V \mathbf{M}^T \frac{\mathbf{k}_1 \mathbf{k}_{r1n}}{\mu_{1n}} \rho_{1n} \Gamma \mathbf{M} dV \tag{6.162}$$

$$\mathbf{W}_{1w} = -\int_V \nabla \mathbf{M}^T \frac{\mathbf{k}_1 \mathbf{k}_{r1w}}{\mu_{1w}} \rho_{1w} \nabla \mathbf{M} dV \tag{6.163}$$

$$\mathbf{W}_{2w} = -\int_V \nabla \mathbf{M}^T \frac{\mathbf{k}_2 \mathbf{k}_{r2w}}{\mu_{2w}} \rho_{2w} \nabla \mathbf{M} dV \tag{6.164}$$

$$\mathbf{W}_{1n} = -\int_V \nabla \mathbf{M}^T \frac{\mathbf{k}_1 \mathbf{k}_{r1n}}{\mu_{1n}} \rho_{1n} \nabla \mathbf{M} dV \tag{6.165}$$

$$\mathbf{W}_{2n} = -\int_V \nabla \mathbf{M}^T \frac{\mathbf{k}_2 \mathbf{k}_{r2n}}{\mu_{2n}} \rho_{2n} \nabla \mathbf{M} dV \tag{6.166}$$

$$\mathbf{L}_{1w} = -\alpha_1 \int_V \mathbf{B}^{\mathbf{T}} \mathbf{m} \mathbf{M} \left(\mathbf{S}_{1w} + \mathbf{p}_{1c} \frac{\partial \mathbf{S}_{1w}}{\partial \mathbf{p}_{1c}} \right) dV \tag{6.167}$$

$$\mathbf{L}_{2w} = -\alpha_2 \int_V \mathbf{B}^{\mathbf{T}} \mathbf{m} \mathbf{M} \left(\mathbf{S}_{2w} + \mathbf{p}_{2c} \frac{\partial \mathbf{S}_{2w}}{\partial \mathbf{p}_{2c}} \right) dV \tag{6.168}$$

$$\mathbf{L}_{1n} = -\alpha_1 \int_V \mathbf{B}^{\mathbf{T}} \mathbf{m} \mathbf{M} \left(1 - \mathbf{S}_{1w} - \mathbf{p}_{1c} \frac{\partial \mathbf{S}_{1w}}{\partial \mathbf{p}_{1c}} \right) dV \tag{6.169}$$

$$\mathbf{L}_{2n} = -\alpha_2 \int_V \mathbf{B}^{\mathbf{T}}\mathbf{m}\mathbf{M}\left(1 - \mathbf{S}_{2w} - \mathbf{p}_{2c}\frac{\partial \mathbf{S}_{2w}}{\partial \mathbf{p}_{2c}}\right) dV \tag{6.170}$$

$$\mathbf{K}_{1w} = \alpha_3 \int_V \mathbf{M}^T(1-n_2)\mathbf{S}_{1w}\rho_{1w}\mathbf{m}^T\mathbf{B}dV \tag{6.171}$$

$$\mathbf{K}_{2w} = \alpha_3 \int_V \mathbf{M}^T(1-n_1)\mathbf{S}_{2w}\rho_{2w}\mathbf{m}^T\mathbf{B}dV \tag{6.172}$$

$$\mathbf{K}_{1n} = \alpha_3 \int_V \mathbf{M}^T(1-n_2)(1-\mathbf{S}_{1w})\rho_{1n}\mathbf{m}^T\mathbf{B}dV \tag{6.173}$$

$$\mathbf{K}_{2n} = \alpha_3 \int_V \mathbf{M}^T(1-n_1)\mathbf{S}_{2n}\rho_{2n}\mathbf{m}^T\mathbf{B}dV$$

$$\mathbf{W}_{11} = \int_V \mathbf{M}^T n_1 \mathbf{S}_{1w}\frac{\rho_{1w}}{K_{1w}}\mathbf{M}dV - \int_V \mathbf{M}^T n_1 \rho_{1w}\frac{\partial \mathbf{S}_{1w}}{\partial \mathbf{p}_{1c}}\mathbf{M}dV + \int_V \mathbf{M}^T$$
$$(1-n_2)\mathbf{S}_{1w}\rho_{1w}\frac{(\alpha_3 - n_1 - n_2)}{K_s}\left(\mathbf{S}_{1w} + \mathbf{p}_{1c}\frac{\partial \mathbf{S}_{1w}}{\partial \mathbf{p}_{1c}}\right)\mathbf{M}dV -$$
$$\int_V \mathbf{M}^T\mathbf{S}_{1w}\rho_{1w}n_1 n_2 K_{f1}^{-1}\left(\mathbf{S}_{1w} + \mathbf{p}_{1c}\frac{\partial \mathbf{S}_{1w}}{\partial \mathbf{p}_{1c}}\right)\mathbf{M}dV \tag{6.174}$$

$$\mathbf{W}_{12} = \int_V \mathbf{M}^T n_1 \frac{\partial \mathbf{S}_{1w}}{\partial \mathbf{p}_{1c}}\rho_{1w}\mathbf{M}dV + \int_V \mathbf{M}^T(1-n_2)\mathbf{S}_{1w}\rho_{1w}$$
$$\frac{(\alpha_3 - n_1 - n_2)}{K_s}\left(1 - \mathbf{S}_{1w} - \mathbf{p}_{1c}\frac{\partial \mathbf{S}_{1w}}{\partial \mathbf{p}_{1c}}\right)\mathbf{M}dV -$$
$$\int_V \mathbf{M}^T\mathbf{S}_{1w}\rho_{1w}n_1 n_2 K_{f1}^{-1}\left(1 - \mathbf{S}_{1w} - \mathbf{p}_{1c}\frac{\partial \mathbf{S}_{1w}}{\partial \mathbf{p}_{1c}}\right)\mathbf{M}dV \tag{6.175}$$

$$\mathbf{W}_{13} = \int_V \mathbf{M}^T(1-n_2)\mathbf{S}_{1w}\rho_{1w}\frac{(\alpha_3 - n_1 - n_2)}{K_n s^*}\left(\mathbf{S}_{2w} + \mathbf{p}_{2c}\frac{\partial \mathbf{S}_{2w}}{\partial \mathbf{p}_{2c}}\right)$$
$$\mathbf{M}dV + \int_V \mathbf{M}^T\mathbf{S}_{1w}\rho_{1w}(1-n_2)n_2 K_{f2}^{-1}\left(\mathbf{S}_{2w} + \mathbf{p}_{2c}\frac{\partial \mathbf{S}_{2w}}{\partial \mathbf{p}_{2c}}\right)\mathbf{M}dV \tag{6.176}$$

$$\mathbf{W}_{14} = \int_V \mathbf{M}^T(1-n_2)\mathbf{S}_{1w}\rho_{1w}\frac{(\alpha_3 - n_1 - n_2)}{K_n s^*}\left(1 - \mathbf{S}_{2w} - \mathbf{p}_{2c}\frac{\partial \mathbf{S}_{2w}}{\partial \mathbf{p}_{2c}}\right)$$
$$\mathbf{M}dV + \int_V \mathbf{M}^T\mathbf{S}_{1w}\rho_{1w}(1-n_2)n_2 K_{f2}^{-1}\left(1 - \mathbf{S}_{2w} - \mathbf{p}_{2c}\frac{\partial \mathbf{S}_{2w}}{\partial \mathbf{p}_{2c}}\right)\mathbf{M}dV \tag{6.177}$$

$$\mathbf{W}_{21} = \int_V \mathbf{M}^T n_1 \frac{\partial \mathbf{S}_{1w}}{\partial \mathbf{p}_{1c}}\rho_{1n}\,\mathbf{M}dV + \int_V \mathbf{M}^T(1-n_2)\mathbf{S}_{1n}\rho_{1n}$$
$$\frac{(\alpha_3 - n_1 - n_2)}{K_s}\left(\mathbf{S}_{1w} + \mathbf{p}_{1c}\frac{\partial \mathbf{S}_{1w}}{\partial \mathbf{p}_{1c}}\right)\mathbf{M}dV - \int_V \mathbf{M}^T$$

$$\mathbf{S}_{1n}\rho_{1n}n_1n_2K_{f1}^{-1}\left(\mathbf{S}_{1w}+\mathbf{p}_{1c}\frac{\partial\mathbf{S}_{1w}}{\partial\mathbf{p}_{1c}}\right)\mathbf{M}dV \tag{6.178}$$

$$\mathbf{W}_{22}=\int_V\mathbf{M}^Tn_1\mathbf{S}_{1n}\frac{\rho_{1n}}{K_{1nw}}\mathbf{M}dV-\int_V\mathbf{M}^Tn_1\frac{\partial\mathbf{S}_{1w}}{\partial\mathbf{p}_{1c}}\rho_{1n}\mathbf{M}dV+$$

$$\int_V\mathbf{M}^T(1-n_2)\mathbf{S}_{1n}\rho_{1n}\frac{(\alpha_3-n_1-n_2)}{K_s}\left(1-\mathbf{S}_{1w}-\mathbf{p}_{1c}\frac{\partial\mathbf{S}_{1w}}{\partial\mathbf{p}_{1c}}\right)$$

$$\mathbf{M}dV-\int_V\mathbf{M}^T\mathbf{S}_{1n}\rho_{1n}n_1n_2K_{f1}^{-1}\left(1-\mathbf{S}_{1w}-\mathbf{p}_{1c}\frac{\partial\mathbf{S}_{1w}}{\partial\mathbf{p}_{1c}}\right)\mathbf{M}dV \tag{6.179}$$

$$\mathbf{W}_{23}=\int_V\mathbf{M}^T(1-n_2)\mathbf{S}_{1n}\rho_{1n}\frac{(\alpha_3-n_1-n_2)}{K_ns^*}\left(\mathbf{S}_{2w}+\mathbf{p}_{2c}\frac{\partial\mathbf{S}_{2w}}{\partial\mathbf{p}_{2c}}\right)$$

$$\mathbf{M}dV+\int_V\mathbf{M}^T\mathbf{S}_{1n}\rho_{1n}(1-n_2)n_2K_{f2}^{-1}\left(\mathbf{S}_{2w}+\mathbf{p}_{2c}\frac{\partial\mathbf{S}_{2w}}{\partial\mathbf{p}_{2c}}\right)\mathbf{M}dV \tag{6.180}$$

$$\mathbf{W}_{24}=\int_V\mathbf{M}^T(1-n_2)\mathbf{S}_{1n}\rho_{1n}\frac{(\alpha_3-n_1-n_2)}{K_ns^*}\left(1-\mathbf{S}_{2w}-\mathbf{p}_{2c}\frac{\partial\mathbf{S}_{2w}}{\partial\mathbf{p}_{2c}}\right)$$

$$\mathbf{M}dV+\int_V\mathbf{M}^T\mathbf{S}_{1n}\rho_{1n}(1-n_2)n_2K_{f2}^{-1}\left(1-\mathbf{S}_{2w}-\mathbf{p}_{2c}\frac{\partial\mathbf{S}_{2w}}{\partial\mathbf{p}_{2c}}\right)\mathbf{M}dV \tag{6.181}$$

$$\mathbf{W}_{31}=\int_V\mathbf{M}^T(1-n_1)\mathbf{S}_{2w}\rho_{2w}\frac{(\alpha_3-n_1-n_2)}{K_s}\left(\mathbf{S}_{1w}+\mathbf{p}_{1c}\frac{\partial\mathbf{S}_{1w}}{\partial\mathbf{p}_{1c}}\right)$$

$$\mathbf{M}dV+\int_V\mathbf{M}^T\mathbf{S}_{2w}\rho_{2w}(1-n_1)n_1K_{f1}^{-1}\left(\mathbf{S}_{1w}+\mathbf{p}_{1c}\frac{\partial\mathbf{S}_{1w}}{\partial\mathbf{p}_{1c}}\right)\mathbf{M}dV \tag{6.182}$$

$$\mathbf{W}_{32}=\int_V\mathbf{M}^T(1-n_1)\mathbf{S}_{2w}\rho_{2w}\frac{(\alpha_3-n_1-n_2)}{K_s}\left(1-\mathbf{S}_{1w}-\mathbf{p}_{1c}\frac{\partial\mathbf{S}_{1w}}{\partial\mathbf{p}_{1c}}\right)$$

$$\mathbf{M}dV+\int_V\mathbf{M}^T\mathbf{S}_{2w}\rho_{2w}(1-n_1)n_1K_{f1}^{-1}\left(1-\mathbf{S}_{1w}-\mathbf{p}_{1c}\frac{\partial\mathbf{S}_{1w}}{\partial\mathbf{p}_{1c}}\right)\mathbf{M}dV \tag{6.183}$$

$$\mathbf{W}_{33}=\int_V\mathbf{M}^Tn_2\mathbf{S}_{2w}\frac{\rho_{2w}}{K_{2w}}\mathbf{M}dV-\int_V\mathbf{M}^Tn_2\frac{\partial\mathbf{S}_{2w}}{\partial\mathbf{p}_{2c}}\rho_{2w}\mathbf{M}dV+\int_V\mathbf{M}^T$$

$$(1-n_1)\mathbf{S}_{2w}\rho_{2w}\frac{(\alpha_3-n_1-n_2)}{K_ns^*}\left(\mathbf{S}_{2w}+\mathbf{p}_{2c}\frac{\partial\mathbf{S}_{2w}}{\partial\mathbf{p}_{2c}}\right)\mathbf{M}dV-$$

$$\int_V\mathbf{M}^T\mathbf{S}_{2w}\rho_{2w}n_1n_2K_{f2}^{-1}\left(\mathbf{S}_{2w}+\mathbf{p}_{2c}\frac{\partial\mathbf{S}_{2w}}{\partial\mathbf{p}_{2c}}\right)\mathbf{M}dV \tag{6.184}$$

$$\mathbf{W}_{34}=\int_V\mathbf{M}^Tn_2\frac{\partial\mathbf{S}_{2w}}{\partial\mathbf{p}_{2c}}\rho_{2w}\mathbf{M}dV+\int_V\mathbf{M}^T(1-n_1)\mathbf{S}_{2w}\rho_{2w}$$

$$\frac{(\alpha_3 - n_1 - n_2)}{K_n s^*}\left(1 - \mathbf{S}_{2w} - \mathbf{p}_{2c}\frac{\partial \mathbf{S}_{2w}}{\partial \mathbf{p}_{2c}}\right)\mathbf{M}dV -$$

$$\int_V \mathbf{M}^T \mathbf{S}_{2w}\rho_{2w}n_1 n_2 K_{f2}^{-1}\left(1 - \mathbf{S}_{2w} - \mathbf{p}_{2c}\frac{\partial \mathbf{S}_{2w}}{\partial \mathbf{p}_{2c}}\right)\mathbf{M}dV \tag{6.185}$$

$$\mathbf{W}_{41} = \int_V \mathbf{M}^T(1 - n_1)\mathbf{S}_{2n}\rho_{2n}\frac{(\alpha_3 - n_1 - n_2)}{K_s}\left(\mathbf{S}_{1w} + \mathbf{p}_{1c}\frac{\partial \mathbf{S}_{1w}}{\partial \mathbf{p}_{1c}}\right)$$

$$\mathbf{M}dV + \int_V \mathbf{M}^T \mathbf{S}_{2n}\rho_{2n}(1 - n_1)n_1 K_{f1}^{-1}\left(\mathbf{S}_{1w} + \mathbf{p}_{1c}\frac{\partial \mathbf{S}_{1w}}{\partial \mathbf{p}_{1c}}\right)\mathbf{M}dV \tag{6.186}$$

$$\mathbf{W}_{42} = \int_V \mathbf{M}^T(1 - n_1)\mathbf{S}_{2n}\rho_{2n}\frac{(\alpha_3 - n_1 - n_2)}{K_s}\left(1 - \mathbf{S}_{1w} - \mathbf{p}_{1c}\frac{\partial \mathbf{S}_{1w}}{\partial \mathbf{p}_{1c}}\right)$$

$$\mathbf{M}dV + \int_V \mathbf{M}^T \mathbf{S}_{2n}\rho_{2n}(1 - n_1)n_1 K_{f1}^{-1}\left(1 - \mathbf{S}_{1w} - \mathbf{p}_{1c}\frac{\partial \mathbf{S}_{1w}}{\partial \mathbf{p}_{1c}}\right)\mathbf{M}dV \tag{6.187}$$

$$\mathbf{W}_{43} = \int_V \mathbf{M}^T n_2 \frac{\partial \mathbf{S}_{2w}}{\partial \mathbf{p}_{2c}}\rho_{2n}\mathbf{M}dV + \int_V \mathbf{M}^T(1 - n_1)\mathbf{S}_{2n}\rho_{2n}$$

$$\frac{(\alpha_3 - n_1 - n_2)}{K_n s^*}\left(\mathbf{S}_{2w} + \mathbf{p}_{2c}\frac{\partial \mathbf{S}_{2w}}{\partial \mathbf{p}_{2c}}\right)\mathbf{M}dV -$$

$$\int_V \mathbf{M}^T \mathbf{S}_{2n}\rho_{2n}n_1 n_2 K_{f2}^{-1}\left(\mathbf{S}_{2w} + \mathbf{p}_{2c}\frac{\partial \mathbf{S}_{2w}}{\partial \mathbf{p}_{2c}}\right)\mathbf{M}dV \tag{6.188}$$

$$\mathbf{W}_{44} = \int_V \mathbf{M}^T n_2 \mathbf{S}_{2n}\frac{\rho_{2n}}{K_{2nw}}\mathbf{M}dV - \int_V \mathbf{M}^T n_2 \frac{\partial \mathbf{S}_{2w}}{\partial \mathbf{p}_{2c}}\rho_{2n}\mathbf{M}dV + \int_V \mathbf{M}^T$$

$$(1 - n_1)\mathbf{S}_{2n}\rho_{2n}\frac{(\alpha_3 - n_1 - n_2)}{K_n s^*}\left(1 - \mathbf{S}_{2w} - \mathbf{p}_{2c}\frac{\partial \mathbf{S}_{2w}}{\partial \mathbf{p}_{2c}}\right)\mathbf{M}dV -$$

$$\int_V \mathbf{M}^T \mathbf{S}_{2n}\rho_{2n}n_1 n_2 K_{f2}^{-1}\left(1 - \mathbf{S}_{2w} - \mathbf{p}_{2c}\frac{\partial \mathbf{S}_{2w}}{\partial \mathbf{p}_{2c}}\right)\mathbf{M}dV \tag{6.189}$$

$$\mathbf{Q}_{1w} = \int_{A_e} \mathbf{M}\mathbf{q}_{1w}dA_e \tag{6.190}$$

$$\mathbf{Q}_{1n} = \int_{A_e} \mathbf{M}\mathbf{q}_{1n}dA_e \tag{6.191}$$

$$\mathbf{Q}_{2w} = \int_{A_e} \mathbf{M}\mathbf{q}_{2w}dA_e \tag{6.192}$$

$$\mathbf{Q}_{2n} = \int_{A_e} \mathbf{M}\mathbf{q}_{2n}dA_e \tag{6.193}$$

where K_{f1}^{-1} and K_{f2}^{-1} are defined in Eq. (3.175) from Chapter 3.

Eq. (6.160) represents the spatial discretization of the domain but is continuous in time. The solution at various discretized times may be obtained via finite difference algorithms. Using a fully implicit time-stepping scheme, yields

$$\begin{cases} \dot{\mathbf{u}}^{t+\Delta t} = \frac{1}{\Delta t}\left(\mathbf{u}^{t+\Delta t} - \mathbf{u}^{t}\right) \\ \dot{\mathbf{p}}_{1w}^{t+\Delta t} = \frac{1}{\Delta t}\left(\mathbf{p}_{1w}^{t+\Delta t} - \mathbf{p}_{1w}^{t}\right) \\ \dot{\mathbf{p}}_{2w}^{t+\Delta t} = \frac{1}{\Delta t}\left(\mathbf{p}_{2w}^{t+\Delta t} - \mathbf{p}_{2w}^{t}\right) \\ \dot{\mathbf{p}}_{1n}^{t+\Delta t} = \frac{1}{\Delta t}\left(\mathbf{p}_{1n}^{t+\Delta t} - \mathbf{p}_{1n}^{t}\right) \\ \dot{\mathbf{p}}_{2n}^{t+\Delta t} = \frac{1}{\Delta t}\left(\mathbf{p}_{2n}^{t+\Delta t} - \mathbf{p}_{2n}^{t}\right). \end{cases} \tag{6.194}$$

Using the expression in Eq. (6.194) to substitute the time derivatives in Eq. (6.160), and positioning all presumed known values on the right-hand side of the equation, Eq. (6.160) can be reformulated as

$$\frac{1}{\Delta t}\begin{bmatrix} \mathbf{K}_e & \mathbf{L}_{1w} & \mathbf{L}_{1n} & \mathbf{L}_{2w} & \mathbf{L}_{2n} \\ \mathbf{K}_{1w} & (\mathbf{W}_{1w}+\mathbf{E}_w)\Delta t + \mathbf{W}_{11} & \mathbf{W}_{12} & -\mathbf{E}_w\Delta t + \mathbf{W}_{13} & \mathbf{W}_{14} \\ \mathbf{K}_{1n} & \mathbf{W}_{21} & (\mathbf{W}_{1n}+\mathbf{E}_n)\Delta t + \mathbf{W}_{22} & \mathbf{W}_{23} & -\mathbf{E}_n\Delta t + \mathbf{W}_{24} \\ \mathbf{K}_{2w} & -\mathbf{E}_w\Delta t + \mathbf{W}_{31} & \mathbf{W}_{32} & (\mathbf{W}_{2w}+\mathbf{E}_w)\Delta t + \mathbf{W}_{33} & \mathbf{W}_{34} \\ \mathbf{K}_{2n} & \mathbf{W}_{41} & -\mathbf{E}_n\Delta t + \mathbf{W}_{42} & \mathbf{W}_{43} & (\mathbf{W}_{2n}+\mathbf{E}_n)\Delta t + \mathbf{W}_{44} \end{bmatrix}\begin{Bmatrix} \mathbf{u} \\ \mathbf{p}_{1w} \\ \mathbf{p}_{1n} \\ \mathbf{p}_{2w} \\ \mathbf{p}_{2n} \end{Bmatrix}^{\tau^*}$$

$$= \frac{1}{\Delta t}\begin{bmatrix} \mathbf{K}_e & \mathbf{L}_{1w} & \mathbf{L}_{1n} & \mathbf{L}_{2w} & \mathbf{L}_{2n} \\ \mathbf{K}_{1w} & \mathbf{W}_{11} & \mathbf{W}_{12} & \mathbf{W}_{13} & \mathbf{W}_{14} \\ \mathbf{K}_{1n} & \mathbf{W}_{21} & \mathbf{W}_{22} & \mathbf{W}_{23} & \mathbf{W}_{24} \\ \mathbf{K}_{2w} & \mathbf{W}_{31} & \mathbf{W}_{32} & \mathbf{W}_{33} & \mathbf{W}_{34} \\ \mathbf{K}_{2n} & \mathbf{W}_{41} & \mathbf{W}_{42} & \mathbf{W}_{43} & \mathbf{W}_{44} \end{bmatrix}\begin{Bmatrix} \mathbf{u} \\ \mathbf{p}_{1w} \\ \mathbf{p}_{1n} \\ \mathbf{p}_{2w} \\ \mathbf{p}_{2n} \end{Bmatrix}^{t} + \begin{Bmatrix} \dot{\mathbf{F}} \\ \mathbf{Q}_{1w} \\ \mathbf{Q}_{1n} \\ \mathbf{Q}_{2w} \\ \mathbf{Q}_{2n} \end{Bmatrix}^{\tau^*} \tag{6.195}$$

where $\tau^* = (t + \Delta t)$ and there are seven explicit unknowns at the nodal level, i.e., displacements, u_x, u_y, u_z, wetting phase pressures p_{1w}, p_{2w}; and nonwetting phase pressures p_{1n}, p_{2n}. With seven degrees of freedom per node, the eight-node brick 3-D element results in a total of 56 equations in the final system, at the element level.

Four secondary unknowns are saturations of the wetting phase, S_{1w} and S_{2w}, and of the nonwetting phase, S_{1n} and S_{2n} for matrix blocks and fractures, respectively. They are related to the primary unknowns (i.e., pressures) via the relationships between capillary pressure and saturation. In general, the matrix capillary pressure is greater than the fracture capillary pressure (Kazemi et al. 1976).

For dual-porosity media, four auxiliary variables are relative permeabilities of the wetting phase, k_{r1w} and k_{r2w}, and of the nonwetting phase, k_{r1n} and k_{r2n}, respectively. They are related to the secondary unknowns via the permeability versus saturation relationship (Mattax and Dalton 1990).

The iterative steps needed to solve the nonlinear coupled system of equations are as follows:

(1) At the initial time $t = t_0$, input initial values of pressures, saturations, relatively permeabilities, and capillary pressures.

(2) Calculate the nonlinear coefficients in the final system of equations using the implicit iterative scheme described earlier. In doing so, all nonlinear parameters are updated by using the current values of fluid pressures within each iterative step.

(3) Solve for the unknowns, u_x, u_y, u_z, p_{1w}, p_{1n}, p_{2w}, and p_{2n} at each node.

(4) With updated values of unknowns, return to step (2) to update the nonlinear coefficients until the convergence condition is satisfied at each nodal point.

(5) Go to the next time-step and repeat steps (2) through (5).

6.5 MODEL VALIDATION

Model validation can be performed by comparing the numerical calculation and an appropriate analytical solution. Since the availability of analytical solutions is typically limited, especially for the coupled processes discussed here, rigorous validation of numerical models is restricted to certain simple configurations such as those of lower spatial dimensions.

6.5.1 Analytical Solution of 1-D Consolidation

A detailed derivation of 1-D consolidation behavior provides insight into the physical response of this simple system.

As demonstrated in Chapter 2, the strain-stress relationship for poroelastic behavior is expressed generally as

$$2G\varepsilon_{ij} = \sigma_{ij} - \frac{\nu}{1+\nu}\sigma_{kk}\delta_{ij} + \frac{\alpha(1-2\nu)}{1+\nu}p\delta_{ij} \tag{6.196}$$

where σ_{kk} is the total stress, which can be rewritten as

$$\sigma_{kk} = 3K\varepsilon_{kk} - 3\alpha p \tag{6.197}$$

where

$$K = \frac{E}{3(1-2\nu)} = \frac{2G(1+\nu)}{3(1-2\nu)}. \tag{6.198}$$

From Eqs. (6.197) and (6.198), Eq. (6.196) can be reformulated as

$$\sigma_{ij} + \alpha p\delta_{ij} = 2G\varepsilon_{ij} + \frac{2G\nu}{1-2\nu}\varepsilon_{kk}\delta_{ij}. \tag{6.199}$$

For 1-D uniaxial strain,

$$\sigma_{ij} = \sigma_{xx}, \quad \varepsilon_{ij} = \varepsilon_{kk} = \varepsilon_{xx}. \tag{6.200}$$

Substituting Eq. (6.200) into Eq. (6.199), yields

$$\sigma_{xx} = \frac{2G(1-\nu)}{1-2\nu}\varepsilon_{kk} - \alpha p. \tag{6.201}$$

For the initial undrained condition, Skempton's effect (Skempton 1954) can be expressed as

$$\frac{\sigma_{xx} + \sigma_{yy} + \sigma_{zz}}{3} = -\frac{p}{B} \tag{6.202}$$

where B is the Skempton coefficient.

From the theory of elasticity (Timoshenko 1934), it is known that under hydrostatic loading,

$$\sigma_{yy} = \frac{\nu_u}{1-\nu_u}\sigma_{xx} \tag{6.203}$$

and

$$\sigma_{zz} = \frac{\nu_u}{1-\nu_u}\sigma_{xx}. \tag{6.204}$$

Substituting Eqs. (6.203) and (6.204) into Eq. (6.202), gives:

$$p = -\frac{B}{3}\left(\frac{1+\nu_u}{1-\nu_u}\right)\sigma_{xx} \tag{6.205}$$

where ν_u is the undrained Poisson's ratio.

The diffusion equation, considering the effect of body strain (Biot 1941), can now be written as

$$\frac{\partial p}{\partial t} - c_{dif}\frac{\partial^2 p}{\partial x^2} = -\frac{B}{3}\left(\frac{1+\nu_u}{1-\nu_u}\right)\frac{d\sigma_{xx}}{dt} \tag{6.206}$$

where c_{dif} is the diffusivity coefficient, which is expressed as (Rice and Cleary 1976)

$$c_{dif} = \frac{2kG(1-\nu)(\nu_u-\nu)}{\alpha^2\mu(1-2\nu)^2(1-\nu_u)}. \tag{6.207}$$

Since the axial load applied at the upper boundary, σ_{xx} is a constant, Eq. (6.206) reduces to

$$\frac{\partial p}{\partial t} - c_{dif}\frac{\partial^2 p}{\partial x^2} = 0. \tag{6.208}$$

Assuming the following boundary condition:

$$\begin{cases} p = 0, & x = 0 \\ \dfrac{\partial p}{\partial x} = 0, & x = L_c \end{cases} \tag{6.209}$$

where L_c is the the column height, and initial condition:

$$p_0 = -\frac{B}{3}\left(\frac{1+\nu_u}{1-\nu_u}\right)\sigma_{xx}, \qquad t = 0 \tag{6.210}$$

Eq. (6.208) can be solved using "separation of variables" as

$$p(x,t) = \frac{4p_0}{\pi}\sum_{m=1,3,\ldots}^{\infty}\frac{1}{m}\sin\left(\frac{m\pi x}{2L_c}\right)\exp\left(-\frac{m^2\pi^2 c_{dif}t}{4L_c^2}\right). \tag{6.211}$$

Differentiating Eq. (6.201) with respect to x, and converting the strain to displacement, results in

$$\frac{\partial\sigma_{xx}}{\partial x} = \frac{2G(1-\nu)}{1-2\nu}\frac{\partial^2 u_x}{\partial x^2} - \alpha\frac{\partial p}{\partial x}. \tag{6.212}$$

Because σ_{xx} is a constant, Eq. (6.212) reduces to

$$\frac{2G(1-\nu)}{1-2\nu}\frac{\partial^2 u_x}{\partial x^2} - \alpha\frac{\partial p}{\partial x} = 0. \tag{6.213}$$

Eq. (6.201) can be rewritten as

$$(\sigma_{xx} + \alpha p) = \frac{2G(1-\nu)}{1-2\nu}\frac{\partial u_x}{\partial x}. \tag{6.214}$$

Integrating with respect to x in Eq. (6.214), and assuming the boundary condition

$$u_x = 0, \qquad x = L_c \tag{6.215}$$

yields the solution from Eq. (6.214) as

$$u_x = \frac{(1-2\nu)}{2G(1-\nu)}(\sigma_{xx} + \alpha p)(x - L_c). \tag{6.216}$$

At the initial time, $t = 0$, the pore pressure has no influence on the solid displacement. As a result, with substitution of ν with ν_u, Eq. (6.216) can be written as

$$u_x|_{t=0} = u_x^0 = -\frac{\sigma_{xx} L_c (1-2\nu_u)}{2G(1-\nu_u)}\left(1 - \frac{x}{L_c}\right). \tag{6.217}$$

The total displacement should be expressed as

$$u_x = u_x^0 + \Delta u_x. \tag{6.218}$$

As $t \to \infty$, the displacement reaches its maximum value, which was derived by Biot (1941) as

$$u_{max}^{\infty} = \frac{(1-2\nu)L_c p_0}{2G(1-\nu)}. \tag{6.219}$$

Eq. (6.219) can also be deduced from physical intuition; i.e., at $t \to \infty$, pore pressure should have no effect on u_{max}^{∞}. In addition, u_{max}^{∞} must occur at the location where $x = 0$. The maximum stress σ_{xx}^{max} should be equal to the initial pore pressure p_0, but with opposite signs. With this intuition, Eq. (6.219) can be deduced from Eq. (6.216).

Substituting Eq. (6.210) into Eq. (6.219) gives

$$u_{max}^{\infty} = -\frac{(1-2\nu)(1+\nu_u)L_c B \sigma_{xx}}{6G(1-\nu)(1-\nu_u)}. \tag{6.220}$$

Skempton's constant B can be expressed as (Bai et al. 1993; Detournay and Cheng 1993)

$$B = \frac{3(\nu_u - \nu)}{\alpha(1-2\nu)(1+\nu_u)}. \tag{6.221}$$

Where $\alpha = 1$ (Terzaghi's 1-D consolidation), B reduces to

$$B = \frac{3(\nu_u - \nu)}{(1-2\nu)(1+\nu_u)}. \tag{6.222}$$

Substituting B in Eq. (6.222) into Eq. (6.220), one has

$$u_{max}^{\infty} = -\frac{(\nu - \nu_u)L_c \sigma_{xx}}{2G(1-\nu)(1-\nu_u)}. \tag{6.223}$$

The asymptotic solution that offers the transition in solution from the undrained to drained conditions can be written from Eq. (6.218) as

$$u_x = u_x^0 + \Delta u_x = u_x^0 + u_{max}^{\infty} F(x,t) \tag{6.224}$$

where $F(x,t)$ can be obtained from the solution of Eq. (6.213), incorporating the solution of p in Eq. (6.211) as

$$F(x,t) = \frac{8}{\pi^2} \sum_{m=1,3,\ldots}^{\infty} \frac{1}{m^2} \cos\left(\frac{m\pi x}{2L_c}\right) \left[1 - \exp\left(-\frac{m^2\pi^2 c_{dif} t}{4L_c^2}\right)\right]. \tag{6.225}$$

Two limiting cases can be deduced from Eq. (6.225):

$$\begin{cases} F(x,0) = 0, & t = 0 \\ F(x,\infty) = 1 - \frac{x}{L_c}, & t \to \infty \end{cases} \tag{6.226}$$

The subsequent two limiting cases from Eq. (6.224) are

$$u_x(x, t=0) = u_x^0 \tag{6.227}$$

and

$$u_x(x, t \to \infty) = u_x^0 + u_{max}^{\infty}\left(1 - \frac{x}{L_c}\right) = -\frac{\sigma_{xx} L_c (1-2\nu)}{2G(1-\nu)}\left(1 - \frac{x}{L_c}\right). \tag{6.228}$$

Eq. (6.228) is equal to Eq. (6.217) if $\nu_u = \nu$. In the simulation, additional poroelastic parameters may be provided, that are not necessary as input parameters (i.e., they are not independent parameters). However, they can be used to check the physical compatibilities among all the dependent and independent parameters.

The solutions in Eqs. (6.211) and (6.224) will be compared with the numerical results in the validation of the finite element model.

6.5.2 Comparative Analysis

A comparative study is conducted to compare the analytical and numerical solutions. In the latter solution, finite element models are related to the coupled single-phase fluid flow in both single-porosity and dual-porosity deformable media. A 1-D column is selected for the consolidation analysis.

6.5.2.1 Single-Porosity Media

A column length, L_c, is chosen as 1 m. At the initial time ($t = 0^+$ sec.), it is subjected to a uniform compressive load of $\sigma_{xx} = 1$ MPa on the top surface (Figure 6-1), and only vertical displacement is allowed within the column. No-flow lateral boundary conditions prevail in the column, except on the top surface, where fluid may exit freely from the column. 2-D plane strain finite elements are used to simulate this 1-D situation. The basic modeling parameters are listed in Table 6.1.

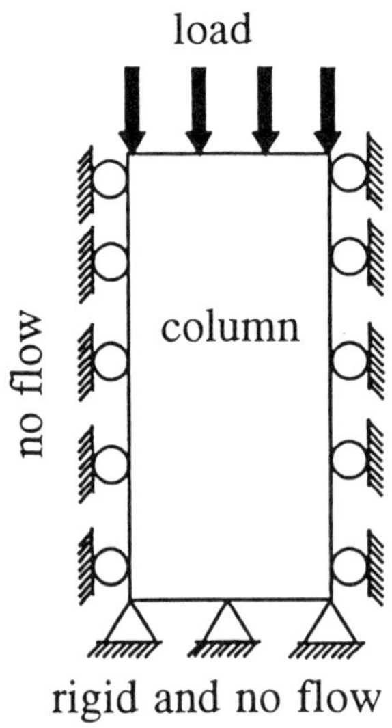

Figure 6-1. Schematic of 1-D column consolidation.

Table 6.1. Parameters for the single-porosity model.

Parameter	Definition	Magnitude	Unit
E	modulus of elasticity	2.4	MN/m^2
ν	Poisson's ratio	0.2	-
c_{dif}	diffusivity	1	m^2/s
M	Biot's modulus	4.1813	MN/m^2
α	Biot coefficient	0.8929	-
K_f	fluid bulk modulus	0.1183	MN/m^2
n_1	porosity	0.02	-
k/μ	mobility	0.5381	$m^4/(MN{\cdot}s)$
K_s	solid grain bulk modulus	12.445	MN/m^2
B	Skempton coefficient	0.8	-
ν_u	undrained Poisson's ratio	0.4	-
σ_{xx}	loading stress	1.0	MN/m^2

It may be noted in Table 6.1 that certain parameters may not be needed in the numerical simulation, such as Biot's modulus, M. However, such a parameter can be used to verify the compatibility of the poroelastic system through the following relationships:

$$\begin{cases} G = \dfrac{E}{2(1+\nu)}, \quad \alpha = \dfrac{3(\nu_u - \nu)}{B(1-2\nu)(1+\nu_u)} \\ M = \dfrac{2G(\nu_u - \nu)}{\alpha^2(1-2\nu_u)(1-2\nu)}, \quad K = \dfrac{E}{3(1-2\nu)} \\ K_s = \dfrac{K}{1-\alpha}, \quad \dfrac{1}{M} - \dfrac{\alpha - n}{K_s} + \dfrac{n}{K_f} \\ \dfrac{k}{\mu} = c_{dif} \dfrac{\alpha^2(1-2\nu)^2(1-\nu_u)}{2G(1-\nu)(\nu_u - \nu)} \end{cases} \tag{6.229}$$

where n is the porosity.

Comparisons between the analytical solutions (solid lines) and numerical results (scattered points) are made for the temporal evolution of displacement at the top surface of the column (Figure 6-2). The excellent match between the analytical and numerical results is apparent.

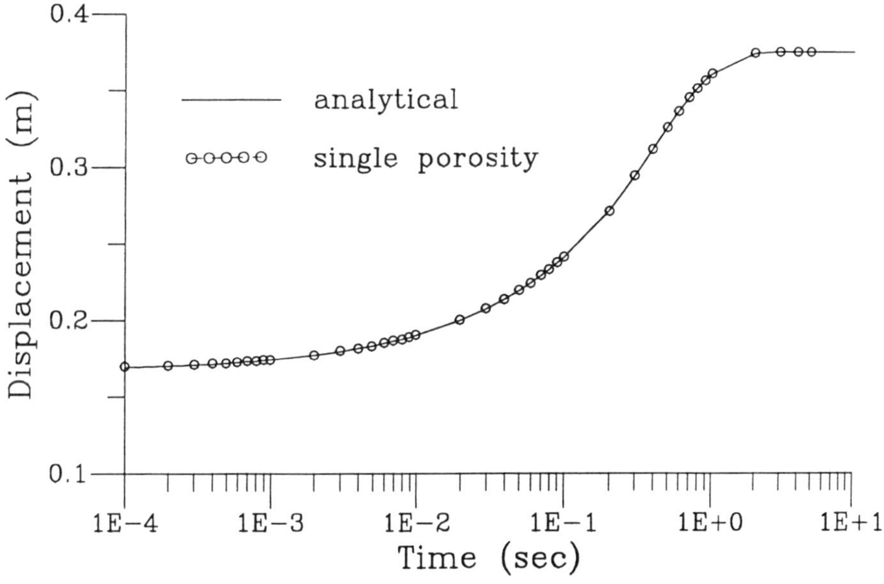

Figure 6-2. Temporal displacement for a single-porosity medium.

6.5.2.2 Dual-Porosity Media

The primary purposes of validating the dual-porosity model are not only to verify that the solutions match the analytical ones under similar conditions, but also to identify the differences and physical implications between the single-porosity and dual-porosity models. Some basic parameters for the dual-porosity model are listed in Table 6.2.

Table 6.2. Parameters for the dual-porosity model.

Parameter	Definition	Magnitude	Unit
E	modulus of elasticity	2.4	MN/m^2
ν	Poisson's ratio	0.2	-
α_1	matrix pressure factor	1	-
α_2	fracture pressure factor	1	-
K_f	fluid bulk modulus	2×10^3	MN/m^2
n_1	matrix porosity	0.02	-
n_2	fracture porosity	0.002	-
k_1/μ	matrix mobility	0.375	$m^4/(MN{\cdot}s)$
k_2/μ	fracture mobility	3.75	$m^4/(MN{\cdot}s)$
K_s	solid grain bulk modulus	1×10^{15}	MN/m^2

This analysis assumes that the bulk modulus of the solid grains is much larger than that of the porous or fractured medium. As a result, both α_1 and α_2 are equal to 1. For a very large fracture spacing ($s^* = 1 \times 10^7$ m), and very large fracture normal stiffness ($K_n = 1 \times 10^7$ MPa/m), the response of a dual-porosity medium should resemble that of a single-porosity medium. Under such a condition, Figures 6-3 and 6-4 indeed indicate a good match between the analytical and dual-porosity calculations for temporal displacement and spatial pressure along the depth of column, respectively.

For reduced spacings between fractures ranging from 0.1 m to 1 m, and a reduced fracture normal stiffness $K_n = 120$ MPa/m, the temporal displacements are compared between the single-porosity and dual-porosity models, as shown in Figure 6-5. The result from the dual-porosity model more closely approximates that from the single-porosity model for greater fracture spacing. This is sensible since the fractured medium is transformed to a non-fractured medium when the fracture spacing is sufficiently large. Another observation from Figure 6-5 is the delayed progression of solid displacement as the spacing is increased. This is also physically meaningful since, at larger spacing, interporosity mass transfer between the matrix blocks and fractures is slowed. This process consequently delays the temporal variation of the displacement for dual-porosity media.

As shown in Figure 6-6, the temporal pressures close to the surface (5% of depth from the top) are compared between the single-porosity and the dual-porosity models. In the dual-porosity model, the fracture spacing is chosen as $s^* = 0.1$ m. Due to the similar nature of solid and flow properties in each of these examples, the matrix pressure approaches the pressure in the single-porosity model. However, they are not identical due to the fluid interchange

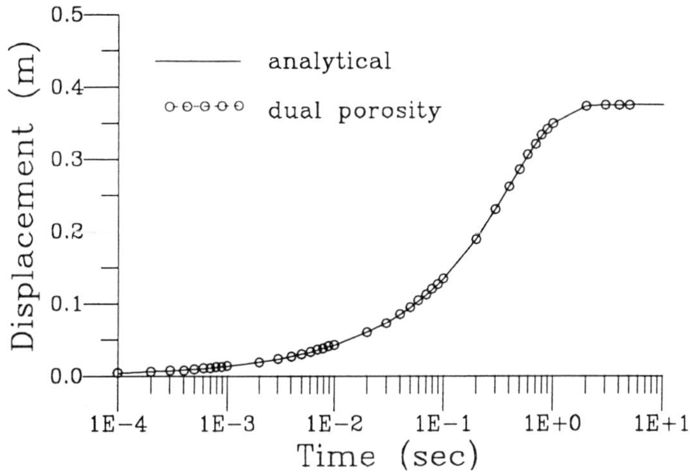

Figure 6-3. Temporal displacement for a dual-porosity medium.

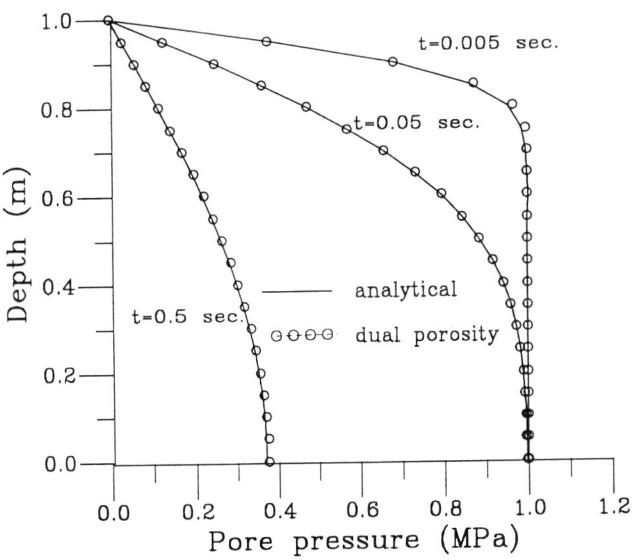

Figure 6-4. Pore pressures between analytical solution and the FEM dual-porosity model.

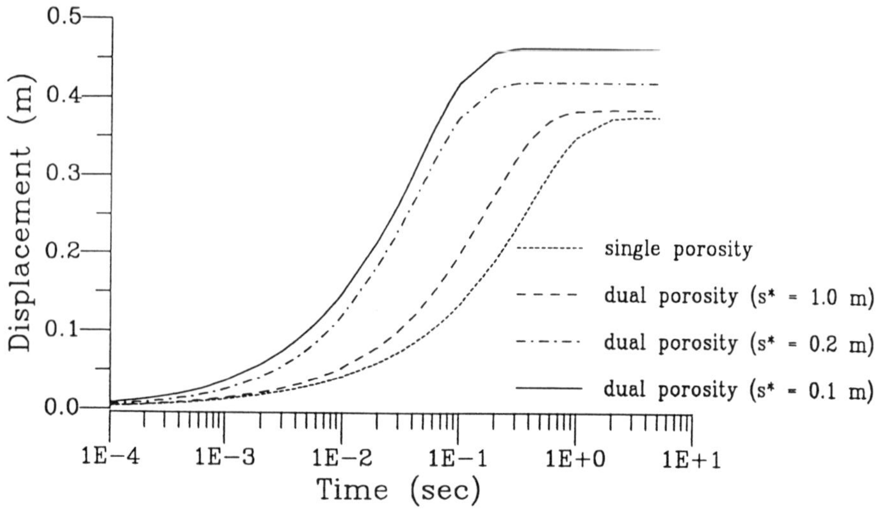

Figure 6-5. Displacements between single-porosity and dual-porosity models.

between the matrix blocks and the fractures. The fracture pressure is quite different from the matrix pressure. The much lower fracture pressure than the matrix pressure reveals that fluid initially flows from the matrix blocks to the fractures as a result of the drop in pressure. Eventually, the fluid pressures of the matrix blocks and the fractures reach equilibrium at late times. Comparison of temporal pressures in the single-porosity system and the pressures of matrix blocks with different fracture spacings is shown in Figure 6-7. Again, the larger the spacing, the closer the correspondence between the pressure responses of the single-porosity and dual-porosity models. It may be noted that there is a transition at early times when all results show very similar fluid pressure response.

The effects of pore pressure factors for matrix blocks and fractures (α_1 and α_2) may also be investigated. Figure 6-8 depicts the comparison of pore pressures when $\alpha_1 = \alpha_2 = 1$ (dotted lines) and when $\alpha_1 = 1$ but $\alpha_2 = 0.89$ (solid lines). The discrepancies are mainly attributed to the difference between the bulk modulus of the porous medium and that of fractured medium. One interesting observation is that the fracture pressure of the solid line has an early time reversal of the non-diffusive behavior, similar to the well-known Mandel effect (Mandel 1953).

The comparisons of spatial pressure distributions along the column, between the single-porosity and the dual-porosity models (pressure of matrix blocks), of various spacings, from early time to late time are shown in Figures

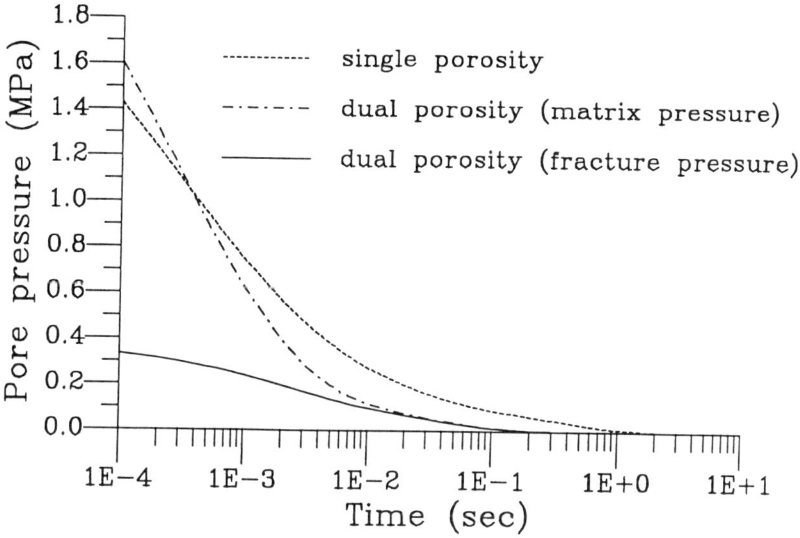

Figure 6-6. Pressures between single-porosity and dual-porosity. models with spacing of 0.1 m

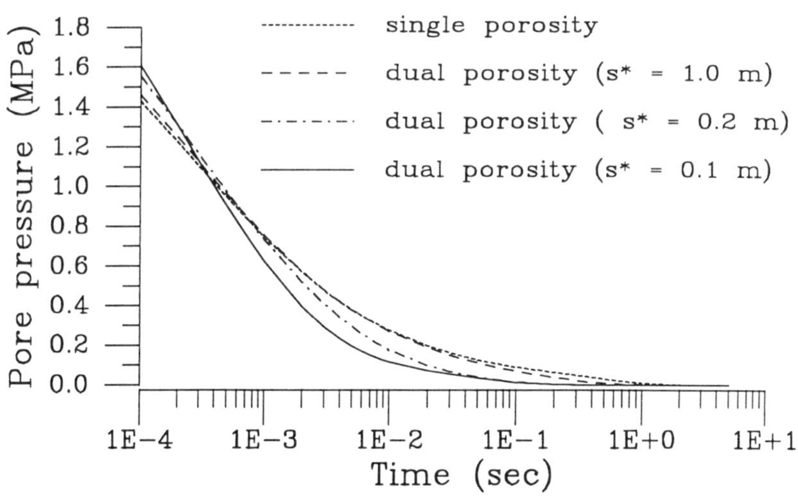

Figure 6-7. Comparative pressures between the single-porosity. model and dual-porosity model

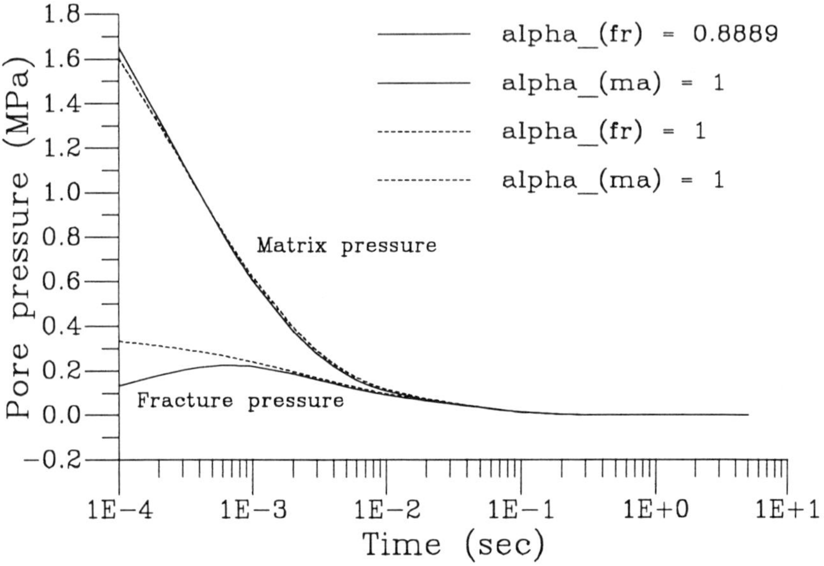

Figure 6-8. Effect of pressure factor (α).

6-9, 6-10, and 6-11, respectively. The difference is most significant when the fracture spacing is the smallest (i.e., $s^* = 0.1$ m). In contrast, the comparative results are almost identical when the fracture spacing is the largest (i.e., $s^* = 10$ m), especially beyond early time (i.e., $t = 0.05$ sec. and 0.5 sec.).

Figure 6-12 compares the spatial pressure distributions in the dual-porosity model between the matrix blocks and the fractures at various times for a fracture spacing of 0.1 m. As time elapses, the difference between matrix pressures and the fracture pressures are reduced.

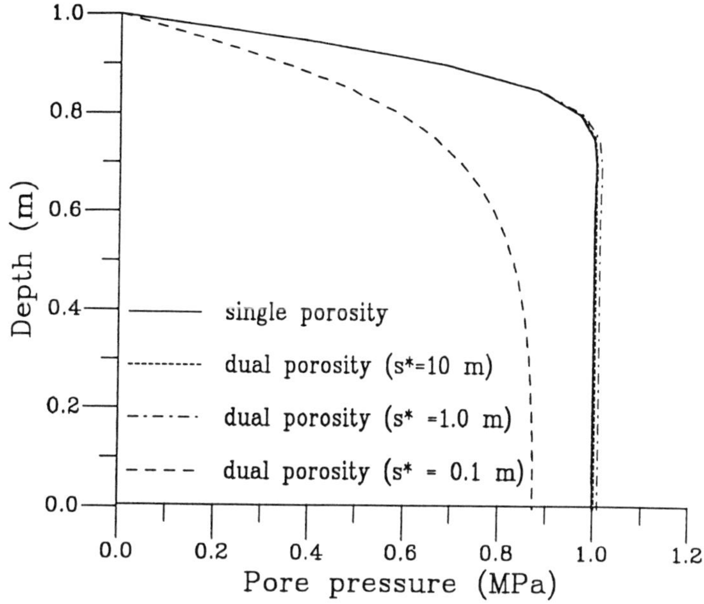

Figure 6-9. Comparative pressures at $t = 0.005$ sec.

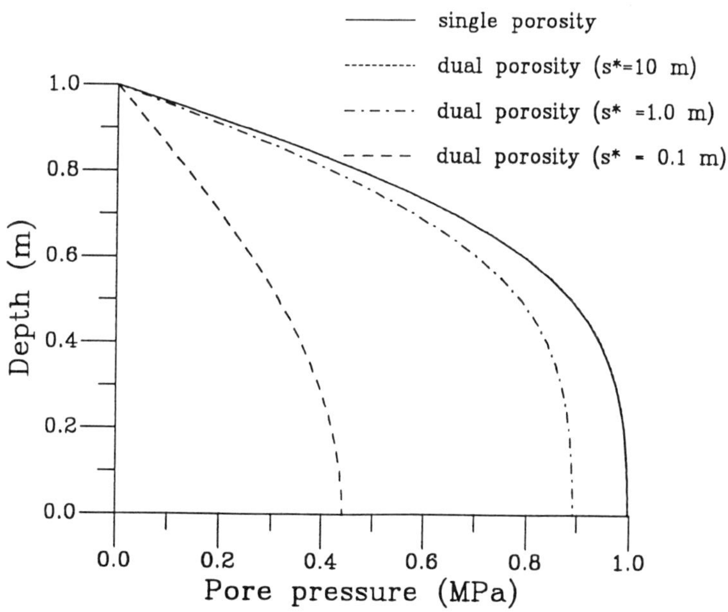

Figure 6-10. Comparative pressures at $t = 0.05$ sec.

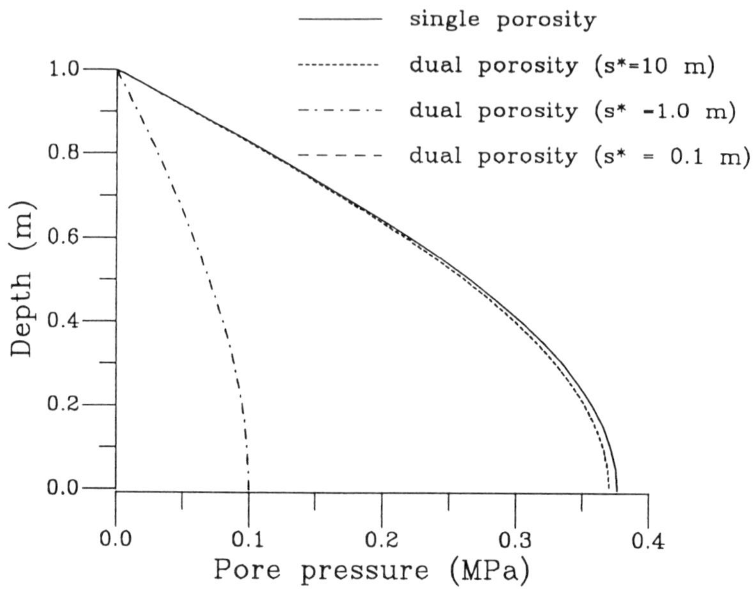

Figure 6-11. Comparative pressures at $t = 0.5$ sec.

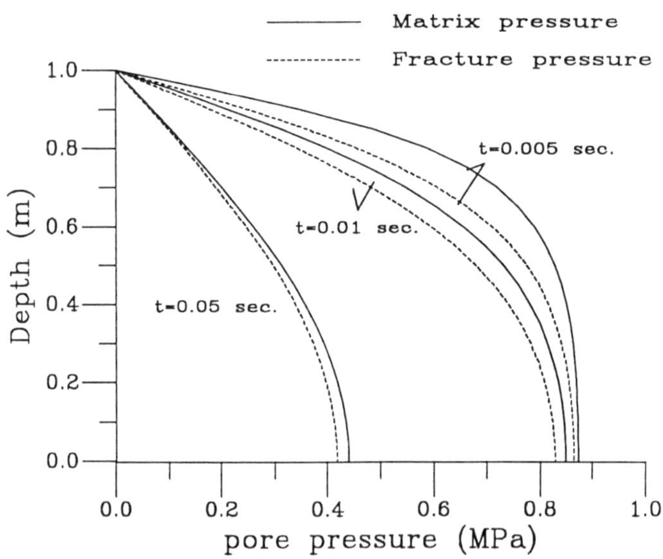

Figure 6-12. Comparative pressures between matrix blocks and fractures.

Chapter 7

APPLICATION

7.1 INTRODUCTION

The concepts, formulations, and solutions presented in the previous chapters provide a framework to address a variety of applications in science and engineering. In reality, many important stability, flow, and transport problems may be adequately represented by neglecting the couplings with other phenomena. For example, for groundwater flow in porous media, the normal assumption is that total stresses in the system remain constant, and that the release of fluid from storage may be considered in isolation from stress reorientation effects. In the majority of flow applications, this approach is sufficient; however, for others, the coupling is the essence of the behavior itself. For example, slope instability may not be confined to the effects of deformation alone, but may be concurrently affected by the effect of seepage due to groundwater flow. Similarly, the production of hydrocarbons may not be restricted to the consequence of mass balance due to fluid flow, but may be influenced significantly by stress variations due to force equilibrium, especially near the wellbore, where there is a corresponding feedback on permeability magnitudes.

The relative importance of various coupled phenomena is illustrated in the following applications to subsurface problems.

7.2 TUNNEL SUBSIDENCE

In this civil engineering application, surface subsidence is evaluated as a result of fluid discharge or recharge into a tunnel.

7.2.1 Problem Definition

The proposed site for a tunnel is located in fractured and karsified limestones and dolomites, present beneath alluvium (Aboulseiman et al. 1995). Maintaining tunnel stability during the service period of the tunnel may be difficult as a result of the impact of consolidation due to (a) seasonal precipitation; (b) the presence of cavities in the karsified limestone; (c) subsidence; (d) storage and conductance interactions between fractures and matrix blocks; and, (e) the activation of tectonic stresses. Common concerns should also consider such aspects as the permeability and durability of the tunnel lining, ground and water pressures, grouting configuration, and mechanical properties of the surrounding rock mass. Gravitational and tectonic stress states, as well as shear stresses mobilized by faulting in the fractured rock mass, may also greatly affect tunnel stability. The numerical procedure adopted in the following is as described in Chapter 6.

7.2.2 Numerical Modeling

The primary objective of the numerical modeling is to identify the key factors that influence the stability of a pair of waterway tunnels, separated 30 m laterally, and situated in fractured carbonate rock masses at a depth of 60 m below the ground surface. The tunnels, each about 3 m in diameter, are subject to both gravitational loads and pore pressure due to combined seepage and running-water forces as a result of inflow and outflow from the tunnels, respectively. Some field and modeling parameters are described in Table 7.1. In the analysis, displacement and time have been normalized against maximum values to aid the presentation.

Table 7.1. Modeling parameters.

Parameter	Symbol	Fluid	Fracture	Matrix	Unit
Gravitational load	F_g			1.5083	MN/m^2
Pore pressure load	F_p	0.6033			MN/m^2
Elastic modulus	E			200	GN/m^2
Poisson ratio	ν			0.3	-
Porosity	n_1, n_2		0.1	0.4	-
Permeability	k_1, k_2		10^{-3}	10^{-5}	m^4/(MN s)
Fracture stiffness	K_n		2.0		MN/m^2/m
Fluid bulk modulus	K_f	1			MN/m^2
Fracture spacing	s^*		0.5		m
Elapsed time	t	10			min

Quarter symmetry is used to reduce the number of unknowns in the solution, as shown in the finite element mesh of Figure 7-1.

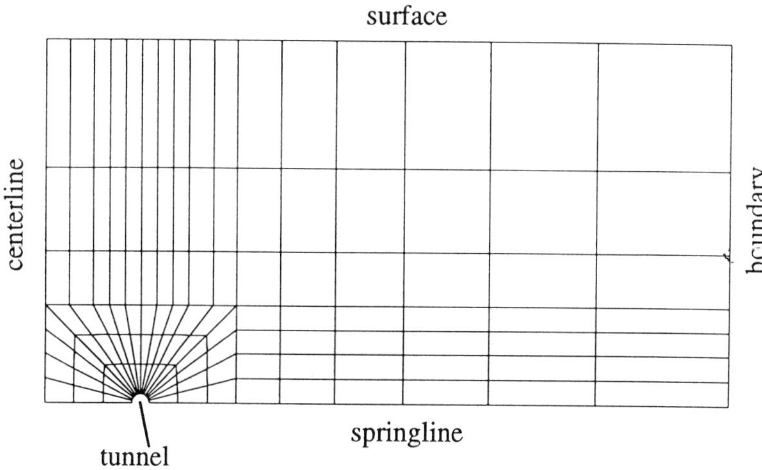

Figure 7-1. Finite element mesh layout.

Figure 7-2 compares the magnitudes of subsidence, measured 3.5 m above the tunnel, for five different scenarios: (a) conventional elastic media, (b) fractured porous media with outflow from the tunnel when the tunnel fills with water, (c) fractured media with reduced outflow, (d) poroelastic media without flow within the tunnel, and (e) fractured media with natural inflow. As a consequence of the presence of a rock pillar between the two tunnels, the profiles of the subsidence curves are not symmetric about the tunnel centerline. Subsidence immediately above the tunnel roof varies substantially between the differing physical assumptions. Fluid flow counteracts the development of rock deformations, apparent in the most conservative subsidence magnitude where outflow from the tunnel is considered. An apparent increase in maximum subsidence is observed for the instance with reduced outflow, and this is greater for the case of poroelastic media without any fluid flow from the tunnel as a representation of drained conditions. Where natural inflow occurs at the tunnel edge, and where the consolidation process is most pronounced, maximum subsidence results. The subsidence of conventional elastic media falls part way between that for the situation with outflow and that with reduced outflow.

This analysis is based on the dual-porosity poroelastic formulation, for which the tunnel configuration appears to be an important factor. The depth of the tunnel controls vertical stress magnitudes; deeper tunnels generally experience larger stress concentrations and greater potential for distress, while shallower tunnels induce greater subsidence and the potential for loss of inter-

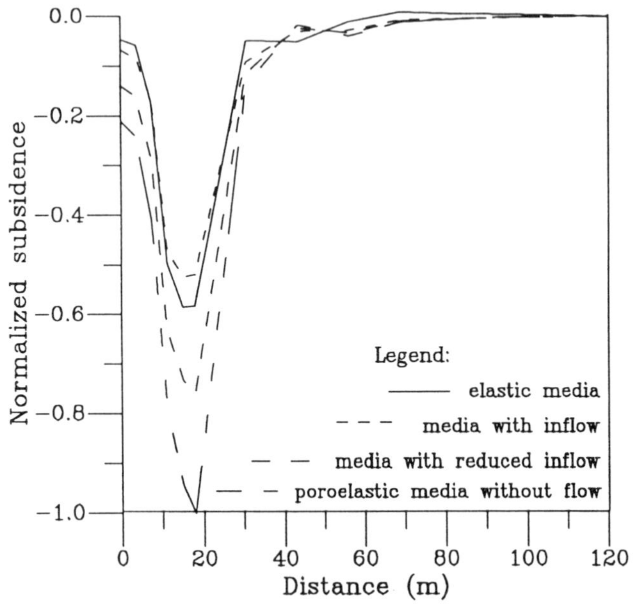

Figure 7-2. Spatial subsidence 3.5 m above the tunnel.

lock and arching in the roof. Shallow tunnels may be subject to the intrusion of surface water from precipitation, penetrating into the transported water, with the potential for its contamination. Consequently, the optimal design must account for these many interacting factors to reach the most desired compromise.

7.2.3 Concluding Remarks

The transient poroelastic effects are most prominent at the stage of initial fluid recharge or discharge. Conventional models, such as those based on either linear elasticity or diffusive fluid flow, frequently fail to quantify the changes in rock stress and pore pressure status. As a result, conventional analysis may provide an erroneous estimation of the rate of water transportation, tunnel performance, and durability. This shortcoming can be avoided using the present method. The conditions of either water inflow into, or outflow from, the tunnels appear to be important factors in regulating the rate of water transportation, as well as evaluating the structural stability of the tunnel.

7.3 SLOPE STABILITY

In this civil/mining engineering application, slope stability is evaluated in a surface mine.

7.3.1 Problem Definition

The prediction of surface movements at a petroleum factory near the north slope of the Fushun west open pit in China is considered as a tutorial example (Liu et al. 1992). Because the excavation is in the complex geological conditions comprising the north slope, deformation and damage resulted in the development of a landslide. This, in turn, caused large surface deformations at the factory and affected its normal operation. With the cooperation of a number of research institutes, study resulted in the resolution of the problem (Ma et al. 1990; Liu et al. 1992; Yang et al. 1995).

Calculations performed by traditional uncoupled numerical methods (solid mechanics only) did not conform to the field measurements. Alternatively, analysis based on poroelastic behavior (Biot 1941), especially dual-porosity poroelasticity theory, did match the field observations. In dual-porosity poroelasticity, changes in stresses will interactively affect the change of fluid pressure in both fractures and matrix blocks. This numerical model enables the simulation of coupled flow-deformation phenomena. The objective of the present study is to demonstrate the capability of the numerical model to match field observations. The numerical model and relevant details of the formulation, together with some solutions, are previously available in Chapter 6.

7.3.2 Finite Element Simulation

The north slope of the Fushun open pit contains a large area, extending about 2 km along strike, where a series of strata movements are evident, and where the stability of nearby buildings has been seriously affected. The finite element program used in this analysis was developed by Elsworth and Bai (1992) and by Bai et al. (1993). From the geological cross-section, the finite element mesh arrangement and the material distribution are shown in Figures 7-3 and 7-4.

The finite element model represents 2-D plane strain behavior, with the mesh representing a section 1,800 m long and 800 m deep. The model base and lateral boundaries are fully restrained and impermeable. The overburden is fully saturated with fluid infiltrating due to precipitation. To simulate surface excavation, stress-release tractions are added to represent the release of strain energy. In this method, the calculations are divided into two steps: first, the strata movement of the slope, when it developed from the position in 1988 to that in 1991, is evaluated; second, strata movements are predicted when the slope reached the design boundary limit. In the finite element calculation, there are nine parameters for each of the materials: modulus of elasticity

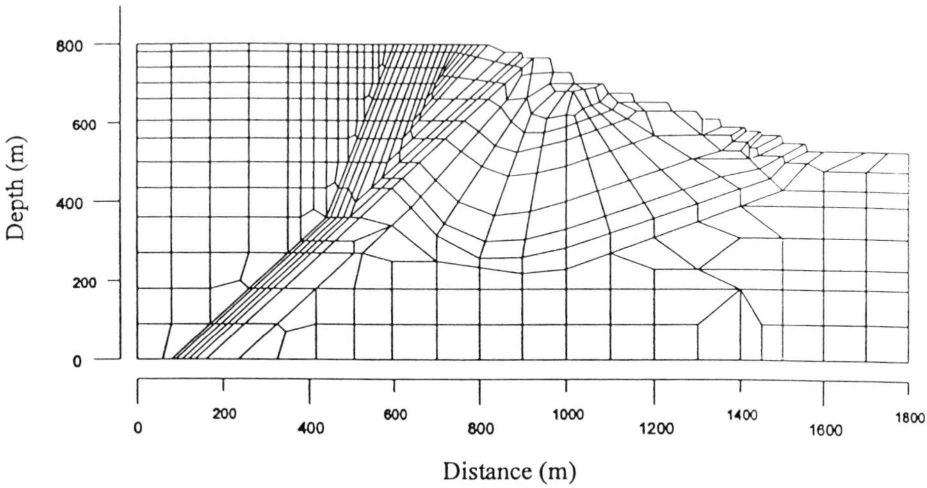

Figure 7-3. Finite element layout.

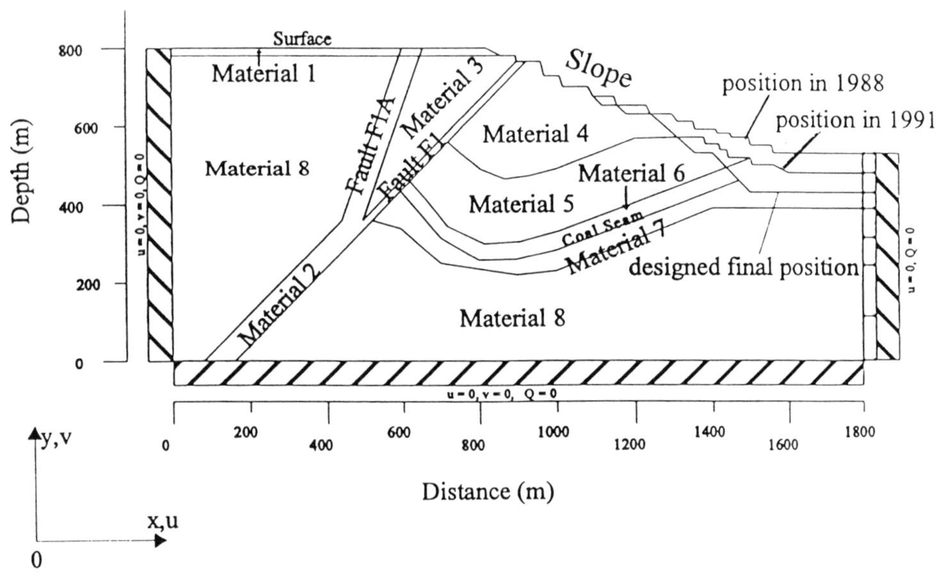

Figure 7-4. Material distribution.

(E), Poisson's ratio (ν), fracture normal stiffness (K_n), fluid bulk modulus (K_f), matrix porosity (n_1), fracture porosity (n_2), matrix permeability (k_1/μ), fracture permeability (k_2/μ), and fracture spacing (s^*). Poroelastic parameters are selected based on the experimentally determined rock characteristics (e.g., those of the solid phase), or are estimated (e.g., those of the fluid phase). As illustrated in Figure 7-4, eight different materials are selected.

7.3.3 Case Analysis

The horizontal strata movement induced by the surface excavation is depicted in Figure 7-5, where the maximum value is present near the excavation level, which may cause slip failure in the open pit wall. With the release of the rock weight, upward vertical displacements may be observed in the overburden, beneath the open pit, with the maximum value located at the center of the pit. This is in contrast to the downward vertical displacement adjacent to the fault, reflecting large fault movement. The maximum horizontal displacement occurs close to where the vertical displacement changes its orientation from upward to downward. The calculations reflect the possibility of a step subsidence over the surface as a result of mining, which has been detected in the field. Progressive mining activity may increase the potential to cause large landslides where stress concentrations are present.

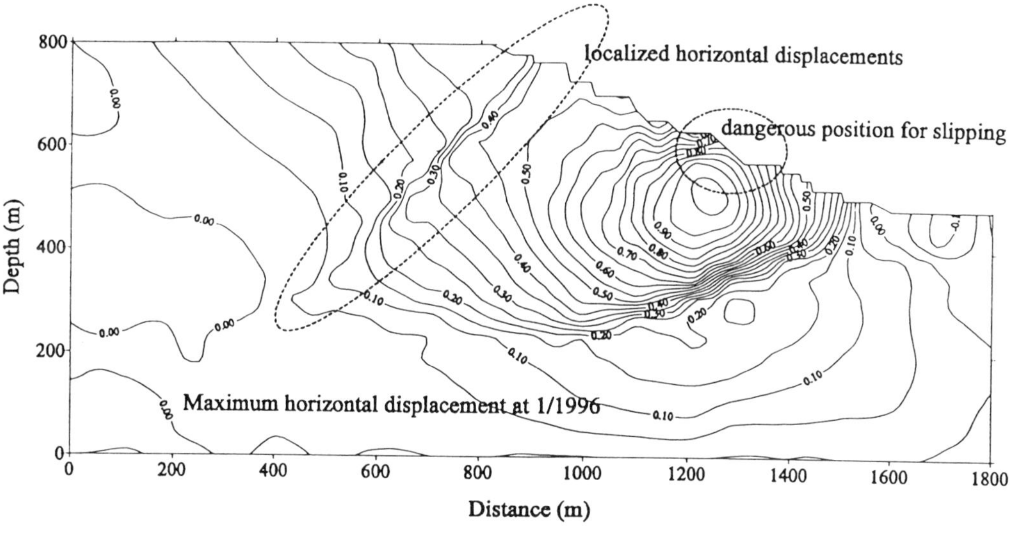

Figure 7-5. Predicted horizontal displacements.

Subject to either external or internal changes of load, the interactions within dual-porosity media are twofold and result from: (a) rock deformations and pore pressure variations, and (b) interporosity flow between fractures and matrix blocks. In the former case, macroscopic fluid flow in the overburden is induced by regional strata movements. In the latter case, microscopic fluid flow between fractures and matrix is the consequence of rock deformation at the pore scale. For this reason, fluid flow can be considered as being introduced from internal sources. From field measurements (Figure 7-6), it is known that strata movements are also due to external fluid flow such as seasonal rainfall and subsequent fluid discharge from the site.

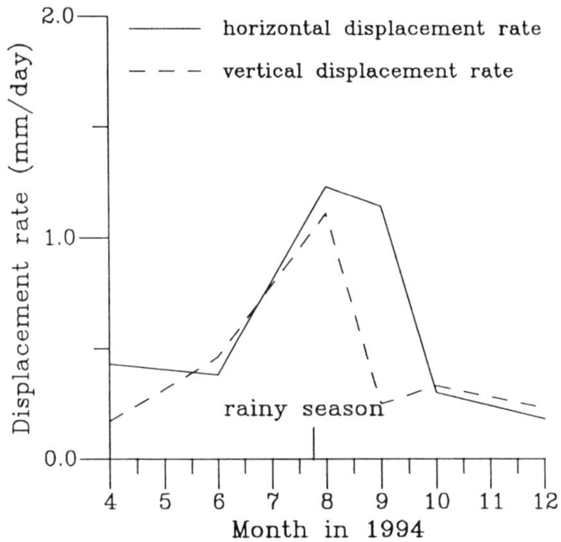

Figure 7-6. Measured rate of subsidence.

Based on past analyses (Ma et al. 1990), dewatering was initiated in the north slope of the mine to prevent potential landslides. Field measurements indicated that in situ pore pressure declined significantly. The present study reveals that the step subsidence near the fault is related to the dewatering activity. Among other reasons, the dramatic reduction of pore pressure within the fault, which may activate the fault movement due to the uniform reduction of shearing resistance in the faulting zone, appears to be directly responsible for the step subsidence.

With the lack of field hydrogeological data, the analysis is based on assumed flow boundary conditions, since the actual locations of pumping are ill constrained. The numerical results indicate that the magnitude of subsidence increases in proportion to an increase in pumping, as shown in Figure 7-7. The step subsidence is aggravated due to the existence of the fault.

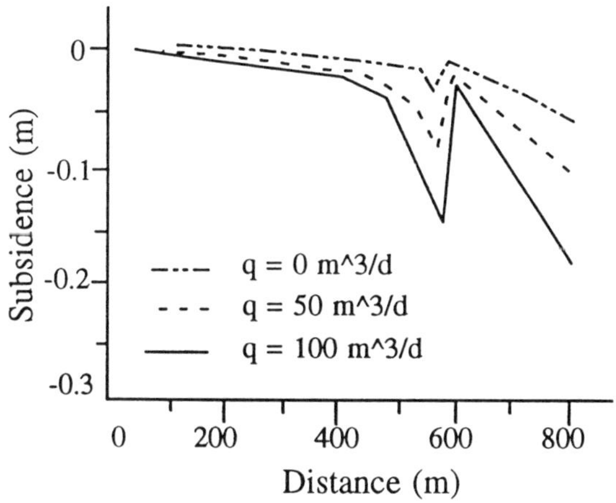

Figure 7-7. Flow rate-dependent subsidence.

Previously, the calculation of step subsidence was unsuccessfully attempted using the theory of elasticity. The modeling was successful in simulating the step subsidence, however, only by applying poroelastic theory coupled with dual-porosity behavior. The predicted surface subsidence is compared with available field measurements (Yang et al. 1995 and Zhou 1995), as depicted in Figure 7-8. The present calculation indicates that the step subsidence is mainly due to the influence of fluid pressure dissipation. It is shown that good agreement has been reached with most of the measuring locations within the open pit. The discrepancy may be the result of differences in boundary and pumping conditions between the calculation and the actual field conditions.

The effect of pumping is related directly to the surface subsidence. The maximum subsidence occurs close to the pumping location. The calculations indicate that pumping accounts for up to 75% of the total subsidence. The pore pressure evolution within the fractures and the matrix blocks reflects the decrease of pore pressure which is the result of pumping, thereby increasing displacements.

7.3.4 Concluding Remarks

The present analysis indicates that poroelasticity theory may be applicable to assess slope stability and associated surface subsidence in the Fushun open pit in China. Field measurements reveal that the strata movements in the

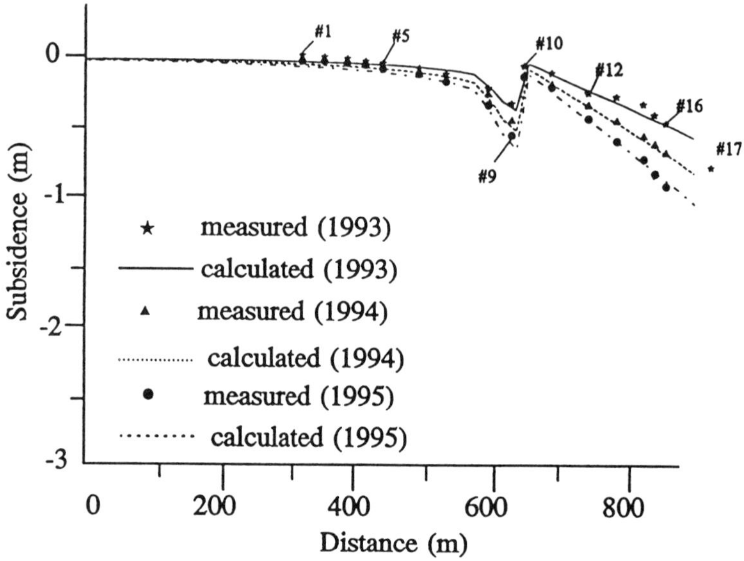

Figure 7-8. Comparison of calculation and measurement.

mining region are strongly related to the regional fluid flow. A comparison between numerical calculations based on dual-porosity poroelasticity, and the field measurements, shows a good agreement. This is in clear contrast to previous comparisons based on elastic theory that failed to predict the observed step subsidence. Although fault movement in the nearby formation may contribute partially to the irregular surface subsidence profile, it is determined through the comparative analysis that regional fluid flow plays a major role. As a result, control of the fluid flow in the area may constitute a key element to minimize the influence of surface subsidence.

7.4 PERMEABILITY DETERMINATION

In this application, numerical approaches are applied in the determation of formation permeability. The full mathematical formulation is documented in Chapter 3 while associated numerical solutions can be found in Chapter 6.

7.4.1 Unstressed Condition

Permeability is one of the most important properties of reservoir rocks. This property is reflected by the permeability tensor that provides a mathematical explanation of anisotropy or preferred flow pathways. In addition, heterogeneous porous media may possess a spatially variable characteristic as a result of point-to-point variation in permeability. In contrast, directional permeability changes are primarily attributed to formation anisotropy (Greenkorn et al. 1965).

The significance of studying methods of permeability determination relates to determining reservoir production, predicting aquifer recharge or yield, and in assessing the rate and penetration of subsurface contaminants. In the study of coupled processes, permeability is the parameter that governs the impact of fluid flow on a deforming system (Bai and Elsworth 1994). For a flow system, permeability uniquely defines fluid transmission through the porous domain.

Formation permeability is commonly determined from laboratory flow experiments in jacketed samples. During the tests, either flow rate or pore pressure may be controlled over the injection area. At the observation area, either pressure or flow rate may be measured. Since the flow cross-sectional area through the specimen is not uniform, a factor normally termed the "geometric factor," must be deduced to link the calculated rate-pressure relationship to the analytical expression via Darcy's law.

7.4.1.1 Linear Relation Between Pore Pressure and Flow Rate

For steady-state flow in anisotropic media, the governing equation can be written as (Zienkiewicz 1983):

$$\frac{\partial}{\partial x}\left(\frac{k_x}{\mu}\frac{\partial p}{\partial x}\right)+\frac{\partial}{\partial y}\left(\frac{k_y}{\mu}\frac{\partial p}{\partial y}\right)+\frac{\partial}{\partial z}\left(\frac{k_z}{\mu}\frac{\partial p}{\partial z}\right)+Q=0 \tag{7.1}$$

where k_x, k_y, and k_z are the permeabilities in the x, y, and z directions, and μ is the fluid dynamic viscosity. In the present analysis, μ is assumed to be 0.5 cp (1 cp $= 1 \times 10^{-3}$ Pa $\cdot$ s). For simple geometries and initial and boundary conditions, an analytical solution can be obtained from Eq. (7.1) (Carslaw and Jaeger 1959). For more complicated boundary conditions, such as flow through cylindrical specimens with small fluid injection and withdrawal areas, a numerical method such as the finite element method must be applied when solving Eq. (7.1).

Examining a steady-state flow situation in a 2-D rectangular domain with isotropic permeability, i.e., $-a_c < x < a_c$ and $-b_c < y < b_c$ (Carslaw and Jaeger 1959), the flow equation can be written as

$$k\frac{\partial^2 p}{\partial x^2}+\frac{k}{\eta_c^2}\frac{\partial^2 p}{\partial y^2}+Q=0 \tag{7.2}$$

where η_c is a constant and Q is the flow rate at the geometric center of the rectangular domain. The boundaries $x = \pm a_c$ and $y = \pm b_c$ are kept at zero fluid pressure. The solution for the above equation under the given boundary conditions can be expressed as

$$p = \frac{Q}{k} F(x, y) \tag{7.3}$$

where

$$F(x, y) = \frac{a_c^2 - x^2}{2} - \frac{16a_c^2}{\pi^3} \sum_{n=0}^{\infty} \frac{(-1)^n \cos\left[\frac{(2n+1)\pi x}{2a_c}\right] \cosh\left[\frac{(2n+1)\pi \eta_c y}{2a_c}\right]}{(2n+1)^3 \cosh\left[\frac{(2n+1)\pi \eta_c b_c}{2a_c}\right]}. \tag{7.4}$$

At any fixed location (x, y) with uniform domain boundaries (i.e., a_c and b_c are constants), $F(x, y)$ also becomes a constant. For a constant permeability, k, the above solution demonstrates that the relationship between pore pressure p and flow rate Q is linear.

7.4.1.2 Determination of Geometric Factor

To relate the numerical solution with a non-uniform flow cross-sectional area to the analytical solution with a uniform area, a conversion factor termed a "geometric factor," may be added to the pressure-rate relationship, as described in Eq. (7.4). Derivation of this factor is described below.

It is assumed that the subscripts "t" and "a" indicate the numerically calculated (true) and the analytically calculated (approximated) quantities, respectively. Using Darcy's law, these quantities can be expressed as

$$q_t = -A_t \frac{k_t}{\mu} \frac{\Delta p_t}{L_f} \tag{7.5}$$

$$q_a = -A_a \frac{k_a}{\mu} \frac{\Delta p_a}{L_f} \tag{7.6}$$

where q is the flow rate, A is the flow cross-sectional area, k is the specimen's permeability, μ is the fluid dynamic viscosity, Δp is the pressure difference across the length of the specimen, and L_f is the linear flow length.

Since $q_t = q_a$ and $k_t = k_a$, the following relation exists:

$$G_f = \frac{A_a}{A_t} = \frac{\Delta p_t}{\Delta p_a} \tag{7.7}$$

where G_f is the so-called geometric factor, which is a constant for the present scenario. Even though A_t is indeterminate, in general, $A_a \leq A_t$ so $G_f \leq 1$.

From the expression in Eq. (7.7), Eq. (7.5) can be rewritten as

$$q_t = -A_t \frac{k_t}{\mu} \frac{\Delta p_t}{L_f} = -G_f \times A_t \frac{k_a}{\mu} \frac{\Delta p_a}{L_f}. \tag{7.8}$$

Since $q_t = q_a$, Eq. (7.6) is readily recovered from Eq. (7.8).

7.4.1.3 Numerical Procedure

Since the steady-state condition prevails, the numerical procedure is straightforward, as outlined in the following:

- Design the finite element mesh for cylindrical specimens applying the symmetric rule (e.g., usually only half of the cylinder needs to be simulated). Employ the finite element model of fluid flow with the prescribed boundary conditions.
- Calculate the pore pressure at the fluid outlet location while a constant flow rate is imposed at the fluid inlet place.
- Determine the change of flow rate using Darcy's law along with the derived permeability, Δk, and pore pressure, Δp, e.g.,

$$\Delta Q = -A_t \frac{\Delta k}{\mu} \frac{\Delta p}{L_f} \tag{7.9}$$

where A_t is the effective flow cross-sectional area.

A_t may be determined through defining a "geometric factor," that relates the numerical solution to the analytical solution and is determined by identifying the difference between the uniform flow cross-sectional area A_a using Darcy's law and the effective cross-sectional area A_t in the numerical computation of fluid flow through a rock specimen. The geometric factor is defined in Eq. (7.7). The injection area is defined as A_a, which can be chosen as the uniform cross-sectional flow area for the analytical solution (Darcy's velocity). Numerical formulation and solutions are given in greater detail in Chapter 6.

7.4.1.4 Flow Through Specimen

Horizontal flow through the rock specimen is used to determine the horizontal permeability of the formation.

Figure 7-9 shows the configuration of flow through the specimen, where the cylindrical specimen is 1.5 in (3.81 cm) in height and 1.0 in (2.54 cm) in diameter. The horizontal flow areas are located across the center of the specimen. The inlet and outlet are circular holes of 0.25 in (0.64 cm) in diameter. The surface of the holes are curved to match the radius of curvature of the specimen. The inlet is kept at constant flow rate, while the pressure at the outlet is measured or calculated. Using the eight-node isoparametric elements, the finite element mesh arrangement is made with total of 378 nodes and 252 elementrs.

In the numerical simulation, the incompressible fluid flows horizontally through a specimen with the length L_f in the flow direction, and with cross-sectional area A_a. With reference to Figure 7-9, Figure 7-10 shows a flow net

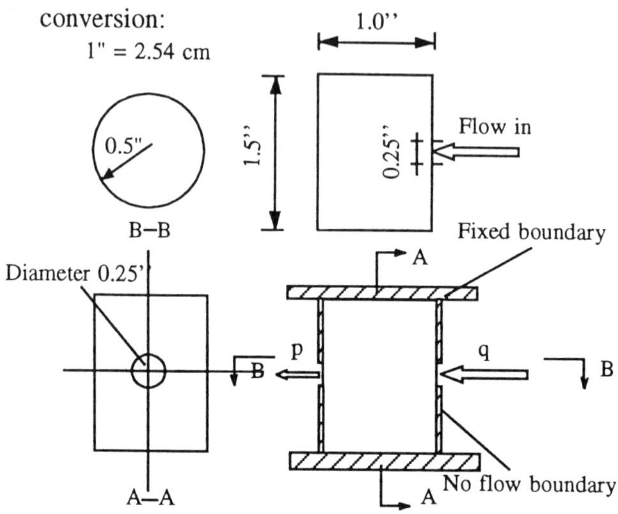

Figure 7-9. Horizontal flow configuration.

along the cross-section (A-A) and the plan-view-section (B-B). The variable flow area is defined as A_t. For comparison with the analytical calculation, a fixed flow area (A'-A') equal to the injection area is assumed as A_a, as also shown in Figure 7-10. The case with a variable flow area is analyzed using the finite element method, where the flow rate is $q_t = -A_t \frac{k_t}{\mu} \frac{\Delta p_t}{L_f}$. Darcy's law can be applied directly for the case with a fixed flow area, where the flow rate is $q_a = -A_a \frac{k_a}{\mu} \frac{\Delta p_a}{L_f}$. Assuming $q_t = q_a$ and $k_t = k_a$, then $A_t \Delta p_t = A_a \Delta p_a$, e.g., $G_f = \frac{A_a}{A_t} = \frac{\Delta p_t}{\Delta p_a}$, where G_f is the geometric factor, $A_a = \pi r^2$, and r is the fixed flow (injection) radius.

For the isotropic porous specimen (e.g., $k_x = k_y = k_z$), Figure 7-11 compares the pressure-rate relationship between the numerical and analytical solutions. The geometric factor can be derived from the ratios of the analytical solution to the numerical solution, implied in Figure 7-11. Figure 7-12 indicates that the geometric factor is independent of the flow rate. In other words, if the geometry of the specimen is fixed, the geometric factor shows a constant value irrespective of different flow rates and permeabilities.

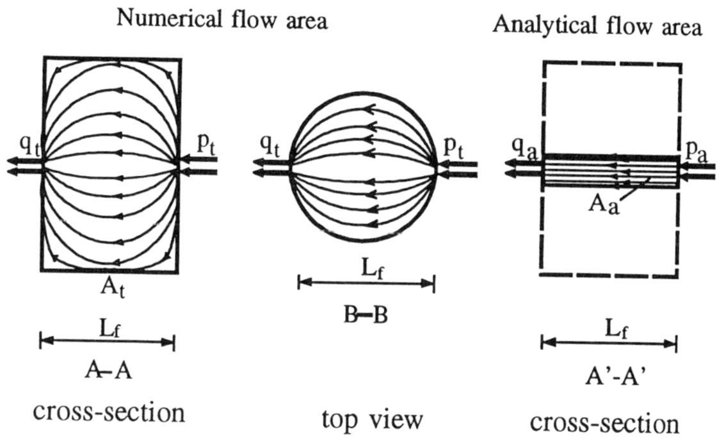

Figure 7-10. Schematic numerical and analytical horizontal flow nets

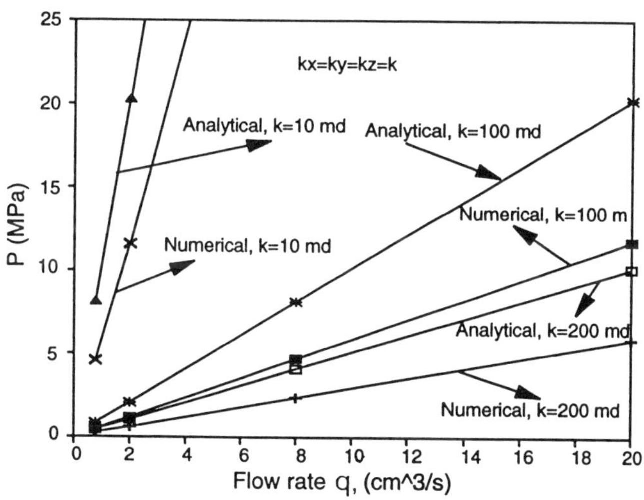

Figure 7-11. Comparison results between numerical and analytical approaches.

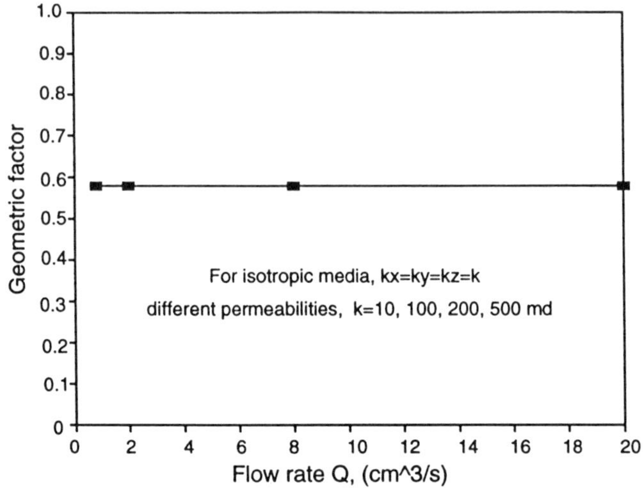

Figure 7-12. Geometric factor-rate relation for an isotropic specimen.

7.4.2 Stressed Condition

Although permeability represents a fundamental property of the porous system, this property can be modified when subjected to variations in stress. Permeability variations may become significant, especially if the porous medium is naturally fractured where the transmissive capacity of highly conductive fractures are extremely sensitive to perturbations in stress.

If permeability depends on position within a geological formation, the formation is heterogeneous. Heterogeneities may result from faulting and fracturing induced during prior tectonism. They also may be present at a variety of length scales, from grain scale of the order of microns to fault zones covering many kilometers. These heterogeneities typically induce a degree of anisotropy that may further control the hydraulic and transport performance of porous and fractured media. The primary cause of anisotropy on a small scale is the orientation of clay minerals in sedimentary rocks and unconsolidated sediments or the alignment of microcracks in indurated materials. Laboratory core samples of clays and shales show horizontal to vertical anisotropy ratios in the range 3:1 to 10:1 (Freeze and Cherry 1979). At a larger scale, field observations indicate a relationship between layered heterogeneity and anisotropy, which may lead to regional anisotropy values on the order of 100:1 or even greater (Maasland 1957; Marcus and Evenson 1961). Snow (1969) showed that fractured rocks behave anisotropically because of the directional variations in fracture aperture and spacing.

Evaluation of anisotropic permeability magnitudes is becoming increasingly recognized as necessary, with core testing techniques providing a viable means to couple this analysis under a varying stress environment (Bai and Meng 1997). In the petroleum industry, interest is motivated from concerns regarding reservoir compaction and the resulting changes in reservoir production that accompany compaction. The present study is motivated by these interests to replicate this field behavior through an investigation of the coupled effect of uniaxial stress on the anisotropic permeabilities of fractured rock specimens in the laboratory.

7.4.2.1 Conceptual Models

The mathematical formulation referenced in this section is included in Chapter 3.

Neglecting turbulent flow and assuming laminar flow within a fracture network, Louis (1969) used the parallel plate analog to evaluate the permeability of fracture networks containing a set of regularly spaced parallel fractures subject to steady state fluid flow, as shown in *mode a* of Figure 7-13.

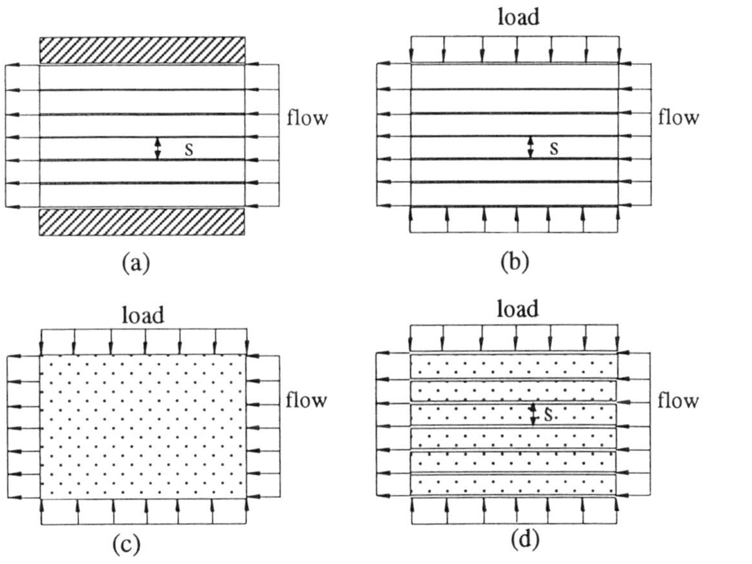

Figure 7-13. Modes a,b,c,d: flow through fractured and intact media.

Applying a uniaxial load to the fracture network, Elsworth (1989) derived the permeability change by assuming the individual fractures are distinctly soft with respect to the porous matrix. This restriction was released by Bai and Elsworth (1994) where the contribution of deformations from both fracture and matrix were distinguished and incorporated (*mode b* of Figure 7-13).

For homogeneous nonfractured porous media, the permeability variations can also be associated with the stress change (*mode c* of Figure 7-13). Instead of a change in aperture, changes in either void space or grain volume are the typical consequence that cause permeability changes. Application of Hertz' (1895) elastic contact theory enables permeability changes for intact media to be evaluated (Bai and Elsworth 1994).

In reality, rock deformation and fluid flow contribute mutually from both matrix blocks and fractures to the system, as depicted schematically in *mode d* of Figure 7-13. The system may be evaluated individually, and the result may be superposed from each contribution [refer to Eq. (3.239) in Chapter 3].

7.4.2.2 Determination of Equivalent Geometric Factor

Determining laboratory permeability magnitudes by conducting steady state flow tests is common. Permeability is derived from the test data using a modified Darcy's law in which a "geometric factor" is used to relate the actual flow geometry to an ideal geometry, as shown in Section 7.4.1 describing unstressed conditions. The cross-sectional area of flow is not constant for the flow geometry defined in simulation. The resulting flow geometry may be related to the specimen dimensions and inlet/outlet areas. The desire is to provide a geometric factor that allows permeability to be calculated directly with reference to the measurable inlet port area alone. The previously described geometric factor becomes an "equivalent" value for the tests involved in the external loading since this conversion factor incorporates not only the geometric variations between actual and analytical flow cross-sectional areas but also the specimen spatial changes as a result of the mechanical loading. The evaluation of correction factors, incorporating both geometric and stress effects, is only possible with the aid of numerical modeling.

Again, assuming that the subscript "t" indicates the quantity from the test or numerical simulation, while the subscript "a" implies the analytically calculated values for a constant flow cross-sectional area, Darcy's law can be written symbolically for each individual situation, as shown in Eqs. (7.5) and (7.6).

If steady state flow conditions prevail for the unstressed condition, one has

$$q_t = q_a, \qquad k_t = k_a. \tag{7.10}$$

However, if the influence of the external load is incorporated, Eq. (7.10) should be modified as

$$q_t^* = q_t - \Delta q_t = q_a, \qquad k_t^* = k_t - \Delta k_t = k_a \tag{7.11}$$

where Δq_t and Δk_t are the changes of flow rate and permeability due to the stress changes.

Let the modified Darcy's law, with reference to the loading environment, be given as

$$q_t^* = -A_t \frac{k_t^*}{\mu} \frac{\Delta p_t}{L_f}. \tag{7.12}$$

Substituting Eq. (7.11) into (7.12), gives

$$q_t^* = -A_t \left(\frac{k_t - \Delta k_t}{\mu} \right) \frac{\Delta p_t}{L_f}. \tag{7.13}$$

The equivalent geometric factor G_e can be obtained by equating Eq. (7.6) to (7.13), which results in (Bai and Meng 1997)

$$G_e = \frac{\Delta p_t}{\Delta p_a} = \frac{A_a k_a}{A_t(k_t - \Delta k_t)}. \tag{7.14}$$

In general, $A_a \leq A_t$, therefore, $G_e \leq 1$. The utility of the equivalent geometric factor, G_e, is to relate laboratory flow rates and pressure drops, measured within a non-standard specimen geometry to invariant magnitudes of permeability, k_a. When there is no change due to mechanical effects, $\Delta k_t = 0$ and the factor relates purely to the mismatch in geometries. In this situation, G_e represents the ratio of pressure drop measured in the cross-specimen test to that measured, say longitudinally within a constant diameter core. It is purely a geometric correction for the selected test geometry where the same volumetric flow rate occurs in each test. Similarly, where mechanical changes in permeability are applied to the specimen, G_e represents the ratios of pressure drops in the actual testing configuration to that for a longitudinal test within a core. Now, stress and geometric effects are included in the result, enabling the anticipated mechanical effect on reservoir permeability to be conveniently determined for any desired specimen configurations. Prior evaluation for magnitudes of G_e enables permeability magnitudes to be deduced directly from laboratory data from non-standard specimens, and extrapolated to define expected magnitudes of permeability change with altered stress fields.

The numerical procedure is similar to the one for the unstressed condition, except that the displacement needs to be calculated and the stress-permeability relation needs to be applied to account for the stress effect.

7.4.2.3 Case Study

The primary purpose of the present analysis is to determine the effects of stress dependency on permeability variations in the four modes depicted in Figure 7-13. In the following, schematic flows through different types of media subjected to uniaxial loading are labeled as mode-a, mode-b, mode-c, and mode-d, respectively (Bai et al 1997b). However, the proposed models are purely conceptual and may not represent the actual flow and loading situations, the real material structures, or the present finite element layout. They are designed simply for the interpolation of stress-permeability relationships. Figure 7-14 reflects the practical aspects of these models, where the desired experiments are projected.

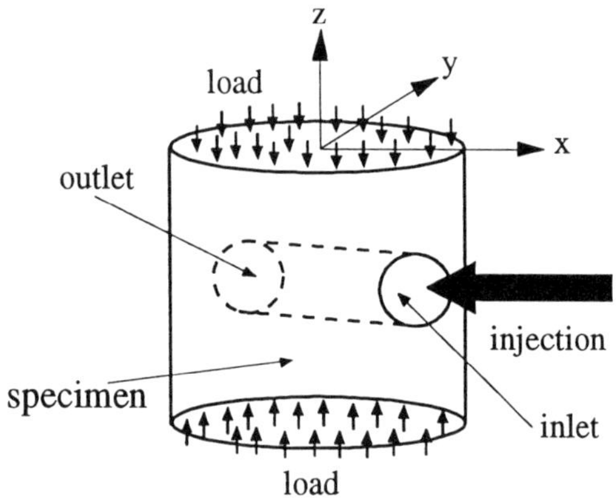

Figure 7-14. Schematic flow and loading in a rock specimen.

Basic parameters for the modeling are listed in Table 7.2. The injection area is confined to represent flow restriction in actual experiments. Flow is maintained at steady state throughout the simulation. Constant rate is applied at the inlet, while the pressure is numerically calculated at the outlet. For simplicity of analysis, the vertical static load is uniaxial on both the top and bottom surfaces of the specimen, free to expand laterally due to loading.

Comparisons of the p-q relationship between mode-b, mode-c (no-loading), mode-c (with loading) and mode-d cases are illustrated in Figure 7-15 for an initial permeability of 100 md for the intact media and 200 md for the fractured media. Different from the impact of loading, there is a significant discrepancy between the fractured and the intact media (e.g., mode-b and mode-c). Comparisons of the equivalent geometric factors for different modes are depicted in Figure 7-16, which shows consistent patterns in the p-q relationship.

These results indicate that the equivalent geometric factors are independent of flow rates. However, these factors may be affected by other influencing variables. Figure 7-17 illustrates that an increase of an equivalent geometric factor is the result of increasing vertical stress, and the resulting equivalent geometric factor is proportional to the applied uniaxial load. This increased equivalent geometric factor can be attributed to the reduced effective cross-sectional area of flow as a result of the incremental increase in the mechanical load. The relationship between the equivalent geometric factor and the applied load is nonlinear.

Table 7.2. Parameters for modeling stress-dependent permeability.

Parameter	Mode-a	Mode-b	Mode-c	Mode-d
Flow rate (cm^3/s)	0.8-2	0.8-2	0.8-2	0.8-2
External load (MN)	-	1e-5	0 or 1e-5	1e-5
Elastic modulus (MN/m^2)	100	100	100	100
Poisson's ratio (-)	0.25	0.25	0.25	0.25
Fracture stiffness (MN/m^2 /m)	100	100	-	100
Matrix porosity (-)	-	-	0.1	0.1
Fracture porosity (-)	0.05	0.05	-	0.05
Matrix permeability (md)	-	-	100	100
Fracture permeability (md)	200	200	-	200
Fluid viscosity (cp)	0.5	0.5	0.5	0.5
Fracture spacing (cm)	0.254	0.254	-	0.254

Note: 1 in = 2.54 cm; 1 md = 9.87×10^{-6} cm^2; 1 cp = 1×10^{-3} Pa · s.

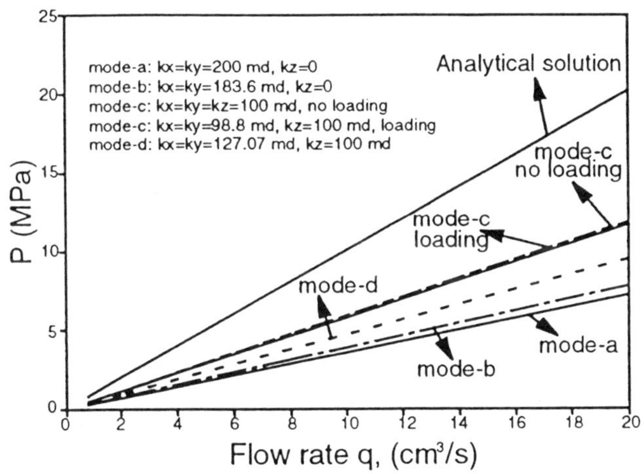

Figure 7-15. The p-q relationship for various cases.

The equivalent geometric factors may also be affected by the specimen size. Figure 7-18 indicates that the reduction of the geometric factors is associated

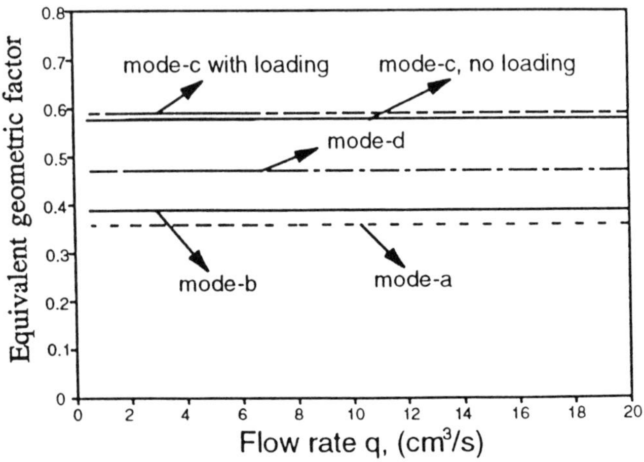

Figure 7-16. Equivalent geometric factors for various cases.

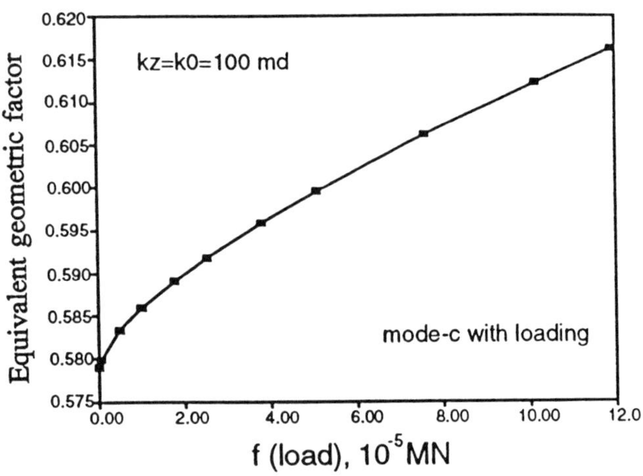

Figure 7-17. Correlation between equivalent geometric factor and loading.

with the increase of core heights for different modes of flow and loading. This increase in core sizes (e.g., heights) leads to an enlarged effective cross-sectional area of flow, and consequently results in a decrease in the equivalent geometric factors.

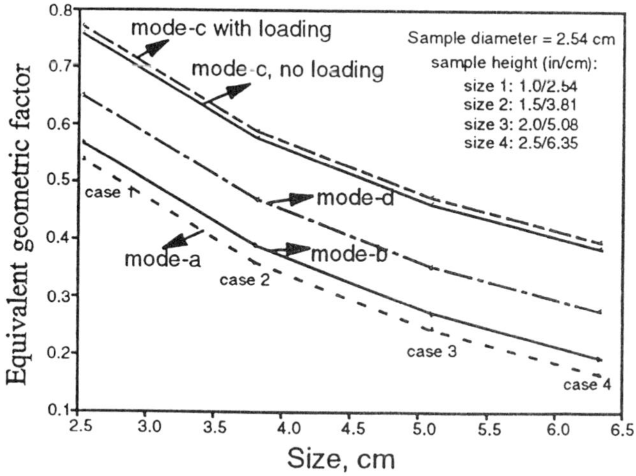

Figure 7-18. Equivalent geometric factor vs. specimen geometry.

7.4.3 Concluding Remarks

Hydrogeologists, geophysicists, and petroleum engineers have made significant efforts to quantify formation permeability, an important parameter in determining rate and magnitude of fluid flow through structured porous media. This quantification is further complicated because the original permeability, a geometric quantity, can be significantly modified by other factors of influence, such as mechanical impacts. Several conceptual models, relating rock permeability to pure fluid flow and to additional mechanical loading in porous and fractured media, are presented. These conceptualizations are incorporated into finite element models. Numerical schemes are subsequently developed to investigate permeability variations that result when specimens are subjected to changes in fluid pressure or combinations of pressure and external loading during laboratory experiments. The permeability and the stress-dependent permeability can be determined through comparison between analytical and numerical solutions via either a "geometric factor" or an "equivalent geometric factor," respectively.

7.5 WELL TESTING

This petroleum engineering application presents transient pressure analyses in naturally fractured reservoirs, practiced in the majority of well tests. The problems are solved using the solutions described previously in Chapter 5.

7.5.1 Flow

This analysis focuses on dual-porosity reservoirs of finite radial dimension. The pressure change due to pumping is conditioned by the following factors: (a) r_0 - the ratio between matrix permeability (k_{ma}) and fracture permeability (k_{fr}), i.e., $r_0 = k_{ma}/k_{fr}$ (defined as the permeability ratio); (b) r_1 - the ratio between fracture compressibility ($1/K_{fr}$) and fluid compressibility ($1/K_f$), i.e., $r_1 = K_f/K_{fr}$ (fracture compressibility ratio for short); and (c) r_2 - the ratio between solid grain compressibility ($1/K_s$) and fluid compressibility ($1/K_f$), i.e., $r_2 = K_f/K_s$ (grain compressibility ratio for short).

An attempt is made to simulate a naturally fractured reservoir with a finite radius ($r_{De} = 10000$). A constant fluid pressure prevails at the outer boundary and a constant pumping rate is maintained at the wellbore. It is assumed that the reservoir is composed of equally-spaced fractures defining uniformly sized porous blocks. The matrix has an average porosity of 0.2, in contrast to an average fracture porosity of 0.02. The pressure analysis focuses on the fractures, although the matrix pressure can be easily calculated, if desired. At a fixed permeability ratio ($r_0 = 0.01$) and grain compressibility ratio ($r_2 = 1$), the dual-porosity behavior is manifest within the certain range of fracture compressibility ratios (r_1), as shown in Figure 7-19. Under the circumstances depicted, the fluid in the fractures drains quickly and the matrix storage resupplies the fractures. Afterwards, fluid flow and fluid pressure between matrix and fractures maintain an equilibrium condition within the total system. This scenario is similar to that observed in the infinite reservoir by Warren and Root (1963).

7.5.2 Flow and Deformation

In this situation, fluid flow is coupled with rock deformation, and solutions of pressure and displacement are obtained sequentially through partial decoupling (refer to Chapter 5).

7.5.2.1 Parametric Relationships

The parameters used in the previous analysis for the fractured-porous rock mass may be classified into four categories: (a) fluid properties, (b) mechanical properties, (c) physical properties, and (d) poromechanical properties. Most fluid and mechanical parameters can be obtained routinely through labora-

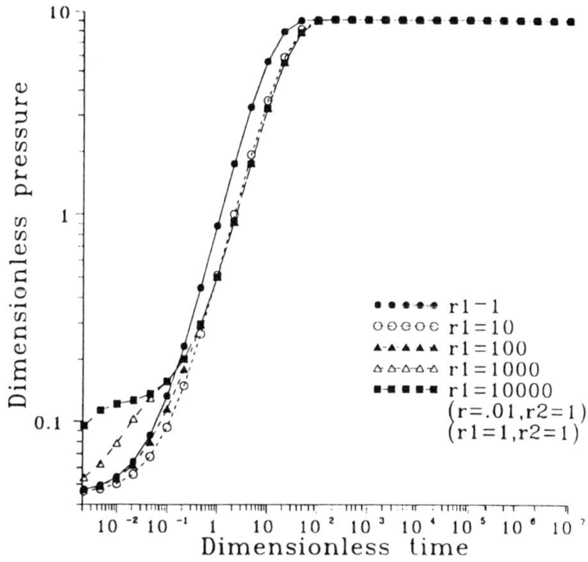

Figure 7-19. Temporal pressure vs. smaller fracture compressibility ratio. (Reprinted from Bai et al. 1994b, with permission from John Wiley and Sons.)

tory tests, in contrast to the physical parameters which are generally acquired via field measurements. More sophisticated macromechanical testing methods with fluid pressure coupling are usually required to determine poromechanical parameters. As an initial approximation, Table 7.3 lists the values or the approximate ranges of general parameters that are used in the calculation (Bai et al. 1995a). These parameters may be re-adjusted to suit individual situations. In particular, the mechanical parameters are determined from laboratory tests and should be adjusted to match in situ magnitudes.

7.5.2.2 Case Study

The basic modeling parameters are listed in Table 7.4. In addition to the parametric ratios defined in Section 7.5.1, (i.e., r_0, r_1 and r_2), the elasticity modulus ratio, r_3, must be defined as the ratio between the compressibility of the porous matrix (i.e., $1/E$ where E is the elastic modulus) and the fluid compressibility ($1/K_f$). $r_3 = K_f/E$ is labeled as the matrix compressibility ratio.

For a fixed permeability ratio ($r_0 = 0.01$) and matrix compressibility ratio ($r_3 = 1$), the pressure change is least sensitive to the change of the fracture compressibility ratio (r_1) and the grain compressibility ratio (r_2) at later times, as shown in Figure 7-20. For a constant fluid compressibility, a decrease in fracture compressibility leads to delayed pressure disturbance. The dual-porosity behavior is observed at large r_1. For larger r_1, the fluid in the fractures drains quickly, and the matrix storage supplies the fractures within a short period of time. Fluid flow and fluid pressure between matrix and fractures are subsequently maintained in equilibrium within the system.

Table 7.3. Referenced material properties.

Parameter	Value	Unit	Type	Note
Elastic modulus (E)	$5 \sim 80$	GPa	sandstone	(1)
Poisson ratio (ν)	$0.07 \sim 0.33$		sandstone, limestone	(1)
Fluid bulk modulus (K_f)	$0.5 \sim 5$	GPa	oil	(2)
	3.3	GPa	water	(3)
Grain bulk modulus (K_s)	$36 \sim 50$	GPa	sandstone, marble	(3)
Fracture stiffness (K_n)	0.1	GPa/cm		(4,5)
Fracture spacing (s^*)	$0.2 \sim 30$	m		(6)
Matrix porosity (n_1)	$0.04 \sim 0.2$		sandstone, limestone	(1)
Fracture porosity (n_2)	$0.0001 \sim 0.01$			(6)
Matrix p factor (α_1)	0.5			(7)
Fracture p factor (α_2)	0.9			(7)
Matrix permeability (k_1)	$10^{-10} \sim 10^{-14}$	cm^2	sandstone	(8)
Fracture permeability (k_2)	$10^{-7} \sim 10^{-11}$	cm^2	meta.* rocks	(8)
Dynamic viscosity (μ)	0.012	cp	oil	(2)
	1.5	cp	water	(2)

Note: (1) Jumikis 1983; (2) Craft and Hawkins 1959; (3) Rice and Cleary, 1976; (4) Iwai 1976; (5) Rosso 1976; (6) Snow 1968; (7) Walsh 1981; (8) Freeze and Cherry 1979; 1 cp = 1×10^{-3} Pa · s; meta.* = metamorphic.

Table 7.4. Parameters in the case study.

Porosity n		Factor α		Coefficient	Radius	
matrix	fractures	matrix	fractures	ν	r_w	r_{De}
0.2	0.02	0.5	0.9	0.25	0.1 m	10000

Note: r_w is the well radius, and r_{De} is the dimensionless radius at the reservoir outer boundary.

For a fixed Poisson ratio, the magnitude of rock elastic modulus represents the stiffness of the rock skeleton, which plays a critical role in the determination of the pressure profiles. This effect is illustrated in Figure 7-21, where the

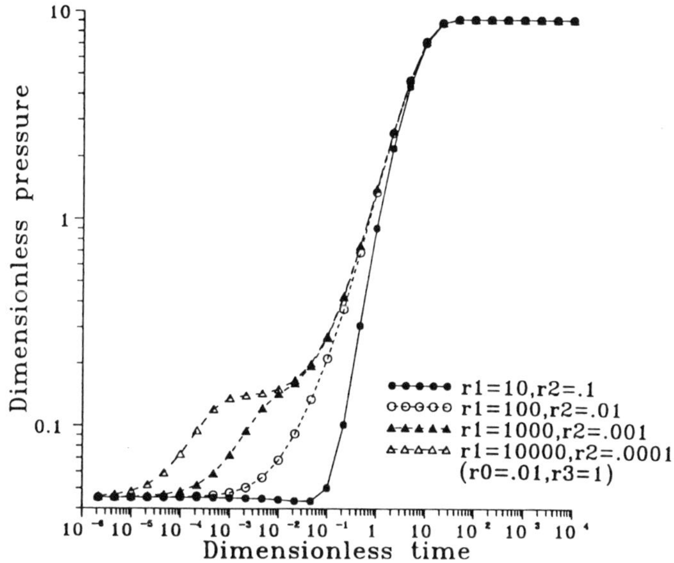

Figure 7-20. Temporal pressure vs. fracture and grain compressibility ratios. (Reprinted from Bai et al. 1995a, with permission from Elsevier Science.)

dual-porosity behavior is apparent only when the rock mass is relatively stiff, e.g., $r_3 = 0.01$. The poromechanical effect is obvious when r_3 falls in the range between 0.1 and 1,000, where it is important to note that the depleting pressure magnitude at early periods appears greater than the reservoir pressure, as depicted in Figure 7-21. This poromechanical impact cannot be observed if a conventional flow model is used alone. Both dual-porosity phenomena and the poromechanical influence diminish when the matrix compressibility ratio (r_3) is equal to or larger than 1,000. Because large r_3 represents relatively soft rock masses, the dual-porosity behavior is least observable for soft media.

Figure 7-22 illustrates the radial pressure, displacement, stress, and strain profiles, assuming $r_0 = 0.1$, $r_1 = r_2 = r_3 = 1$, and at a dimensionless time (t_D) equal to 10. Unlike the uniform pressure decline from the well, displacement, stress, and strain decline dramatically within a dimensionless distance of 600 from the well, signifying a significant local deformation effect as a result of hydrocarbon production. This implies that the impact of poroelastic behavior of the fractured porous media should not be neglected within this region.

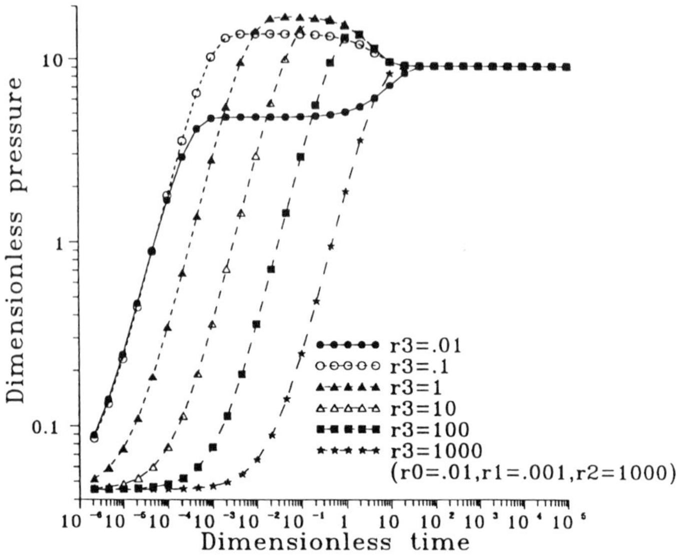

Figure 7-21. Temporal pressure vs. matrix compressibility ratio. (Reprinted from Bai et al. 1995a, with permission from Elsevier Science.)

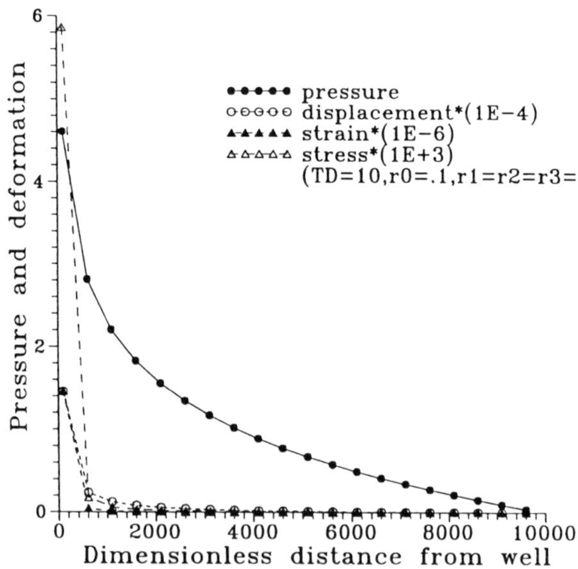

Figure 7-22. Pressure and deformation profiles with radial distance. (Reprinted from Bai et al. 1995a, with permission from Elsevier Science.)

7.5.3 Concluding remarks

Analytical results have been presented to evaluate both flow and poromechanical effects on the fluid pressure response, close to the well, as a result of pumping in a fractured-porous rock mass. In comparison with conventional approaches to dual-porosity behavior, the present model provides a more reasonable description of the transition with regard to the storage interaction between matrix blocks and fractures. The parametric study reveals the critical factors in generating the dual-porosity or poroelastic responses, especially in near-well regions, where the modifications of reservoir initial flow and deformation conditions are dramatic. Change in the flow system in the vicinity of a well-test or a production well may therefore be misinterpreted without considering these important dual-porosity or poromechanical factors.

For a constant fluid compressibility, the dual-porosity behavior of the fractured-porous rock mass is identified in the early transient stage when fracture drainage leads to significant compression of the fractures. Dual-porosity behavior may not be observable if the rock skeleton is relatively soft, even if a significant disparity exists between fracture and grain compressibilities. However, the poromechanical influence is noticeable for soft rock masses, where the pressure magnitude at early time appears greater than the average reservoir pressure, a phenomenon not attainable using traditional flow models. Dual-porosity behavior is most obvious in reservoirs comprising a stiff rock mass with relatively small fracture compressibilities and large grain compressibilities.

7.6 CONTAMINANT TRANSPORT

The analyses of contaminant migration through fractured porous media, subjected to nonisothermal conditions, is presented in this environmental engineering application.

7.6.1 Matrix Diffusion and Matrix Replenishment

Contaminant transport through fractured rock masses has become a very important issue due to the recognition that the fate of a contaminant in a fractured medium can be significantly different from that in a homogeneous porous medium. Serious errors in prediction may be made if this difference is neglected. In comparison with a fracture-dominated medium or a homogeneous porous medium, pollutant migration in a fractured porous medium may be first accelerated by migration along fractures, and later attenuated due to molecular diffusion in the matrix blocks. This, consequently, results in a rate reduction in the concentration change at each tracer sampling location. This scenario is valid only if the concentration in the fractures is greater than in the matrix blocks. If this condition is reversed, contaminants tend to diffuse back into the fractures, which may cause additional pollutant spreading in the primary flow channels. As depicted in Figure 7-23, where c_{ma} and c_{fr} are the concentrations in secondary (matrix blocks) and primary (fractures) flow pathways, respectively, the contaminant is primarily transported through fractures due to their significant conductivity if $c_{ma} = c_{fr}$. However, the rate of transport in the fractures begins to increase if the concentration in the fractures is less than that in the matrix blocks, i.e., $c_{fr} < c_{ma}$. This phenomenon may be called "matrix replenishment." Conversely, the rate of transport in the fractures starts to decrease if the concentration in the matrix blocks is less than that in the fractures, i.e. $c_{fr} > c_{ma}$. This response is usually called "matrix diffusion."

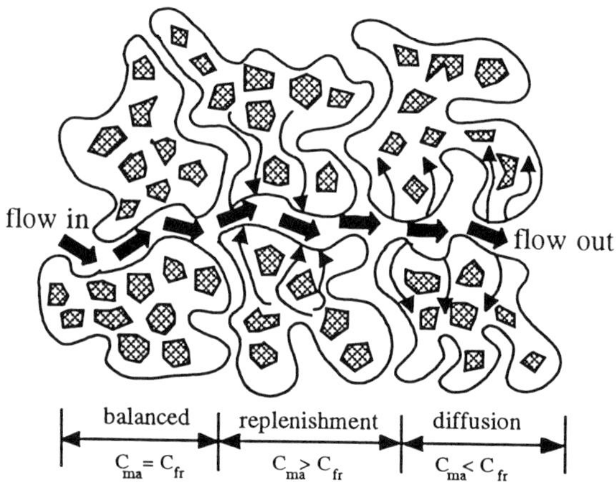

Figure 7-23. Contaminant transport in dual-porosity media.

Figure 7-23 depicts the spatial variation of contaminant concentration in a local region. At a specific location, the temporal variation of the contaminant concentration may exhibit similar behavior, as shown schematically in Figure 7-24. The variable concentration changes in time actually replicate the abnormal breakthrough curves frequently observed in practice where contaminant transport through fractured porous media is examined, as evidenced by Figure 7-25 that depicts the experimental breakthrough curves for tritiated water subjected to different flow rates (Neretnieks 1993).

Similar to fluid low, the transport of contaminants (or components of the fluid phase) in fractured rock masses may occur in both the porous blocks

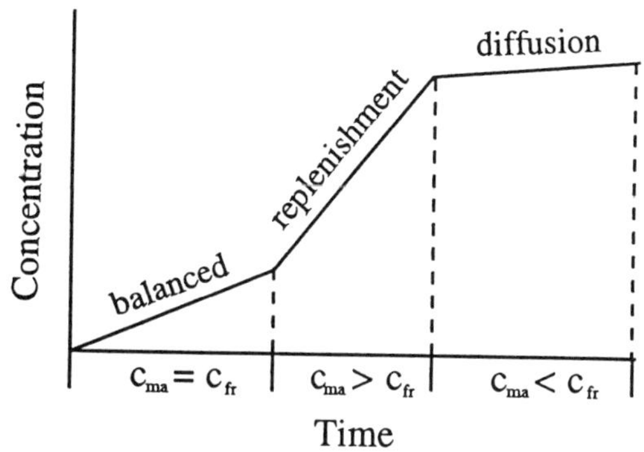

Figure 7-24. Schematic of solute breakthrough.

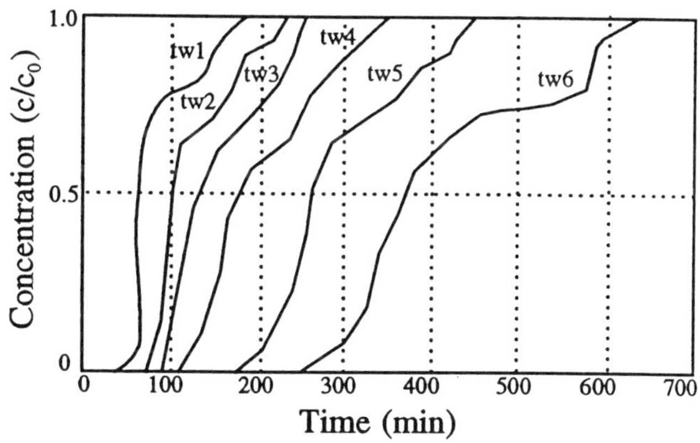

Figure 7-25. Experimental breakthrough curves (after Neretnieks 1993).

and the fracture network. In contrast to solute transport (transport of the mass of a component of a fluid phase), fluid flow can also be envisioned as the transport of mass of a fluid phase (Bear 1993). However, convective mass transport (or bulk flow) due to the effect of change in the velocity field is usually not included in the diffusion-dominated fluid flow, but must be considered in the solute transport. With reference to flow characterization, rock masses generally can be categorized into three different classes: intergranular, fracture dominated, and fracture-matrix mixtures, with the third one being the most complex because of the fracture-matrix interaction. Where the matrix of the rock is porous, the contaminant (e.g., an aqueous component) may move in and out of the porous space, by diffusion. As the groundwater in the matrix pores is primarily stagnant, the dissolved species may thus migrate at a significantly lower speed than the rate of the mobile groundwater (local or regional) in fractures, in addition to being retarded due to its interaction with the solid. The transport process can be further complicated by the presence of existing fractures that often have preferential flow channels for the ground water movement.

Because of significant differences in the characteristics of flow and transport between fractures and rock matrix, the solute transport predicted by the present model is substantially different from the model using a single-porosity approach. The formulations representing transient heat flow and contaminant transport along with steady fluid flow in fractured rock masses are introduced. The solutions are applicable to the analysis of the effect of thermal sweeping used in contaminant site remediation, or modeling contaminant transport under nonisothermal conditions.

7.6.2 Brief Formulation

In solute transport applications, absorption occurs mainly as an imbibition process, mainly in the matrix as characterized by low flow velocity. This retardation process is most significant in the matrix pore space and is generally negligible in the fractures. When the temperature of the saturating fluid phases in a porous medium is not uniform, additional flow may be induced by buoyancy effects. These convective flows depend on density differences due to temperature gradients. An accompanying convective flow is associated with temperature differences due to density gradients. Neglecting chemical reactions, thermal dispersion, as well as the heat transfer by radiation, while incorporating the solute changes due to the coupling of convection as a result of fluid and thermal gradients (Bear 1972), the governing equations for contaminant transport in dual-porosity media formed by rectangular parallelepipeds with an orthogonal fracture network (Warren and Root 1963) can be given as

(Bai and Roegiers 1995)

$$\frac{\partial}{\partial x_i}\left(n_{ma}D_{maij}\frac{\partial c_{ma}}{\partial x_j}\right) - v_i^f\frac{\partial c_{ma}}{\partial x_i} - v_i^h L_h \frac{\partial c_{ma}}{\partial x_i} =$$

$$n_{ma}R_r\frac{\partial c_{ma}}{\partial t} + n_{ma}R_r\lambda_d c_{ma} + \Gamma_c(c_{ma} - c_{fr}) \tag{7.15}$$

$$\frac{\partial}{\partial x_i}\left(n_{fr}D_{fr_{ij}}\frac{\partial c_{fr}}{\partial x_j}\right) - v_i^f\frac{\partial c_{fr}}{\partial x_i} - v_i^h L_h\frac{\partial c_{fr}}{\partial x_i} = n_{fr}\frac{\partial c_{fr}}{\partial t} - \Gamma_c(c_{ma} - c_{fr}) \tag{7.16}$$

where subscripts ma and fr represent matrix blocks and fractures, respectively; c is the concentration of the contaminant; D_{ij} is the hydrodynamic dispersion tensor; v_i^f and v_i^h are the fluid flow velocity (primarily from flow in fractures) and the heat flow velocity (an average quantity from the flow in fractures and in matrix blocks), respectively; L_h is a compatibility factor in terms of fluid and heat fluxes similar to Soret's coefficient (De Groot 1963); R_r is the retardation factor of a solute due to adsorption/desorption; λ_d is a first-order decay coefficient; Γ_c is a concentration exchange coefficient characterizing pollutant exchange between fractures and matrix blocks, which is analogous to the parallelepiped model proposed by Warren and Root (1963); and $\Gamma_c = 60n_{ma}D_{ma}/(s^*)^2$ where s^* is the fracture spacing.

Assuming that the fluid within the matrix blocks is immobile (Coats and Smith 1964), diffusive transport within the pore spaces is less significant due the minimal concentration gradient. As a result, diffusion occurs mainly within the flow path between fractures and matrix blocks. In addition, the thermal flux in the matrix blocks is minimal while considering the fluid as a primary heat carrier. Therefore, Eq. (7.15) can be simplified as a process of quasi-steady matrix transport:

$$-n_{ma}R_r\frac{\partial c_{ma}}{\partial t} = n_{ma}R_r\lambda_d c_{ma} + \Gamma_c(c_{ma} - c_{fr}). \tag{7.17}$$

To recover volatile contaminants in the subsurface, Figure 7-26 depicts a scenario where hot water is injected through a centrally located well. Flow and transport processes are assumed to occur primarily in the radial direction.

Because molecular diffusion becomes apparent only at low flow velocities, which is not the case for fractures, Eq. (7.16), representing transport in fractures, reduces to plane radial flow as

$$\frac{n_{fr}}{r}\frac{\partial}{\partial r}\left(rD_{frr}\frac{\partial c_{fr}}{\partial r}\right) - v_r^f\frac{\partial c_{fr}}{\partial r} - v_r^h L_h\frac{\partial c_{fr}}{\partial r} = n_{fr}\frac{\partial c_{fr}}{\partial t} - \Gamma_c(c_{ma} - c_{fr}) \tag{7.18}$$

where r is the radial distance from the well and D_{frr} is the radial dispersion coefficient. The isotropic fluid velocity can be expressed as

$$v^f = \frac{q_s}{(2\pi h_a r)} \tag{7.19}$$

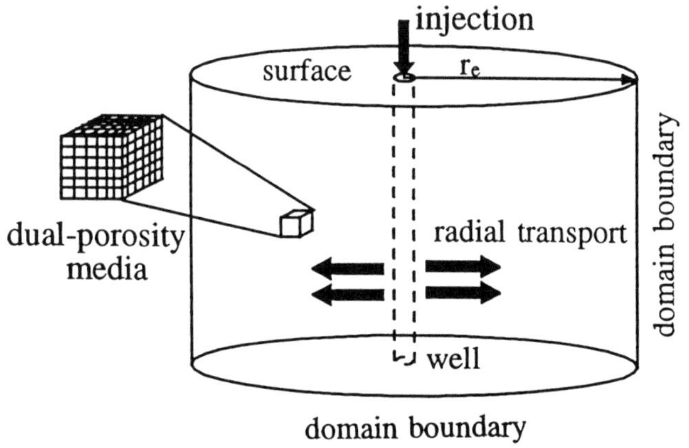

Figure 7-26. Radial transport in fractured porous media.

where q_s is the specific flow rate, h_a is the domain thickness, and the dispersion coefficient can be written as

$$D_{frr} = \alpha_r v^f \tag{7.20}$$

where α_r is the radial dispersivity.

For a steady flow rate (q_s), consider the relations in Eqs. (7.19) and (7.20), the product rD_{frr} becomes a constant, i.e.,

$$rD_{frr} = \frac{q_s \alpha_r}{2\pi h_a}. \tag{7.21}$$

As a result, Eq. (7.18) can be further simplified as

$$n_{fr}\alpha_r v^f \frac{\partial^2 c_{fr}}{\partial r^2} - v^f \frac{\partial c_{fr}}{\partial r} - v^h L_h \frac{\partial c_{fr}}{\partial r} = n_{fr}\frac{\partial c_{fr}}{\partial t} - \Gamma_c(c_{ma} - c_{fr}) \tag{7.22}$$

where the transient heat flow velocity v^h can be determined by applying Fourier's Law of heat conduction, i.e.,

$$v^h = \frac{q_h}{2\pi h_a r} \tag{7.23}$$

where

$$q_h = -K_h^* \frac{\partial T}{\partial r} \tag{7.24}$$

where T is the temperature and K_h^* is the thermal conductivity. The detailed solution of the governing equations (7.17) and (7.22) can be found in Bai and Roegiers (1995).

7.6.3 Simulation

This model can be used in the assessment of the effects of remediation technology applied to nuclear and solid waste disposal sites. The remediation of contaminated aquifers by gas or thermal injection and extraction has shown great promise. Low-level nuclear waste buried in shallow tanks has been found leaking downward toward an underlying aquifer. To clean up sites contaminated by radioactive substances, Hanford and Savannah River (disposal sites of Department of Energy) have developed methodologies using horizontal wells coupled with chemical sweeping (Volk 1994). Two horizontal wells are drilled with the first one used to inject air, methane, or steam into the contaminated zone, while the second one extracts the air or hot liquid containing volatile compounds. When repeated, the process eventually dilutes the radioactive substance below hazardous levels. However, the sweep method using horizontal wells can only be cost-effective if the groundwater flow and fracture development are dominant in the upward direction. For sedimentary rocks where joints and groundwater flow are predominantly horizontal, using vertical wells to inject volatile steam as a means to sweep the contaminated zone appears to be a suitable alternative. In view of modeling, a more complicated transport system may emerge for low permeability matrix rocks such as fractured clay, chalk, or cavernous limestone where dual-porosity phenomena become prevalent. Under this circumstance, traditional methods, using advective-dispersive systems to evaluate a homogeneous medium, may not be applicable, especially when the process is coupled with heat flow. Simplifications in quantifying a real scenario are necessary as a result of the complexity in treating a heterogeneous system.

The goal of using thermal sweeping via a centrally located vertical well (shown in Figure 7-26) is to remediate a contaminated subsurface area where porous rock masses are naturally fractured. The selected modeling parameters are listed in Table 7.5.

In view of the dimensionless concentration versus the distance from the well, Figure 7-27 demonstrates that an extensive concentration change along the radial distance can be found when subjected to various steady-state thermal fluxes. More intensive concentration changes are observed for higher injected thermal fluxes over greater distances. In addition, the profiles of spatial concentration are not similar to the ones anticipated by the single-porosity models. The slope changes in the relative concentration along the radial distance appear to be the result of solute exchange between matrix blocks and fractures.

Figure 7-28 depicts the spatial relative concentration resulting from various exchange coefficients (Γ_c) for more dominant convection in the fractures. The rapid decline of concentration along the radial distance, which deviates from the conventional convection-dominant transport, may be attributed to the aggressive solute interchange between the matrix blocks and the fractures, which occurs when the storage is reduced in the fractures. As mentioned previously, the change in slope of the concentration profile appears to be the result of solute exchange between the matrix and the fractures, where solute replenishment from the matrix to the fractures is enhanced for large exchange coefficients. On the other hand, smaller values of Γ_c represents lower permeability matrix blocks, and thus more dominant solute transport in the fractures. The phenomenon of a conventional convection-dominated transport process is revealed for the case with the smallest Γ_c shown in Figure 7-28.

Table 7.5 Basic modeling parameters.

Parameter	Values	Unit
Porosity (n_{ma}, n_{fr})	0.3, 0.1	-
Retardation factor (R_r)	1.0	-
Decay coefficient (λ_d)	1.0	1/day
Fluid density (ρ_f)	950	kg/m^3
Specific heat (c_h^*)	5000	J/kg ^{o}C
Thermal conductivity (K_h^*)	10000	J/(m day ^{o}C)
Reference temperature (T_0)	250	^{o}C
Well radius (r_w)	0.1	m
Elastic modulus (E)	100	GPa
Poisson ratio (ν)	0.25	-
Thermal expansion coefficient (α_h^*)	10^{-4}	-
Factor (L_h)	$10^{-22} - 1$	m t^2/kg
Domain radius (L_r)	100	m
Domain thickness (h_a)	100	m
Day at calculation (t)	100	day
Distance at calculation (R_c)	10	m
Fluid velocity (v^f)	1 - 50	m/day
Fluid conductivity (K_h)	10^{-5}	m/day

The profiles of spatial concentration can also be significantly affected by the variation in groundwater flow velocities, as depicted in Figure 7-29. In general, the effects of dispersion in the fractures and matrix diffusion (or matrix replenishment) are reduced for the case subjected to larger flow velocities that represent convection-dominated transport in the fractures. Conversely, for transport with less prominent flow velocity, the spatial concentration curve shows diffusive characteristics due to the dominance of dispersion in the fractures, and a variable slope change in the relative concentration as a result

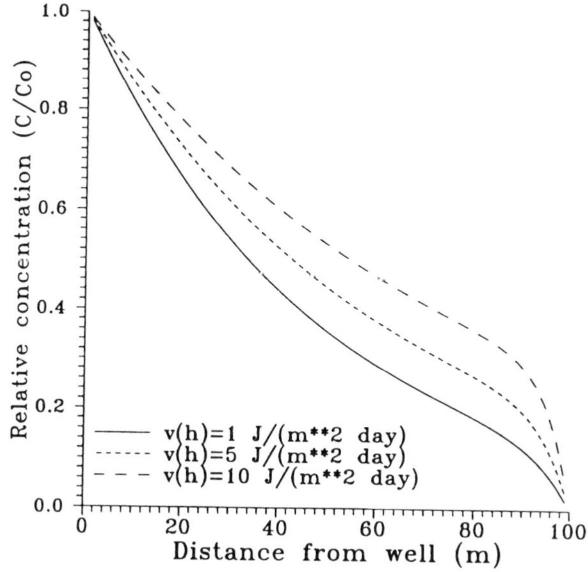

Figure 7-27. Concentration vs. various thermal fluxes.

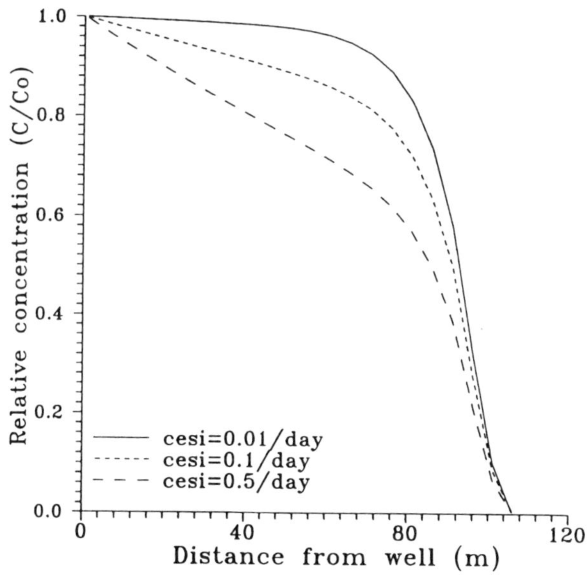

Figure 7-28. Concentration vs. various interchange coefficients.

of the solute interaction between the matrix blocks and the fractures. This general trend is illustrated in Figure 7-29.

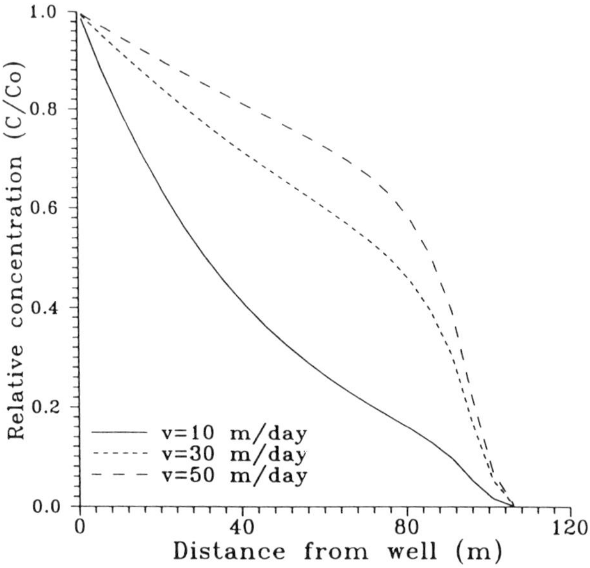

Figure 7-29. Concentration vs. various flow velocities.

The solute breakthrough represents the temporal contaminant occurrence at a specific location from the source. Figure 7-30 indicates the breakthrough curves for various fracture spacings. Larger spacings, which represent relatively more intact rock masses, result in rapid solute breakthrough. However, extensive tailing in the breakthrough curves for the smaller fracture spacings, which represent a more pervasive fracture network, are the result of interporosity transport between the matrix blocks and the fractures. In view of the fracture spacings, the significant difference may be noted for the results when the spacing drops from 10 m to 1 m.

7.6.4 Concluding Remarks

The concept of matrix replenishment, which describes solute diffusion from matrix to fractures, has been proposed. This proposal is important because the existence of the matrix blocks not only yields a temporary reservoir within which the pollutant in the fractures may reside (matrix diffusion, causing contaminant attenuation), but also supplies the pollutant to the fractures from the matrix (matrix replenishment, causing additional contaminant spreading).

Modeling results indicate that the effect of heat flow may be as equally critical as that of fluid flow in view of regulating contaminant migration. The

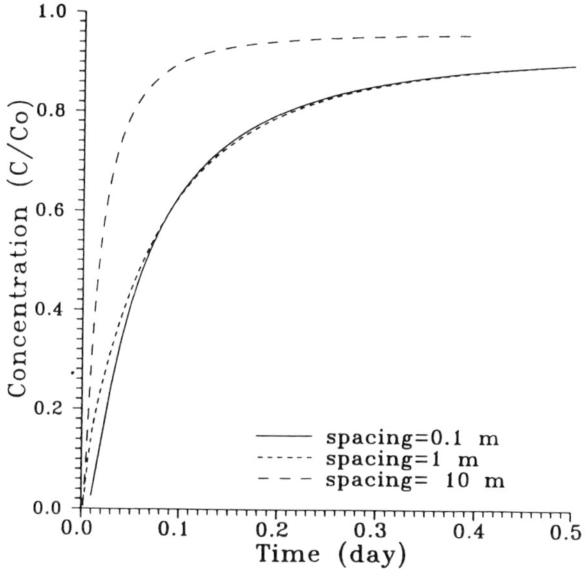

Figure 7-30. Breakthrough for various fracture spacings.

important influences of the thermal effects are identified, and become most apparent near the pollutant and heat sources. In view of the coupling of heat flow in contaminant transport, the proposed model may provide a sensible prediction of the temporal concentration front in fractured porous media when the technique of thermal sweeping is applied. The method appears to be effective in expelling the contaminant to a more remote area, particularly when a constant thermal flux is applied.

REFERENCES

Abdassah, D. and I. Ershaghi. 1986. "Triple-porosity system for representing naturally fractured reservoirs." *Formation Evaluation.* April: 113-127.

Aberg, B. 1992. "Void ratio of noncohesive soils and similar materials." *J. Geotech. Eng.*, ASCE, 118: 1315-1334.

Abousleiman, Y., M. Bai, and J.-C. Roegiers. 1995. "Analysis of tunnel stability in fractured carbonate rocks." *Proc. 3rd Int. Conf. on Computer Methods and Water Resour.*, Beirut, Lebanon: 347-354.

Abousleiman, Y., M. Bai, and J.-C. Roegiers. 1996. "Thermoporoelastic coupling in hydraulic fracturing." *Proc. ISRM Int. Symp. Pred. Perform. Rock Mech. Rock Eng., EUROCK '96*, Torino, Italy: 1387-1394.

Abousleiman, Y., A. H.-D. Cheng, C. Jiang and J.-C. Roegiers. 1993. "A micromechanically consistent poroviscoelasticity theory for rock mechanics applications." *Int. Rock Mech. and Min. Sci. and Geomech. Abstr.* 30: 1177-1180.

Acheson, D. J. 1990. *Elementary Fluid Dynamics.* Oxford, UK: Clarendon Press.

Aifantis, E. C. 1977. "Introducing a multi-porous medium." *Developments in Mechanics* 37: 265-296.

Aifantis, E. C. 1980. "On the problem of diffusion in solids." *Acta Mechanica* 37: 265-296.

Akin, J. E. 1982. *Application and Implementation of Finite Element Methods.* London: Academic Press.

Al-Bemani, A. S. and I. Ershaghi. 1991. "Two-phase flow interporosity effects on pressure transient test response in naturally fractured reservoirs." SPE 22718, *66th Annual Tech. Conf. and Exhib.*, Dallas, TX.

Aly, A., H. Y. Chen and W. J. Lee. 1996. "Development of a mathematical model for multilayer reservoirs with unequal initial pressures." *In Situ* 20: 61-92.

Amadei, B. and E. Pan. 1995. "Estimating in-situ stresses for the selection of the alignment of unlined pressure tunnels." *Proc. 35th U.S. Symp. on Rock Mech.*, Lake Tahoe, NV: 267-272.

Andersson, J. and B. Dverstorp. 1987. "Conditional simulations of fluid flow in three-dimensional networks of discrete fractures." *Water Resour. Res.* 23: 1876-1886.

Atkin, R. J. and R. E. Craine. 1976. "Continuum theories of mixtures: basic theory and historical development." *Q. J. Mech. Appl. Math.* 29: 209-244.
Atkin, R. J. and N. Fox. 1980. *An Introduction to the Theory of Elasticity.* London: Longman.
Atkinson, B. K. ed. 1987. *Fracture Mechanics of Rock.* New York: Academic Press.
Auguilera, R. 1980. *Naturally Fractured Reservoirs.* Petroleum Publishing Company, Tulsa, OK.
Bai, M. 1997. "An efficient algorithm for evaluating coupled processes in radial fluid flow." *J. Computers and Geosciences* 23: 195-202.
Bai, M. 1999. "On equivalence of dual-porosity poroelastic parameters." *J. Geophy. Res.* 104(B5): 10461-10466.
Bai, M. and Y. Abousleiman. 1997. "Thermoporoelastic coupling with application to consolidation." *Int. J. Num. Analy. Meth. Geomech.* 21: 121-132.
Bai, M., A. Bouhroum and D. Elsworth. 1994d. " Some aspects of solving advection dominated flows." Kohle-Erdgas-Petrochemie vereinigt mit Brennstoff-Chemie, *J. Hydrocarbon Tech.* 47: 11-17.
Bai, M. and D. Elsworth. 1993. "Dual porosity poroelastic approach to behavior of porous media over a mining panel." *Trans. Inst. Min. Metall.* 102: A114-124.
Bai, M., and D. Elsworth. 1994. "Modeling of subsidence and stress-dependent hydraulic conductivity of intact and fractured porous media." *Rock Mech. and Rock Eng.* 27: 209-234.
Bai, M. and D. Elsworth. 1995. "On the modeling of miscible flow in multi-component porous media." *Transport in Porous Media* 21: 19-46.
Bai, M., D. Elsworth, H. I. Inyang and J.-C. Roegiers. 1997a. "A semi-analytical solution for contaminant migration with linear sorption in strongly heterogeneous media." *J. of Environ. Eng.*, ASCE 123: 1116-1125.
Bai, M., D. Elsworth and J.-C. Roegiers. 1993. "Multi-porosity/multi-permeability approach to the simulation of naturally fractured reservoirs." *Water Resour. Res.* 29: 1621-1633.
Bai, M., D. Elsworth, J.-C. Roegiers and F. Meng. 1995b. "A three-dimensional dual-porosity poroelastic model." *Proc. of CCMRI - International Mining Tech'95*, Beijing, China: 184-202.
Bai, M., H. I. Inyang, F. Meng, Y. Abousleiman and J.-C. Roegiers. 1998a. "Statistical modeling of particle migration." *Proc. of 4th International Symposium on Environmental Geotechnology and Global Sustainable Development*, Danvers, MA.
Bai, M., A. Khan, F. Meng and J.-C. Roegiers. 1995c. "On the thermoporoelastic properties of fractured reservoir rocks." *Proc. 35th U.S. Symp. on Rock Mech.*, Lake Tahoe, NV: 769-774.
Bai, M., Q. Ma and J.-C. Roegiers. 1994a. "A nonlinear dual-porosity model." *Appl. Math. Modelling* 18: 602-610.
Bai, M., Q. Ma and J.-C. Roegiers. 1994b. "Dual-porosity behavior of naturally

fractured reservoirs." *Int. J. Num. Analy. Meth. Geomech.* 18: 359-376.

Bai, M. and F. Meng. 1997. "Analysis of stress-dependent permeability: homogeneous anisotropic porous media." Report RMC-97-01, University of Oklahoma.

Bai, M., F. Meng, Y. Abousleiman and J.-C. Roegiers. 1998b. "Modeling multiphase fluid flow and rock deformation in fractured porous media." *Proc. Biot Conference on Poromechanics*, Louvain-La-Neuve, Belgium: 333-338.

Bai, M., F. Meng, D. Elsworth, Y. Abousleiman and J.-C. Roegiers. 1999b. "Numerical modeling of flow and deformation in fractured rock specimens." *Int. J. Num. Analy. Meth. Geomech.* 23: 141-160.

Bai, M., F. Meng, D. Elsworth, M. Zaman and J.-C. Roegiers. 1997b. "Numerical modeling of stress-dependent permeability." *Int. J. Rock Mech. Min. Sci.* 34(3/4): 446.

Bai, M. and J.-C. Roegiers. 1994a. "On the correlation of nonlinear flow and linear transport with application to dual-porosity modeling." *J. Petrol. Sci. and Eng.* 11: 63-72.

Bai, M. and J.-C. Roegiers. 1994b. "Fluid flow and heat flow in deformable fractured porous media." *Int. J. Eng. Sci.* 32(10): 1615-1633.

Bai, M. and J.-C. Roegiers. 1995. "Modeling of heat flow and solute transport in fractured rock masses." *Proc. 8th Int. Congress on Rock Mech.*, Tokyo, Japan: 787-793.

Bai, M. and J.-C. Roegiers. 1997. "Triple-porosity analysis of solute transport." *J. of Contaminant Hydrology* 28: 247-266.

Bai, M., J.-C. Roegiers and D. Elsworth. 1994c. "An alternative model of fractured reservoir simulation." *Proc. 8th Int. Conf. Comp. Adv. Geomech.*, Morgantown, WV: 2031-2036.

Bai, M., J.-C. Roegiers and D. Elsworth. 1995a. "Poromechanical response of fractured-porous rock masses." *J. Petrol. Sci. Eng.* 13: 155-168.

Bai, M., J.-C. Roegiers and H. I. Inyang. 1996. "Contaminant transport in nonisothermal fractured porous media." *J. Environ. Eng.*, ASCE 122(5): 416-423.

Bai, M., Z. Shu, J. Cao, M. Zaman, J.-C. and Roegiers. 1999a. "A semi-analytical solution for a two-dimensional capacitance model in solute transport." *J. Petrol. Sci. Eng.* 22: 275-295.

Baker, L. E. 1977. "Effects of dispersion and dead-end pore volume in miscible flooding." *Soc. Petrol. Eng. J.* 17: 219-227.

Bandis, S. C. and N. R. Barton. 1983. "Fundamentals of rock joint deformation." *Int. J. Rock Mech. Min. Sci. and Geomech. Abstr.* 20: 249-268.

Banerjee, P. K. 1976. "Integral equation methods for analysis of piece-wise non-homogeneous three-dimensional elastic solids of arbitrary shape." *Int. J. Mech. Sci.* 18: 293-303.

Banerjee, P. K. and R. Butterfield. 1981. *Boundary Element Methods in Engineering Science.* New York: McGraw-Hill.

Barenblatt, G. E., V. M. Entov and V. M. Ryzhik. 1990. *Theory of Fluid Flows*

Through Natural Rocks. Dordrecht: Kluwer Academic Pub.

Barenblatt, G. E., I. P. Zheltov and I. N. Kochina. 1960. "Basic concept in the theory of homogeneous liquids in fractured rocks." *J. Appl. Math. Mech.* 24(5): 1286-1303.

Barton, N. S. 1976. "Rock mechanics review - the shear strength of rock and rock joints." *Int. J. Rock Mech. Min. Sci. and Geomech. Abstr.* 13: 255-279.

Barton, N. S., S. C. Bandis and K. Bakhtar. 1985. "Strength, deformation and conductivity coupling of rock joints." *Int. J. Rock Mech. Min. Sci. and Geomech. Abstr.* 22: 121-140.

Batchelor, G. K. 1967. *An Introduction to Fluid Dynamics.* Cambridge, UK: Cambridge University Press.

Bear, J. 1972. *Dynamics of Fluids in Porous Media.* New York: American Elsevier.

Bear, J. 1993. "Modeling flow and contaminant transport in fractured rock." In *Flow and Contaminant Transport in Fractured Rock.* Edited by J. Bear, C.-F. Tsang and G. De Marsily. San Diego, CA: Academic Press, Inc.: 1-37.

Bear, J. and Y. Bachmat. 1990. *Introduction to Modeling of Transport Phenomena in Porous Media.* Dordrecht: Kluwer Academic Pub.

Bear, J. and M. Y. Corapcioglu, eds. 1987. *Advances in Transport Phenomena in Porous Media.* Dordrecht: Martinus Nijhoff Pub.

Bear, J., C.-F. Tsang,and G. de Marsily. eds. 1993. *Flow and Contaminant Transport in Fractured Rock.* San Diego, CA: Academic Press.

Bear, J. and A. Verruijt. 1987. *Groundwater Flow and Pollution.* Dordrecht: D. Reidel Publishing Co.

Berryman, J. G., and H. F. Wang. 1995. "The elastic coefficients of double-porosity models for fluid transport in jointed rock." *J. Geophy. Res.* 100: 24611-24627.

Beskos, D. E. and E. C. Aifantis. 1986. "On the theory of consolidation with double porosity - 2." *Int. J. Eng. Sci.* 24: 1697-1716.

Bibby R. 1981. "Mass transport of solutes in dual-porosity media." *Water Resour. Res.* 17: 1075-1081.

Biot, M. A. 1941. "General theory of three-dimensional consolidation." *J. Appl. Phys.* 12: 155-164.

Biot, M. A. 1955. "Theory of elasticity and consolidation for a porous anisotropic solid." *J. Appl. Phys.* 26: 182-185.

Biot, M. A. 1956. "Thermoelasticity and irreversible thermodynamics." *J. Appl. Phys.* 27: 240-253.

Biot, M. A. and D. G. Willis. 1957. "The elastic coefficients of the theory of consolidation." *J. Appl. Mech.* 24: 594-601.

Bishop, A. W. 1955. "The principle of effective stress." *Tekniske Ukeblad* 24: 594-601.

Bishop, A. W. 1973. "The influence of an undrained change in stress on the pore pressure in porous media of low compressibility." *Géotechique* 23: 435-442.

Bishop, A. W. and G. E. Blight. 1963. "Some aspects of effective stress in saturated and partially saturated soils." *Geotéchnique* 13: 177-197.

Bouhroum, A. 1993. "Is tailing of breakthrough curves diffusive? theoretical and experimental evaluation." Paper 13, *5th Saskatchewan Petroleum Conf.*, Regina, Canada.

Bouhroum, A. and M. Bai. 1996. "Experimental and numerical simulation of solute transport in heterogeneous porous media." *Int. J. Num. Analy. Meth. Geomech.* 20: 155-171.

Bouhroum, A. and F. Civan. 1994. "A study of particulate migration in gravel-pack." SPE 27346, *Int. Symp. on Formation Damage Control*, Lafayette, LA: 75-91.

Bourdet, D., J. A. Ayoub and Y. M. Pirard. 1984. "Use of pressure derivative in well test interpretation." SPE 12777, *Cal. Regional Meeting*: 1-17.

Brace, W. F. and D. Byerlee. 1965. "Recent experimental studies of brittle fracture of rocks." Cambridge, MA: MIT.

Brace, W. F. and R. J. Martin, III. 1968. "A test of the law of effective stress for crystalline rocks of low porosity." *Int. J. rock Mech. Min. Sci.* 5: 415-426.

Braester, C. 1984. "Influence of block size on the transition curve for a draw-down test in a naturally fractured reservoir." *Soc. Petrol. Eng. J.* June: 498-504.

Brandt, H. 1955. "A study of the speed of sound in porous granular media." *J. Appl. Mech.* 22: 479-486.

Brebbia, C. A. 1980. *The Boundary Element Method for Engineers.* London: Pentech Press.

Breitenbach, E. A. 1991. "Reservoir simulation: state of the art." *J. Petrol. Tech.* September: 1033-1036.

Butterworth, D. and G. F. Hewitt, eds. 1977. *Two-Phase Flow and Heat Transfer.* Oxford, UK: Oxford University Press.

Carroll, M. M. 1979. "An effective stress law for anisotropic elastic deformation." *J. Geophys. Res.* 84: 7710-7512.

Carslaw, H. S. and J. C. Jaeger. 1959. *Conduction of Heat in Solids.* 2nd Ed. Oxford, UK: Clarendon Press.

Castillo, E., G. M. Karadi and R. J. Krizek. 1972. "Unconfined flow through jointed rock." *Water Resour. Bull.* 8: 226-229.

Charlez, Ph. A. 1991. *Rock Mechanics, vol.1. Theoretical Fundamentals.* Paris: Editions Technip.

Chen, Z.-X. 1989. "Transient flow of slightly compressible fluids through double-porosity, double-permeability systems - a state of the art review." *Transport in Porous Media* 4: 147-184.

Chen, M., M. Bai and J.-C. Roegiers. 1999. "Permeability tensors of anisotropic fracture networks." *Mathematical Geology* 31: 355-373.

Chen, C. C., K. Serra, A. C. Reynolds and R. Raghavan. 1985. "Pressure transient analysis methods for bounded naturally fractured reservoirs." *Soc.*

Petrol. Eng. J. June: 451-464.

Chilingarian, G. V., S. J. Mazzullo and H. H. Rieke. 1992. *Carbonate Reservoir Characterization: A Geologic-Engineering Analysis, Part I.* Amsterdam: Elsevier Sci.

Christie, I., D. F. Griffiths, A. R. Mitchell, and O. C. Zienkiewicz. 1976. "Finite element methods for second order differential equations with significant first derivatives." *Int. J. Num. Meth. Eng.* 10: 1389-1396.

Cleary, M. P. 1977. "Fundamental solutions for a fluid-saturated porous solid." *Int. J. Solids Struct.* 13: 785-806.

Coats, K. H. and B. D. Smith. 1964. "Dead-end pore volume and dispersion in porous media." *Soc. Petrol. Eng. J.* 4: 73-84.

Cook, N. G. W. 1988. "Natural joints in rock: mechanical, hydraulic and seismic behavior and properties under normal stress." *1st Jaeger Memorial Lecture*, Univ. of Minnesota.

Correa, A. C., K. K. Pande and H. J. Ramey. 1987. "Prediction and interpretation of miscible displacement performance using a transverse matrix dispersion model." SPE 16756, *62nd Annual Tech. Conf. and Exhib.*, Dallas, TX.

Coussy, O. 1995. *Mechanics of Porous Continua.* Chichester: John Wiley and Sons Ltd.

Craft, B. C. and M. F. Hawkins. 1959. *Applied Petroleum Reservoir Engineering.* Englewood Cliff, NJ: Prentice-Hall, Inc.

Crawford, G. E., A. R. Hagedorn and A. E. Pierce. 1976. "Analysis of pressure buildup tests in a naturally fractured reservoir." *J. Petrol. Tech.* November: 1295-1300.

Crochet, M. J. and P. M. Naghdi. 1966. "On constitutive equations for flow of fluid through an elastic solid." *Int. J. Eng. Sci.* 4: 383-401.

Crouch, S. L. and A. M. Starfield. 1983. *Boundary Element Methods in Solid Mechanics.* London: George Allen and Unwin.

Crump, K. S. 1976. "Numerical inversion of Laplace transforms using Fourier series approximation." *J. Assoc. Comput. Mach.* 23: 89-96.

Cryer, C. W. 1963. "A comparison of the three-dimensional consolidation theories of Biot and Terzaghi." *Q. J. Mech. Appl. Mech.* 16: 401-412.

Cui, L., V. N. Kaliakin, Y. Aboulseiman and A. H.-D. Cheng. 1997. "Finite element formulation and application of poroelastic generalized plane strain problems." *Int. J. rock Mech. Min. Sci.* 34: 953-962.

Curran, J. and J. L. Carvalho. 1987. "A displacement discontinuity model for fluid-saturated porous media." *Proc. 6th Int. Congress on Rock Mech.*, Balkema, Montreal, Canada: 73-78.

Cushman, J. H. ed. 1990. *Dynamics of Fluids in Hierarchical Porous Media.* London: Academic Press.

Dagger, M. A. S. 1997. A Fully Coupled Two-Phase Flow and Rock Deformation Model for Reservoir Rock, Ph.D dissertation, University of Oklahoma.

Davies, B. and B. L. Martin. 1979. "Numerical inversion of Laplace transform:

a critical evaluation and review of methods." *J. Comp. Phys.* 33: 1-32.

Deans, H. A. 1963. "A mathematical model for dispersion in the direction of flow in porous media." *Trans.*, AIME 228: 49-52.

De Groot, S. R. 1963. *Thermodynamics of Irreversible Processes.* Amsterdam: North-Holland Pub. Co.

deSwaan-O., A. 1976. "Analytical solutions for determining naturally fractured reservoir properties by well testing." *Trans.*, AIME 261, *Soc. Petrol. Eng. J.* June: 117-22.

Detournay, E. and A. H-D. Cheng. 1988. "Poroelastic response of a borehole in a non-hydrostatic stress field." *Int. J. Rock Mech. Min. Sci. and Geomech. Abstr.* 25: 171-182.

Detournay, E. and A. H-D. Cheng. 1993. "Fundamentals of poroelasticity." In *Comprehensive Rock Engineering.* Edited by J. A. Hudson. Pergamon Press 2: 113-171.

De Wiest, R. J. M. ed. 1969. *Flow Through Porous Media.* New York: Academic Press.

Duguid, J. O. and P. C. Y. Lee. 1977. "Flow in fractured porous media." *Water Resour. Res* 13: 558-566.

Dykhuizen, R. C. 1990. "A new coupling term for dual-porosity models." *Water Resour. Res* 26: 351-356.

Economides, M. J. and K. G. Nolte. 1987. *Reservoir Stimulation.* Houston, TX: Schlumberger Educational Services.

Elliot, G. M., E. T. Brown, P. I. Boodt and J. A. Hudson. 1985. "Hydromechanical behaviour of joints in Carnmenellis granite." *Proc. Int. Symp. on Fundamental Rock Joints*, Björkliden: 249-258.

Elsworth, D. 1986. "A hybrid boundary element - finite element analysis procedure for fluid flow simulation in fractured rock masses." *Int. J. Num. Analy. Meth. Geomech.* 10: 569-584.

Elsworth, D. 1987. "A boundary element - finite element procedure for porous and fractured media flow." *Water Resour. Res.* 23: 551-560.

Elsworth, D. 1989. "Thermal permeability enhancement of blocky rocks: one dimensional flows." *Int. J. Rock Mech. Min. Sci. and Geomech. Abstr.* 26: 329-339.

Elsworth, D. 1993. "Computational methods in fluid flow." In *Comprehensive Rock Engineering.* Edited by J. A. Hudson. Pergamon Press 2: 173-189.

Elsworth, D. and M. Bai. 1992. "Coupled flow-deformation response of dual porosity media." *J. Geotech. Eng.*, ASCE 118: 107-124.

Elsworth, D. and J. Xiang. 1989. "A reduced degree of freedom model for thermal permeability enhancement in blocky rock." *Geothermics* 18: 691-709.

Ershaghi, I. and R. Aflaki. 1985. "Problems in characterization of naturally fractured reservoirs from well test data." *Soc. Petrol. Eng. J.* June: 445-449.

Evans, D. D. and T. J. Nicholson. eds. 1987. *Flow and Transport Through Unsaturated Fractured Rock.* Geophysical Monograph 42. Washington DC:

American Geophysical Union.

Fillunger, P. 1930. "Auftrieb und Unterdruck in Staumauern." *Proc. 2nd Trans. World Power Conf.*, Berlin 9: 323-329.

Firoozabadi, A. and L. K. Thomas. 1990. "Sixth SPE comparative solution project: dual-porosity simulators." *J. Petrol. Tech.* June: 710-763.

Fletcher, C. A. J. 1984. *Computational Galerkin Methods.* New York: Springer-Verlag.

Flügge, W. 1967. *Viscoelasticity.* Waltham, MA: Blaisdell Publication.

Freeze, R. A., and J. A. Cherry. 1979. *Groundwater.* Englewood Cliffs, NJ: Prentice-Hall, Inc.

Gale, J. E. 1982. "The effects of fracture type (induced vs. natural) on the stress-fracture closure-fracture permeability relationship." *23rd U.S. Rock Mech. Symp.*, Univ. of California, Berkeley: 290-298.

Gangi, A. F. 1978. "Variation of whole and fractured porous rock permeability with confining pressure." *Int. J. Rock Mech. Min. Sci. and Geomech. Abstr.* 15: 249-257.

Garg, S. K. and A. Nur. 1973. "Effective stress laws for fluid-saturated porous rocks." *J. Geophys. Res.* 78: 5911-5921.

Gassmann, F. 1951. "Über die Elastizität poröser Medien." *Vierteljahrsschrift der Naturforschenden Gesellschaft in Zürich* 96: 1-23.

Gee, G. W., C. T. Kincaid, R. J. Lenhard and C. S. Simmons. 1991. "Recent studies of flow and transport in the vadose zone." *Reviews of Geophysics*, supplement: 227-239.

Geertsma, J. 1957. "The effect of fluid pressure decline on volumetric changes of porous rocks." *Petrol. Trans.*, AIME: 331-340.

Gerke, H. H. and M. T. van Genuchten. 1993. "A dual-porosity model for simulating the preferential movement of water and solutes in structured porous media." *Water Resour. Res.* 29: 305-319.

Ghaboussi, J. and E. L. Wilson. 1973. "Flow of compressible fluid in porous elastic media." *Int. J. Num. Methods in Eng.* 5: 419-442.

Ghafouri, H. R. and R. W. Lewis. 1996. "A finite element double porosity model for heterogeneous deformable porous media." *Int. J. Num. Anal. Meth. Geomech.* 20: 831-844.

Ghanem, R. and S. Dham. 1998. "Stochastic finite element analysis for multiphase flow in heterogeneous porous media." *Transport in Porous Media* 32: 239-262.

Gilman, J. R. and H. Kazemi. 1983. "Improvements in simulation of naturally fractured reservoirs." *Soc. Petrol. Eng. J.* 23: 695-706.

Gladwell, I. and R. Wait. 1979. *A Survey of Numerical Methods for Partial Differential Equations.* Oxford, UK: Oxford Univ. Press: 195-211.

Goodier, J. N. 1936. "The thermal stresses in a strip." *Phys.* 7: 156.

Goodman, R. E. 1976. *Methods of Geological Engineering in Discontinuous Rocks.* New York: West.

Gottardi, G. and M. Venutelli. 1994. "One-dimensional moving finite-element

model of solute transport." *Ground Water* 32: 645-649.
Green, D. H. and H. F. Wang. 1986. "Fluid pressure response to undrained compression in saturated sedimentary rock." *Geophysics* 51: 948-958.
Green, D. H. and H. F. Wang,. 1990. "Specific storage as a poroelastic coefficient." *Water Resour. Res.* 26: 1631-1637.
Greenkorn, R. A. 1983. *Flow Phenomena in Porous Media.* New York: Marcel Dekker, Inc.
Greenkorn, R. A., C. R. Johnson and R. E. Haring. 1965. "Miscible displacement in a controlled natural system." *Petrol. Trans.*, AIME 234: 1329.
Gringarten, A. C. 1984. "Interpretation of tests in fissured and multilayered reservoirs with double-porosity behavior: theory and practice." *J. Petrol. Tech.* April: 549-564.
Grisak, G. E. and J. A. Cherry. 1975. "Hydrologic characteristics and response of fractured till and clay confining a shallow aquifer." *Canadian Geotechnical Journal* 12: 23-43.
Guymon, G. L. 1970. "A finite element solution of the one-dimensional diffusion-convection equation." *Water Resour. Res.* 6: 204-210.
Gwo, J. P., P. M. Jardine, G. V. Wilson and G. T. Yeh. 1995. "A multiple-pore-region concept to modeling mass transfer in subsurface media." *J. Hydrology* 164: 217-237.
Gwo, J. P., P. M. Jardine, G. V. Wilson and G. T. Yeh. 1996. "Using a multiregion model to study the effects of advective and diffusive mass transfer on local physical nonequilibrium and solute mobility in a structured soil." *Water Resour. Res.* 32: 561-570.
Handin, J. 1958. "Effects of pore pressure on the experimental deformation of some sedimentary rocks." *Bull. Geol. Soc. Am.* 69: 1576-1577.
Handin, J., R. V. Hager, Jr., M. Friedman and J. N. Feather. 1963. "Experimental deformation of sedimentary rocks under confining pressure: pore pressure tests." *Bull. Amer. Ass. Petrol. Geol.* 47: 717-755.
Harrison, B., E. A. Sudicky and J. A. Cherry. 1992. "Numerical analysis of solute migration through fractured clay deposits into underlying aquifers." *Water Resour. Res.* 28: 515-526.
Heinrich, J. C., P. S. Huyakorn and O. C. Zienkiewicz. 1977. "An upwind finite element scheme for two dimensional convective transport equations." *Int. J. Num. Meth. Eng.* 11: 131-143.
Hencher, S. R. 1987. "The implications of joints and structures for slope stability." In *Slope Stability.* Edited by M. G. Anderson and K. S. Richards. New York: John Wiley and Sons: 145-186.
Hertz, H. 1895. *Collected Works.* English translation by D. E. Jones and J. T. Walley, London: Macmillian and Co., Ltd.: 179-195.
Hetsroni, G. 1982. *Handbook of Multiphase Systems.* Washington, DC: Hemisphere.
Hill, J. M. and E. C. Aifantis. 1980. "On the theory of diffusion in media with double diffusivity. 2. boundary value problems." *Q. J. Mech. Appl. Math.*

33: 23-41.

Hoek, E. and J. W. Bray. 1981. *Rock Slope Engineering.* London: Institute of Mining and Metallurgy.

Hoek, E. and E. T. Brown. 1980. *Underground Excavation in Rock.* London: Institute of Mining and Metallurgy.

Hoffman, O. 1928. *Permeazoni d'Acqua e loro Effeti nei Muri di Ritenuta.* Milan: Hoepli.

Houi, D. and R. Lenormand. 1986. "Particle accumulation at the surface of a filter." *Filtration Separation* 23: 238-241.

Houseworth, J. E. 1988. "Characterizing permeability heterogeneity in core samples from standard miscible displacement experiments." SPE 18329, *63rd SPE Annual Tech. Conf. and Exhib.*, Houston, TX.

Hsieh, P. A. and S. P. Neuman. 1985. "Field determination of the three-dimensional hydraulic conductivity tensor of anisotropic media, 1. Theory." *Water Resour. Res.* 21: 1655-1665.

Hsieh, P. A., S. P. Neuman, G. K. Stiles and E. S. Simpson. 1985. "Field determination of the three-dimensional hydraulic conductivity tensor of anisotropic media, 2. methodology and application to fractured rocks." *Water Resour. Res.* 21: 1667-1676.

Hu, S. S. and W. E. Schiesser. 1981. "An adaptive grid method in the numerical method of lines." In *Adv. in Computer Methods for Partial Differential Equations IV.* Edited by R. Vichnevetsky and R. S. Stepleman. Bethlehem, PA: Publ. MACS.

Hubbert M. K. 1940. "The theory of ground water motion." *J. Geol.* 48: 785-944.

Hubbert, M. K. and W. W. Rubey. 1959. "Role of fluid pressure in mechanics of overthrust faulting." *Bull. Geol. Soc. Am.* 70: 115-205.

Hubbert, M. K. and W. W. Rubey. 1960. "Role of fluid pressure in mechanics of overthrust faulting: a reply." *Bull. Geol. Soc. Am.* 71: 617-628.

Huyakorn, P. S., B. H. Lester and C. R. Faust. 1983. "Finite element techniques for modeling groundwater flow in fractured aquifers." *Water Resour. Res.* 19: 1019-1035.

Huyakorn, P. S. and G. Pinder. 1983. *Computational Methods in Subsurface Flow.* New York: Academic Press.

Ichikawa, S. 1982. "Numerical processing of the two-dimensional inverse Laplace transform and its application to equation of heat conduction." *IEE J. Jpn.* IP-82: 231-248.

Imdakm, A. O. and M. Sahimi. 1991. "Computer simulation of particle transport processes in flow through porous media." *Chemical Eng.* 46: 1977-1993.

Istok, J. 1989. *Groundwater Modeling by the Finite Element Method.* Monograph 13. Washington, DC: AGU.

Iwai, K. 1976. Fundamental Studies of Fluid Flow Through a Single Fracture. Ph.D. Dissertation, University of California, Berkeley.

Jaeger, J. C. and N. G. W. Cook. 1979. *Fundamentals of Rock Mechanics*. 3rd Ed. London: Chapman and Hall.

Javandel, I., C. Doughty and C.-F. Tsang. 1984. *Groundwater Transport: Handbook of Mathematical Models*. Monograph 10. Washington, DC: AGU.

Jelmert, T. A. 1993. "Double-porosity response of horizontal wells during linear flow period." *J. Petrol. Sci. Eng.* 9: 125-137.

Johns, R. A. and P. V. Roberts. 1991. "A solute transport model for channelized flow in a fracture." *Water Resour. Res.* 27: 1797-1808.

Jones, F. O. 1975. "A laboratory study of the effects of confining pressure on fracture flow and storage capacity in carbonate rocks." *J. Petrol. Technol.* January: 21-27.

Joy, D. M. and N. Kouwen. 1991. "Particulate transport in a porous medium under non-linear flow conditions." *J. Hydraulic Res.* 29: 373-385.

Joy, D. M., W. C. Lennox and N. Kouwen. 1993. "Stochastic model of particulate transport." *J. Hydraulic Eng.* 119: 846-861.

Jumikis, A. R. 1983. *Rock Mechanics*, 2nd Ed. Houston, TX: Trans. Tech. Publications.

Kamal, M. M. 1983. "Interference and pulse testing - a review." *Soc. Petrol. Eng. J.* December: 2257-2270.

Kanwal, R. P. 1971. *Linear Integral Equations*. New York: Academic Press.

Kaviany, M. 1991. *Principles of Heat Transfer in Porous Media*. New York: Springer-Verlag.

Kazemi, H. 1969. "Pressure transient analysis of naturally fractured reservoirs with uniform fracture distribution." *Trans.*, AIME 246, *Soc. Petrol. Eng. J.* December: 451-461.

Kazemi, H., J. R. Gilman and A. M. Elsharkawy. 1992. "Analytical and numerical solution of oil recovery from fractured reservoirs with empirical transfer functions." *SPE Reservoir Engineering*. May: 219-227.

Kazemi, H., L. S. Merril, Jr., K. L. Porterfield and P. R. Zeman. 1976. "Numerical simulation of water-oil flow in naturally fractured reservoirs." *Soc. Petrol. Eng. J.* 16: 317-326.

Khaled, M. Y., D. E. Beskos and E. C. Aifantis. 1984. "On the theory of consolidation with double porosity." *Int. J. Numer. Anal. Methods Geomech.* 8: 101-123.

Koenders, M. A. and A. F. Williams. 1992. "Flow equations of particle fluid mixtures." *Acta Mechanica* 92: 91-116.

Kozeny, J. 1927. *Wiener Akad.* Abstr. 2a, 136: 271.

Kranz, R. L., A. D., Frankel, T. Engelder and C. H. Scholz. 1979. "The permeability of whole and jointed Barre granite." *Int. J. Rock Mech. Min. Sci. and Geomech. Abstr.* 16: 225-234.

Krumbein, W. C. and G. D. Monk. 1943. "Permeability as a function of the size parameters of unconsolidated sand." *Trans. Am. Inst. Min. Metall. Petrol. Eng.* 151: 153-163.

Kucuk, F. and W. K. Sawyer. 1980. "Transient flow in naturally fractured reservoirs and its application to Devonian gas shales." Paper SPE 9397, *SPE Annual Technical Conference and Exhibition*, Dallas, TX: 21-24.

Kümpel, H.-J. 1991. "Poroelasticity: parameters reviewed." *Geophys. J. Int.* 105: 783-799.

Lade, P. V. and R. De. Boer. 1997. "The concept of effective stress for soil, concrete and rock." *Géotechnique* 47: 61-78.

Lai, W. M., D. Rubin and E. Krampl. 1993. *Introduction to Continuum Mechanics.* Oxford, UK: Pergamon Press.

Lamas, L. N. 1995. "An experimental study of the hydromechanical properties of granite joints." *Proc. 8th Int. Congress on Rock Mech.*, Tokyo, Japan: 733-738.

Lamb, H. 1932. *Hydrodynamics.* 6th Ed. New York: Cambridge University Press.

Lapidus, L. and G. F. Pinder. 1982. *Numerical Solution of Partial Differential Equations in Science and Engineering.* New York: John Wiley and Sons.

Leismann, H. M. and E. O. Frind. 1989. "A symmetric-matrix time integration scheme for the efficient solution of advection-dispersion problems." *Water Resour. Res.* 25: 1133-1139.

Lekhnitskii, S. G. 1981. *Theory of Elasticity of an Anisotropic Body.* Moscow: Mir.

Leo, C. J. and J. R. Booker. 1993. "Boundary element analysis of contaminant transport in fractured porous media." *Int. J. Numer. Anal. Methods Geomech.* 17: 471-492.

Lewis, R. W. and H. R. Ghafouri. 1997. "A novel finite element double porosity model for multiphase flow through deformable fractured porous media." *Int. J. Num. Analy. Meth. Geomech.* 21: 789-816.

Lewis, R. W. and B. A. Schrefler. 1987. *The Finite Element Method in the Deformation and Consolidation of Porous Media.* New York: John Wiley and Sons.

Li, X. Y. and X. W. Li. 1992. "On the thermoelasticity of multicomponent fluid-saturated reacting porous media." *Int. J. Eng. Sci.* 30: 891-912.

Li, X., O. C. Zienkiewicz and Y. M. Xie. 1990. "A Numerical model for immiscible two-phase fluid flow in a porous medium and its time domain solution." *Int. J. for Numer. Meth. in Eng.* 30: 1195-1212.

Liggett, J. A. and P. L. F. Liu. 1983. *The Boundary Integral Equation Method for Porous Media Flow.* London: George Allen and Unwin.

Lim, K. T. and K. Aziz. 1995. "Matrix-fracture transfer shape factors for dual-porosity simulators." *J. Petrol. Sci. Eng.* 13: 169-178.

Lin, D. and C. Fairhurst. 1991. " The topological structure of fracture system in rock." *Proc. of the 32nd U.S. Symp. on Rock Mechanics*, Norman, OK: 1155-1163.

Liu, J., D. Elsworth and B. H. Brady. 1999. "Linking stress-dependent effective porosity and hydraulic conductivity fields to RMR." *Int. J. Rock Mech. Min. Sci.* 36: 581-596.

Liu, X. and Z. Chen. 1987. "Exact solution of double-porosity, double-permeability systems including wellbore storage and skin effect." SPE 16849, *66th Annual Tech. Conf. and Exhib.*, Dallas, TX.

Liu, T., Ma, X., W. Zhong and B. Xu. 1992. "Prediction of surface movements and deformations in Fushun Petroleum Factory." CCMRI Report, Beijing, China (in Chinese).

Long, J. C. S., J. S. Remer C. R. Wilson and P. A. Witherspoon. 1982. "Porous media equivalents for networks of discontinuous fractures." *Water Resour. Res.* 18: 645-658.

Louis., C. A. 1969. "Groundwater flow in rock masses and its influence on stability." Rock Mech. Research Report, 10, Imperial College, UK.

Louis, C. A. 1974. "Rock hydraulics." In *Rock Mechanics.* Edited by L. Muller. Vienna: Springer Verlag: 299-382.

Love, A. E. H. 1927. *Treatise on the Mathematical Theory of Elasticity.* 4th Ed. New York: Cambridge University Press.

Lubinski, A. 1954. "The theory of elasticity for porous bodies displaying a strong pore structure." *Proc. 2nd US Nat. Congr. Appl. Mech.*, Ann Arbor, MI.

Luckner, L. and W. M. Schestakow. 1991. *Migration Processes in the Soil and Groundwater Zone.* Chelsea, MI: Lewis Publishers, Inc.

Lunardi, P., P. Froldi and E. Fornari. 1994. "Rock mechanics investigations for rock slope stability assessment." *Int. J. Rock Mech. Min. Sci. and Geomech. Abstr.* 31: 323-345.

Luxmoore, R. J., P. M. Jardine, G. V., Wilson, J. R. Jones and L. W. Zelazny. 1990. "Physical and chemical controls of preferred path flow through a forest hillslope." *Geoderma* 46: 139-154.

Ma, X., Q. Huo, J. Wang and Z. Jiang. 1990. "Report on north slope stability analysis of Fushun open pit." CCMRI Report 01-04, Beijing, China (in Chinese).

Maasland, M. 1957. "Soil anisotropy and land drainage." In *Drainage of Agricultural Lands.* Edited by. J. N. Luthin, Madison, WI: ASA: 216-285.

Mandel, J. 1953. "Consolidation des sols." *Géotechnique* 3: 287-299.

Mannhardt, K. and H. A. Nasr-El-Din. 1994. "A review of one-dimensional convection-dispersion models and their applications to miscible displacement in porous media." *In Situ* 18: 277-345.

Marcus, H. and D. E. Evenson. 1961. "Directional permeability in anisotropic porous media." *Water Resour. Center Contrib.*, No. 31, University of California, Berkeley.

Mathematical Handbook. 1979. *People's Education Pub. Co.*, Beijing, China (in Chinese).

Mattax, C. C. and R. L. Dalton. 1990. *Reservoir Simulation.* SPE Monograph 13. Henry L. Doherty Memorial Fund of AIME, SPE, Richardson, TX.

Mavor, M. J. and H. Cinco-Ley. 1979. "Transient pressure behavior of naturally fractured reservoirs." SPE 7977, *J. Petrol. Tech.* 259: 1290-1306.

McDonald, M. G. and A. W. Harbaugh. 1988. "A modular three-dimensional finite-difference ground-water flow model." *Tech. of Water Resour. Invest. Book 6.* U.S. Geological Survey, Chap. A1.

McKibbin, R. 1985. "Thermal convection in layered and anisotropic porous media: a review." *Proc. CSIRO/DSIR Seminar on Convective Flows in Porous Media*, Wellington, New Zealand: 113-127.

McNamee, J. and R. E. Gibson. 1960. "Displacement functions and linear transforms applied to diffusion through porous elastic media." *Q. J. Mech. Appl. Math* 13: 99-111.

Melan, E. Z. 1940. ("Correction." 1932. *Angew. Math. Mech.*, 20, 368) "Der Spannungszustand der durch eine einzelkraft im innern beanspruchten halbscheibe." *Angew. Math. Mech.* 12: 343-346.

Meng, F. 1998. Three-Dimensional Finite Element Modeling of Two-Phase Fluid Flow in Deformable Naturally Fractured Reservoirs. Ph.D. dissertation, University of Oklahoma.

Meng, F. and M. Bai. 1998. "Two-phase fluid flow in deformable fractured formations: finite element approach." Report RMRC-98-01, University of Oklahoma.

Mitchell, A. R. and D. F. Griffiths. 1980. *Finite Difference Method in Partial Differential Equations.* New York: John Wiley and Sons.

Mo, H., M. Bai, D. Lin and J.-C. Roegiers. 1998. "Study of flow and transport in fracture network using percolation theory." *Applied Math. Modelling* 22: 277-291.

Moench, A. F. 1984. "Double-porosity models for a fissured groundwater reservoir with fracture skin." *Water Resour. Res.* 20: 831-846.

Mohsen, M. F. N. 1984. "Numerical experiments using adaptive finite element methods with collocation." *Finite Elements in Water Resour.*, Burlington, VT: 45-61.

Morita, N. 1993. "Numerical evaluation of gravel-pack damage." SPE 25434, *Prod. Oper. Symp.*, Oklahoma City, OK: 287-298.

Murrell, S. A. F. 1963. "A criterion for brittle fracture of rocks and concrete under triaxial stress, and the effect of pore pressure on the criterion." In *Proc. 5th Symp. Rock Mech.* Edited by C. Fairhurst. New York: Pergamon Press: 563-577.

Muskat, M. 1937. *The Flow of Homogeneous Fluids Through Porous Media.* New York: McGraw-Hill.

Najurieta, H. L. 1980. "A theory for pressure transient analysis in naturally fractured reservoirs." *Trans.*, AIME 269, *J. Petrol. Tech.* July: 1241-1250.

Nanba, T. 1991. "Numerical simulation of pressure transients in naturally fractured reservoirs with unsteady-state matrix-fracture flow." SPE 22719, *66th Annual Tech. Conf. and Exhib.*, Dallas, TX.

Nash, D. 1987. "A comparative review of limit equilibrium methods of stability analysis." In *Slope Stability.* Edited by M. G. Anderson and K. S. Richards. New York: John Wiley and Sons: 11-76.

National Research Council. 1996. *Rock Fractures and Fluid Flow*. Washington, DC: National Academy Press.

Neretnieks, I. 1993. "Solute transport in fractured rock - applications to radionuclide waste repositories." In *Flow and Contaminant Transport in Fractured Rock*. Edited by J. Bear et al. San Diego, CA: Academic Press Inc.: 39-128.

Nguyen, T. S. and A. P. S. Selvadurai. 1994. "Thermo-poroelastic response of a fractured geological medium." *Proc. 8th Int. Confer. Computer Meth. Adv. Geomech.*, Morgantown, WV: 2031-2036.

Nilson, R. H. and K. H. Lie. 1990. "Double-porosity modeling of oscillatory gas motion and contaminant transport in a fractured porous medium." *Int. J. Numer. Analy. Methods Geomech.* 14: 565-585.

Noltimier, H. C. 1971. "A model for grain dispersion and magnetic anisotropy in sedimentary rocks." *J. Geophy. Res.* 76: 3990-4002.

Noorishad, J., C.-F. Tsang, P. Perrochet and A. Musy. 1992. "A perspective on the numerical solution of convection-dominated transport problems: a price to pay for the easy way out." *Water Resour. Res.* 28: 551-561.

Nowacki, W. 1962. *Thermoelasticity*. Oxford, UK: Pergamon Press.

Nur, A. and J. D. Byerlee. 1971. "An exact effective stress law for elastic deformation of rock with fluids." *J. Geophys. Res.* 76: 6414-6419.

Oda, A. 1985. "Permeability tensor for discontinuous rock masses." *Géotechnique* 35: 483-495.

Oda, M. 1986. "An equivalent continuum model for coupled stress and fluid flow analysis in jointed rock masses." *Water Resour. Res.* 22: 1845-1856.

Odeh, A. S. 1965. "Unsteady-state behavior of naturally fractured reservoirs." *Soc. Petrol. Eng. J.* March: 60-66.

Ogata, A. 1964. "Mathematics of dispersion with linear adsorption isotherm." Geological Professional Paper. Washington, DC: U.S. Govt. Print. Off.: 411-H.

Ogata, A. 1970. "Theory of dispersion in a granular medium." Geological Professional Paper. Washington, DC: U.S. Govt. Print. Off.: 411-I.

Ogata, A. and R. B. Banks. 1961. "A solution of the differential equation of longitudinal dispersion in porous media." Geological Professional Paper. Washington, DC: U.S. Govt. Print. Off.: 411-A.

O'Neil, P. V. 1983. *Advanced Engineering Mathematics*. Belmont, CA: Wadsworth Publishing Co.

Ouyang, Z. and D. Elsworth. 1989. "An adaptive characteristics method for advective-diffusive transport." *Appl. Math. Modelling* 13: 682-692.

Ouyang, Z. and D. Elsworth. 1993. "Evaluation of groundwater flow into mined panels." *Int. J. Rock Mech. Min. Sci. and Geomech. Abstr.* 30: 71-80.

Passioura, J. B. 1971. "Hydrodynamic dispersion in aggregated media: 1. theory." *Soil Sci.* 111: 339-344.

Passioura, J. B. and D. A. Rose. 1971. "Hydrodynamic dispersion in aggregated media: 2. effects of velocity and aggregate size." *Soil Sci.* 111: 345-

351.
Peaceman, D. W. 1977. *Fundamentals of Numerical Reservoir Simulation.* Amsterdam: Elsevier Sci.
Piessens, R. and R. Huysmans. 1984. "Algorithm 619 automatic numerical inversion of the Laplace transform [D5]." *ACM Trans. Math. Software* 10: 348-353.
Piggott, A. R. and D. Elsworth. 1993. "Laboratory assessment of the equivalent apertures of a rock fracture." *Geophys. Res. Let.* 20: 1387-1390.
Pinder, G. E. and A. Shapiro. 1979. "A new collocation method for the solution of the convection dominated transport equation." *Water Resour. Res.* 15: 1177-1182.
Piquemal, J. 1992. "On the modeling of miscible displacements in porous media with stagnant fluid." *Transport in Porous Media* 8: 243-262.
Piquemal, J. 1993. "On the modeling conditions of miscible displacements in porous media presenting capacitance effects by a dispersion-convection equation for the mobile fluid and a diffusion equation for the stagnant fluid." *Transport in Porous Media* 10: 271-283.
Prandtl, L. B. 1952. *Essentials of Fluid Dynamics.* New York: Hafner Pub. Co.
Prat, G. Da. 1990. *Well Test Analysis for Fractured Reservoir Evaluation.* Amsterdam: Elsevier Sci.
Price, H. S., J. C. Cavendish and R. S. Varga. 1968. "Numerical methods of higher-order accuracy for diffusion-convection equations." *Soc. Petrol. Eng. J.* 8: 293-303.
Raghavan, R., R. Prijambodo and A. C. Reynolds. 1985. "Well test analysis for well producing layered reservoirs with crossflow." *Soc. Petrol. Eng. J.* June: 380-396.
Reiss, L. H. 1980. *The Reservoir Engineering Aspects of Fractured Formations.* Institut Francais Du Petrole. Paris: Editions Technip.
Reynolds, A. C., W. L. Chang, N. Yeh and R. Raghavan. 1985. "Wellbore pressure response in naturally fractured reservoirs." *J. Petrol. Tech.* May: 908-920.
Rice, J. R. and M. P. Cleary. 1976. "Some basic stress-diffusion solutions for fluid saturated elastic porous media with compressible constituents." *Rev. Geophys. Space Phys.* 14: 227-241.
Richtmyer, R. D. and K. W. Morton. 1967. *Difference Methods for Initial-Value Problems.* New York: Interscience.
Robin, P. Y. F. 1973. "Note on effective pressure." *J. Geophys. Res.* 78: 2434-2437.
Robinson, L. H. Jr. 1959. "The effect of pore and confining pressure on the failure process in sedimentary rock." *Colorado School of Mines Quart.* 54: 177-199.
Rosso, R. S. 1976. "A comparison of joint stiffness measurements in direct shear, triaxial compression, and in situ." *Int. J. Rock Mech. Min. Sci. and*

Geomech. Abstr. 13: 167-172.
Rowe, R. K. and J. R. Booker. 1990. "Contaminant migration in a regular two- or three-dimensional fractured network: reactive contaminants." *Int. J. Numer. Analy. Methods Geomech.* 14: 401-425.
Rudnicki, J. W. 1980. "Fracture mechanics applied to the earth's crust." *Ann. Rev. Earth Planet. Sci.* 8: 489-525.
Sagar, B. and A. Runchal. 1982. "Permeability of fractured rock: effect of fracture size and data uncertainties." *Water Resour. Res.* 18: 266-274.
Sahimi, M. 1993. "Flow phenomena in rocks: from continuum models to fractals, percolation, cellular automata, and simulated annealing." *Rev. Mod. Phys.* 65: 1393-1534.
Sandhu, R. S. and E. L. Wilson. 1969. "Finite element analysis of seepage in elastic media." *J. Eng. Mech.*, ASCE 95: 641-652.
Sardin, M. and D. Schweich, F. J. Leij and van M. T. Genuchten. 1991. "Modeling the nonequilibrium transport of linearly interacting solutes in porous media: a review." *Water Resour. Res.* 27: 2287-2307.
Scheidegger, A. E. 1957. *The Physics of Flow Through Porous Media.* New York: Macmillan Company.
Schestakow, W. M. 1979. "Modeli perenosa v neodnorodnyh plastah i porodah." In Wiss. Konferenz der Techn. Univ. Dresden zur Simulation der Migrationsprozesse im Boden-und Grundwasser, Bd. II. Dresden: Techn. Univ.: 56-61.
Schiffman, R. L. 1970. "The stress components of a porous medium." *J. Geophys. Res.* 75: 4035-4038.
Schiffman, R. L., A. T.-F. Chen and J. C. Jordan. 1969. "An analysis of consolidation theories." *J. Soil Mech. Found. Div.*, ASCE 95: 285-312.
Segall, P. 1985. "Stress and subsidence resulting from subsurface fluid withdrawal in the epicentral region of the 1983 Coalinga earthquake." *J. Geophys. Res.* B8: 6801-6816.
Serafim, J. L. 1964. "The 'uplift area' in plain concrete in the elastic range." *Proc. 8th Congr. on Large Dams*, Edinburge: 5-17.
Serra, K., A. C. Reynolds and R. Raghavan. 1983. "New pressure transient analysis methods for naturally fractured reservoirs." *J. Petrol. Tech.* December: 2271-2283.
Shapiro, A. M. 1987. "Transport equations for fractured porous media." In *Advances in Transport Phenomena in Porous Media.* Edited by J. Bear and M. Y. Corapcioglu. Dordrecht: Martinus Nijhoff Publishers.
Shimo, M. and J. C. S. Long. 1987. "A numerical study of transport parameters in fracture networks." In *Flow and Transport Through Unsaturated Fractured Rock.* Edited by D. D. Evans and T. J. Nicholson. Geophysical Monograph 42. Washington DC: American Geophysical Union: 121-131.
Sih, G. C., R. P. Wei and F. Erdogan, eds. 1975. *Linear Fracture Mechanics.* Lehigh, PA: Envo Publishing Co.
Simon, B. R., O. C. Zienkiewicz and D. R. Paul. 1984. "An analytical solution

for the transient response of saturated porous elastic solids." *Int. J. Num. Analy. Methods Geomech.* 8: 381-398.

Simoni, L. and B. A. Schrefler. 1987. "Mapped infinite elements in soil consolidation." *Int. J. Num. Meth. in Eng.* 24: 513-527.

Skempton, A. W. 1954. "The pore pressure coefficients A and B." *Géotechnique* 4: 143-147.

Skempton, A. W. 1960. " Effective stress in solid, concrete and rock." In *Pore Pressure and Suction in Soils.* London: Butterworths.

Skempton, A. W. and A. W. Bishop. 1954. *Soils, in Building Materials, Their Elasticity and Inelasticity.* Edited by M. Reiner. Amsterdam, The Netherlands: North Holland Publishing Co.: 417-482.

Smith, G. D. 1978. *Numerical Solution of Partial Differential Equations: Finite Difference Methods.* 2nd Ed. Oxford, UK: Clarendon Press.

Smith, I. M. and D. F. Griffiths. 1988. *Programming the Finite Element Method.* Chichester: John Wiley and Sons.

Smith, L. and F. W. Schwartz. 1984. "An analysis of fracture geometry on mass transport in fractured media." *Water Resour. Res.* 20: 1241-1252.

Snow, D. T. 1968. "Rock fracture spacings, openings, and porosities." *J. Soil Mech. and Found. Div. Proc.* ASCE 94: 73-91.

Snow, D. T. 1969. "Anisotropic permeability of fractured media." *Water Resour. Res.* 5: 1273-1289.

Stehfest, H. 1970. "Algorithm 358 - Numerical inversion of Laplace transforms." *Communications of ACM* 13: 47-49.

Streltsova, T. D. 1983. "Well pressure behavior of a naturally fractured reservoir." *Trans.* AIME 275, *Soc. Petrol. Eng. J.* October: 769-780.

Streltsova, T. D. 1988. *Well Testing in Heterogeneous Formations.* New York: John Wiley and Sons.

Streltsova-Adams, T. D. 1978. "Well hydraulics in heterogeneous aquifer formations." *Adv. Hydrosci.* 11: 357-423.

Sudicky, E. A. 1989. "The Laplace transform Galerkin technique: a time-continuous finite element theory and application to mass transport in groundwater." *Water Resour. Res.* 25: 1833-1846.

Sudicky, E. A. and P. S. Huyakorn. 1991. "Contaminant migration in imperfectly known heterogeneous groundwater systems." *Review of Geophysics*, supplement: 240-253.

Sudicky, E. A. and R. G. McLaren. 1992. "The Laplace transform Galerkin technique for large-scale simulation of mass transport in discretely fractured porous formations." *Water Resour. Res.* 28: 499-514.

Suklje, L. 1969. *Rheological Aspects of Soil Mechanics.* New York: Wiley Interscience.

Sun, Y. and R. Sterling. 1994. "A three-dimensional model for transient fluid flow through deformable fractured porous media." *8th Int. Confer. Computer Meth. Adv. Geomech.*, Morgantown, WV: 1275-1280.

Swenson, D. and M. Beikmann. 1992. "An implicitly coupled model of fluid

flow in jointed rock." *Proc. 33nd U.S. Symp. on Rock Mech.*, Santa Fe, NM: 649-658.
Tang, D. H., E. O. Frind and E. A. Sudicky. 1981. "Contaminant transport in fractured porous media: analytical solution for a single fracture." *Water Resour. Res.* 17: 555-564.
Terzaghi, K. 1923. "Die Berechung der Durchassigkeitsiffer des Tones aus dem Verlauf der hydrodynamischen Spannungserscheinungen." Sitzungsber. Akad. Wiss. *Wien Math Naturwiss.*, Kl. Abstr. 2A, 132: 105.
Terzaghi, K. 1943. *Theoretical Soil Mechanics.* New York: John Wiley and Sons.
Terzaghi, K. 1945. "Stress conditions for the failure of saturated concrete and rock." *Proc. Am. Soc. Testing Materials* 45: 777-801.
Thompson, M. and J. R. Willis. 1991. "A reformation of the equations of anisotropic poroelasticity." *J. Appl. Mech.*, ASME 58: 612-616.
Timoshenko, S. 1934. *Theory of Elasticity.* 1st Ed. New York: McGraw-Hill Book Co.
Timoshenko, S. and J. N. Goodier. 1970. *Theory of Elasticity.* New York: McGraw-Hill Book Co.
Todd, A. C., J. E. Somerville and G. Scott. 1984. "The application of depth of formation damage measurements in predicting water injectivity decline." SPE 12498, Denver, CO.
Tritton, D. J. 1988. *Physical Fluid Dynamics.* Oxford, UK: Oxford University Press.
Tuncay, K. and M. Y. Corapcioglu. 1995. "Effective stress principle for saturated fractured porous media." *Water Resour. Res.* 31(12): 3103-3106.
Turner, J. S. 1973. *Buoyancy Effects in Fluids.* Cambridge, UK: Cambridge University Press.
Udey, N. and T. J. T. Spanos. 1993. "The equations of miscible flow with negligible molecular diffusion." *Transport in Porous Media* 10: 1-41.
Valliappan, S. and N. Khalili-Naghadeh. 1990. "Flow through fissured porous media with deformable matrix." *Int. J. Numer. Meth. Eng.* 29: 1079-1094.
van Genuchten, M. T. 1982. "*U.S. Dept of Agric. Tech. Bull.*." 1663: 144.
van Genuchten, M. T. and W. J. Alves. 1982. Analytical Solutions of the One-Dimensional Convective-Dispersive Solute Transport Equation. U.S. Dept. of Agric. Tech. Bull. 1661.
van Genuchten, M. T. and G. F. Pinder. 1977. "Simulation of two dimensional contaminant transport with isoparametric hermitian finite elements." *Water Resour. Res.* 13: 451-458.
Vaziri, H. H. and P. M. Byrne. 1990. "Analysis of stress, flow, and stability around deep wells." *Géotechnique* 40: 63-77.
Verruijt, A. 1969. "An elastic storage of aquifers." In *Flow Through Porous Media.* New York: Academic Press: 331-376.
Volk, B. 1994. Personal communication.
Wallis, G. B. 1969. *One-Dimensional Two-Phase Flow.* New York: McGraw-

Hill.

Walsh, J. B. 1965. "The effect of cracks on compressibility of rock." *J. Geophys. Res.* 70: 381-389.

Walsh, J. B. 1981. "Effect of pore pressure and confining pressure on fracture permeability." *Int. J. Rock Mech. Min. Sci. and Geomech. Abstr.* 18: 429-435.

Wang, J. S. Y. 1991. "Flow and transport in fractured rocks." *Reviews of Geophysics*, supplement: 254-262.

Warren, J. E. and P. J. Root. 1963. "Behavior of naturally fractured reservoirs." *Trans.*, AIME 228, *Soc. Petrol. Eng. J.* September: 245-55.

Wheatcraft, S. W. and J. H. Cushman. 1991. "Hierarchical approaches to transport in heterogeneous porous media." *Review of Geophysics*, supplement: 263-269.

Wilson, R. K., and E. C. Aifantis. 1982. "On the theory of consolidation with double porosity." *Int. J. Eng. Sci.* 20: 1009-1035.

Witherpoon, P. A. 1981. "New approaches to problems of fluid flow in fractured rock masses." *Proc. of 22nd U.S. Symp. Rock Mech.*, MIT, Cambridge, MA.

Witherspoon, P. A., J. S. Y. Wang, K. Iwai and J. E. Gale. 1980. "Validity of cubic law for fluid flow in a deformable rock fracture." *Water Resour. Res.* 16: 1016-1024.

Yang, X., Y. Zhang, F. Xu and H. Zhang. 1995. "Studies on strata movement observation and its regularity." CCMRI Report 93-179, Beijing, China (in Chinese).

Young, A., P. F. Low and A. S. McLatchie. 1964. "Permeability studies of argillaceous rocks." *J. Geophys. Res.* 69: 4237-4245.

Zhang, R., K. Huang and M. T. van Genuchten. 1993. "An efficient Eulerian-Lagrangian method for solving solute transport problems in steady and transient flow fields." *Water Resour. Res.* 29: 4131-4138.

Zhou, W. 1995. "Study on surface and strata movements in petroleum factory." CCMRI Report, Beijing China (in Chinese).

Zienkiewicz, O. C. 1983. *The Finite Element Method.* 3rd Ed. London: McGraw-Hill.

Zienkiewicz, O. C., C. Humpheson and R. W. Lewis. 1977. "A unified approach to soil mechanics problems." In *Finite Element in Geomechanics.* Edited by G. Gudehus. New York: John Wiley and Sons: 151-177.

Zimmerman, R. W., G. Chen, T. Hadgu and G. S. Bodvarsson. 1993. "A numerical dual-porosity model with semianalytical treatment of fracture/matrix flow." *Water Resour. Res.* 29: 2127-2137.

Zimmerman, R. W., W. Somerton and M. King. 1986. "Compressibility of porous rock." *J. Geophys. Res.* 91: 12765-12777.

Index